Point Set Topology

STEVEN A. GAAL

Department of Mathematics
University of Minnesota

Dover Publications, Inc.
Mineola, New York

Copyright

Bibliographical Note

This Dover edition, first published in 2009, is an unabridged republication of the second printing (1966) of the work originally published by Academic Press, Inc., New York, in 1964.

Library of Congress Cataloging-in-Publication Data

Gaal, Steven A.
Point set topology / Steven A. Gaal. — Dover ed.
p. cm.
Originally published: New York : Academic Press, 1964; in series: Pure and applied mathematics ; 16.
ISBN-13: 978-0-486-47222-5
ISBN-10: 0-486-47222-1
1. Topology. I. Title.

QA611.G33 2009
514—dc22

2009005728

Manufactured in the United States of America
Dover Publications, Inc., 31 East 2nd Street, Mineola, N.Y. 11501

Preface

This book for beginning graduate and advanced undergraduate mathematics students presents point set topology not only as an end in itself, but also as a related discipline to the proper understanding of various branches of analysis and geometry. It starts with the basic concepts of set theory and topological spaces and ends with the beginning of functional analysis. The text and nearly all of the exercises presuppose knowledge of only those concepts defined herein, so that the book is self-contained to accommodate those who wish to study topology on their own. Moreover, it includes additional material and literature which make it valuable as a reference work.

The book contains enough material for a one-year course, and I have found it accessible to juniors majoring in mathematics. By omitting carefully chosen sections it is possible, but not easy, to cover most of the book in a one-semester course. When giving a year's course on the foundations of analysis for graduate students, I have been able to include some additional material, such as differentiable manifolds or abstract harmonic analysis and fixed point theorems.

The first chapter contains the fundamental notions associated with a general topological space, and a systematic discussion of the various practical methods used to define topological spaces. The second and third chapters deal with those additional properties that give a general topological space more resemblance to the primitive, intuitive concept associated with the concept of space. The basic properties of functions defined on topological spaces are collected in the fourth chapter, although some concepts, such as continuous maps and homeomorphisms, were introduced earlier for the sake of clarity. The last chapter contains an exposition of the theory of topological convergence using filters and nets, which is applied to problems of compactness, completion, and compactification. The exercises vary in difficulty, and some provide additional insights or new results. The principal theorems are all included as part of the text. The remarks at the end of each chapter contain pertinent comments which did not seem suitable for inclusion elsewhere. While it is hoped that they may give the reader some historical perspective, they do reflect the author's personal thoughts, and no attempt has been made at complete coverage.

The bulk of the material presented here was developed through lectures at Cornell University and the University of Minnesota. I am very grateful to Professor R. J. Walker and to Miss Madelyn Keady of Cornell University and to my wife for their encouragement and help with the preliminary publication. A great deal of additional work and support is needed to make a first draft develop into a book. I want to thank Professor J. B. Rosser for his kind interest in my project and for giving me a helping hand. In Minnesota I was very fortunate to meet Mr. Glenn Schober, who read the entire manuscript and made many improvements. He also read and corrected the proofs and thus helped eliminate many errors that I had overlooked.

I also want to express my appreciation for the support given to me by the various Federal agencies at several stages of the project. Finally, my sincere thanks to Academic Press for their careful work.

STEVEN A. GAAL

June, 1964

Notation

$\mathscr{B}$	Base for open sets; **33**
$\mathscr{B}$	Filter base; **43**
$\mathscr{C}$	Family of closed sets; **22**
$\mathfrak{C}$	Complex number field; **1**
$\mathscr{C}(x)$	Filter base of closed neighborhoods of x; **43**
E_n	Euclidean space of dimension n; **40**
$\mathscr{F}$	Filter; **43**
I	Diagonal; **6, 46**
N_x	Neighborhood of x; **42**
$\mathscr{N}(x)$	Neighborhood filter of x; **43**
O	Open set; **21**
$\mathscr{O}$	Family of open sets; **21**
$\mathscr{O}(x)$	Filter base of open neighborhoods of x; **43**
$\mathscr{P}(X)$	Power set of X; **6, 7**
R	Field of rationals; **1**
$\mathfrak{R}$	Field of real numbers; **1**
$\mathscr{S}$	Subbase for open sets; **34**
$S_\epsilon[x]$	ϵ-Ball with center at x; **38**
$\mathscr{T}$	Topology; **21**
$\mathscr{T}_+$	Right half-open interval topology; **37**
$\mathscr{T}_-$	Left half-open interval topology; **37**
$\mathscr{T}_R$	Topology determined by the saturated open sets relative to R; **73**
$\mathscr{T}/R$	Quotient topology; **71**
U	Uniformity; **46**
U^{-1}	Inverse uniformity; **46**
$U[x]$	Uniformity U evaluated at x; **48**
$U[A]$	Uniformity U evaluated on the set A; **48**
U_Y	Trace of uniformity U on $Y \times Y$; **57**
$\mathscr{U}$	Uniform structure; **45**
$\mathscr{U}_B$	Structure base; **46**
$\mathscr{U}_S$	Subbase for a uniform structure; **47**
$\mathscr{U}_Y$	Trace of uniform structure $\mathscr{U}$ on set Y; **57**
$\{a\}$	Set consisting of single element a; **2**
$\{a, b\}$	Nonordered pair of a and b; **2**
$\{A_n\}$ $(n = 1, 2, \ldots)$	Countable or denumerable family; **8**
(A_n) $(n = 1, 2, \ldots)$	Sequence, i.e. countable family with a fixed enumeration; **8**
$\{A_i\}$ $(i \in I)$	Family of objects with index set I; **13**
(A_i) $(i \in I)$	Indexed family of objects with partially ordered index set I; **14**
(O_d) $(d \in D)$	Scale of open sets; **109**
A^e	Exterior of A; **24**
A^b	Boundary of A; **25**
A^i	Interior of A; **24**
$\bar{A}$	Closure of A; **25**
A'	Derived set of A; **26**
$-B$	Complement of B; **3**
cB	Complement of B; **3**
$c_A B$	Complement of B relative to A; **3**
$A \underset{r}{\cup} B$	Regular union of A and B; **27**
$A \underset{r}{\cap} B$	Regular intersection of A and B; **27**
$A - B$	Relative complement; **3**
$A \setminus B$	Complement of B relative to A; **3**
$A \triangle B$	Symmetric difference of A and B; **4**
$A \circ B$	Composition of A and B; **46, 176**
S^b_Y	Boundary of S relative to subspace Y; **56**
(a, b)	Ordered pair of a and b, or open interval with end points a and b; **4, 5, 35**

Symbol	Meaning
$[a, b]$	Closed interval with end points a and b; **35**
$(a, b]$	Left half-open interval; **35**
$[a, b)$	Right half-open interval; **35**
$(a, +\infty)$	Improper interval with left end point a; **35**
$(-\infty, b)$	Improper interval with right end point b; **35**
$(a_1, ..., a_n)$	Ordered n-tuple; **5**
$(A \times B)$	Product of the ordered pair (A, B) of sets A, B; generally written as $A \times B$ when it causes no confusion; **5**
$(A_1 \times ... \times A_n)$	Product of the ordered n-tuple $(A_1, ..., A_n)$ of sets $A_1, ..., A_n$; **5**
$A_1 \times ... \times A_n$	Cartesian product on the natural numbers; **14**
$\prod X_s$	Product of the sets X_s or the product of the topological spaces X_s with the product topology; **59**
$\pi_s(A)$	Projection of A into the sth factor; **59**
$Z(s, A_s)$	Cylinder with base A_s in the sth factor; **59**
x^*	Star of a point x; **143**
$d(X)$	Diameter of the metric space X; **133**
$d(a, b)$	Metric function; distance of a and b; **38**
$d(A, B)$	Distance of sets A and B in the metric space having metric d; **141**
$A \leqslant B$	Cardinality of A is at most as large as that of B; **12**
$A < B$	The set A is of smaller cardinality than B; **12**
$\mathscr{T}_1 \leqslant \mathscr{T}_2$	The topology $\mathscr{T}_1$ is coarser than $\mathscr{T}_2$; **22**
$\mathscr{T}_1 < \mathscr{T}_2$	The topology $\mathscr{T}_1$ is strictly coarser than $\mathscr{T}_2$; **22**
$\mathscr{U}_1 \leqslant \mathscr{U}_2$	The uniform structure $\mathscr{U}_1$ is coarser than $\mathscr{U}_2$; **53**
$\mathscr{U}_1 < \mathscr{U}_2$	The uniform structure $\mathscr{U}_1$ is strictly coarser than $\mathscr{U}_2$; **53**
$\mathscr{T}_x \sim \mathscr{T}_y$	The topological spaces X and Y are homeomorphic; **22**
$a\,R\,b$	a is related to b by the relation R; **5**

Symbol	Meaning
$A \sim B$	A and B are equivalent sets; **7**
R_2/R_1	Quotient of the equivalence relations R_1 and R_2; **72**
$f: A \to B$	Function on A with values in B; occasionally function from A in B; **6, 175**
$f \circ g$	Composition of functions f and g; **177**
$f \cup g$	Maximum of f and g; **185**
$f \cap g$	Minimum of f and g; **185**
$f(\mathscr{T})$	Direct image of the topology $\mathscr{T}$ under the map f; **70**
$f^{-1}(A)$	Inverse image of the set A under f; **68**
$f^{-1}(\mathscr{T})$	Inverse image of the topology $\mathscr{T}$ under f; **68**
$F \circ G$	Composition of graphs; **176**
$\text{glb}\{\mathscr{F}_\alpha\}$	Greatest lower bound of the filters $\mathscr{F}_\alpha$; **262**
$\text{glb}\{\mathscr{T}_i\}$	Greatest lower bound of the topologies $\mathscr{T}_i$; **34**
$\text{lub}\{\mathscr{F}_\alpha\}$	Least upper bound of the filters $\mathscr{F}_\alpha$ provided it exists; **262**
$\text{lub}\{\mathscr{T}_i\}$	Least upper bound of the topologies $\mathscr{T}_i$; **34**
$\text{lub}\{\mathscr{U}_i\}$	Least upper bound of the uniform structures $\mathscr{U}_i$; **53**
adh $\mathscr{B}$	Adherence of the filter base $\mathscr{B}$; **259**
adh $\mathscr{F}$	Adherence of the filter $\mathscr{F}$; **259**
lim $\mathscr{B}$	Limit of the filter base $\mathscr{B}$; **260**
lim $\mathscr{F}$	Limit of the filter $\mathscr{F}$; **260**
$\text{adh}(x_d)$	Adherence of the net (x_d); **261**
$\lim(x_d)$	Limit of the net (x_d); **262**
$\emptyset$	Empty set, void set; **1**
$\cup$	Union, maximum; **3, 185**
$\cap$	Intersection, minimum; **3, 185**
$\bigcup$	Union; **3**
$\bigcap$	Intersection; **3**
$\prod$	Cartesian product, product; **14**
$\in$	Membership relation; **1, 2**
$\notin$	Negation of the membership relation; **1**
$\subset$	Proper inclusion; **2**
$\subseteq$	Improper inclusion, subset of ...; **2**
$\not\subset$	Negation of proper inclusion; **2**
$\not\subseteq$	Negation of improper inclusion; **2**

Contents

Point Set Topology

Introduction to Set Theory

1. Elementary Operations on Sets

At the time of this writing the fundamental system in mathematics from which all others are constructed by logical reasonings is the theory of sets. This is a fairly recent discovery, for the concept of an abstract set had been isolated only at the end of the last century by Georg Cantor. His results were published in two famous papers dated 1895 and 1897.

The purpose of this introductory chapter is to help the reader in forming an intuitive concept of an abstract set. Cantor's first memoir begins with the following sentences: "By a set M we understand any collection into a whole of definite and separate objects m of our intuition or our thought. These objects are called the elements of M. In symbols we express this as follows: $M = \{m\}$." Needless to say, this is far from a precise definition and it would be very foolish to build the whole of mathematics on such shaky foundations. However, these are just the opening notes of a magnificent theory and one hardly could begin differently.

Sets will be denoted by the symbols $a, b, c, \ldots; A, B, C, \ldots; \mathscr{A}, \mathscr{B}, \mathscr{C}, \ldots$ or $\mathfrak{A}, \mathfrak{B}, \mathfrak{C}, \ldots$. It is likely that the reader knows a great variety of things which he will correctly recognize as being sets: Collections of common objects, aggregates of people, finite families of natural numbers, the collection of all natural numbers $\{1, 2, 3, \ldots\}$, the set of integers $\mathfrak{Z}$, the rationals R, the reals $\mathfrak{R}$, the complex numbers $\mathfrak{C}$, the Euclidean spaces. These are very valuable examples of sets and we suppose that the reader is familiar with these concepts, but as far as the theory is concerned they are irrelevant. Instead, one stipulates the existence of one particular set which will be called the *void set* or the *empty set* and shall always be denoted by the symbol $\emptyset$.

Some sets are elements of others. If a is an element of A we write $a \in A$ and if not then $a \notin A$. Throughout this book the notations will be chosen in the following manner: If a reasoning involves sets and their elements, then the symbols $A, B, C, \ldots$ and $a, b, c, \ldots$ will be used. If we deal also with sets of sets, then these will be denoted by $\mathscr{A}, \mathscr{B}, \mathscr{C}, \ldots$ and their elements by $A, B, C, \ldots$. On a few occasions we shall go a step further and use a set $\mathfrak{A}$, its elements $\mathscr{A}, \mathscr{B}, \ldots$, the elements $A, B, C, \ldots$ of these, and also the elements $a, b, c, d, \ldots$ of the sets $A, B, C, \ldots$. In the abstract axiomatic theory of sets, the *membership relation* $a \in A$ is a primitive notion and it is further characterized only by the axiomatic statements in which it occurs. One of these states that if a is a set then $a \notin \varnothing$. In other words, the empty set is really void.

If every element of A belongs to B, we say that A is a *subset* of B and write $A \subseteq B$. The symbol $A \nsubseteq B$ will be used to denote that A is not a subset of B. If $A \subseteq B$ but $B \nsubseteq A$, then A is called a *proper subset* of B and one uses the notation $A \subset B$. Two sets A and B are considered identical if they have the same elements, i.e., if $A \subseteq B$ and $B \subseteq A$. If this is the case we write $A = B$. Clearly, $\varnothing \subseteq A$ for every set A and if $A \neq \varnothing$ then $\varnothing \subset A$.

We shall often define sets by specifying their elements. If the set is finite we simply enclose between curly brackets all the symbols which designate these elements. For instance $\{1, 2, 4, 8\}$ or {Conn., Minn., N.J.} or $\{\varnothing, \{1\}, \{2\}, \{1, 2\}\}$. For sets for which this notation would be too cumbersome to handle or if the set to be described is not finite, we shall use other means. For example,

$$\{n : n = k^3 \quad \text{and} \quad k = 1, 2, \ldots\}$$

denotes the set of cubes of natural numbers and

$$\{n : n = x^2 + y^2 \quad \text{and} \quad x, y \in \mathfrak{Z}\}$$

is the set of integers which can be written as the sum of two squares. In terms of this symbolism we have $A = \{a : a \in A\}$ for any set A.

Our intuition suggests that the families described above indeed are sets. In the strict axiomatic treatment it is considerably harder to establish the fact that we are really defining certain sets. For instance, the existence of the natural numbers is not taken for granted but is derived from the axioms. The particular axiom which is needed here states that if a and b are sets then there is a set whose elements are just the sets a and b. This is called the *nonordered pair* of a and b and is denoted by $\{a, b\}$. By choosing $a = b$ we obtain the existence of a set whose sole element is the set a. It is convenient to use the simpler notation $\{a\}$ instead of $\{a, a\}$. For a we can certainly choose the void set $\varnothing$ and get a new set $\{\varnothing\}$. Then $a = \{\varnothing\}$ yields $\{\{\varnothing\}\}$, the set whose only element is a set consisting of the single element $\varnothing$. By continuing in this fashion we end up with a whole class of sets:

$$\{\varnothing\}, \quad \{\{\varnothing\}\}, \ldots, \{\ldots\{\{\varnothing\}\}\ldots\}, \ldots .$$

Special names and symbols are attached to these curious sets: *One*, *two*, *three*, ... and 1, 2, 3, ...

If $\mathscr{A}$ is a set whose elements A are also sets, then we let $\bigcup \mathscr{A}$ denote the set consisting of the elements of the sets A:

$$\bigcup \mathscr{A} = \{a : a \in A \quad \text{and} \quad A \in \mathscr{A}\}.$$

This is called the *union* of $\mathscr{A}$ or the sum of $\mathscr{A}$. Its existence is postulated in one of the axioms. If $\mathscr{A}$ is finite, e.g., if $\mathscr{A} = \{A, B\}$ or $\mathscr{A} = \{A, B, C\}$, then we shall write $A \cup B$ or $A \cup B \cup C$ instead of $\bigcup \mathscr{A}$. We notice that $A \cup B = B \cup A$ and $A \cup (B \cup C) = A \cup B \cup C = (A \cup B) \cup C$. Moreover, $A \cup A = A$ and $A \cup \varnothing = A$ for any set A. Similarly, we define the *intersection* of $\mathscr{A}$ as the set of common elements of the sets A belonging to $\mathscr{A}$:

$$\bigcap \mathscr{A} = \{a : a \in A \quad \text{for every} \quad A \in \mathscr{A}\}.$$

In the finite case we use the notation $A \cap B$ or $A \cap B \cap C$, etc. We have $A \cap B = B \cap A$ and $A \cap (B \cap C) = A \cap B \cap C = (A \cap B) \cap C$ and also $A \cap A = A$ and $A \cap \varnothing = \varnothing$.

The operations $A \cup B$ and $A \cap B$ are meaningful for any pair of sets A, B and yield new sets. Moreover, as we have seen, these operations follow the *commutativity* and *associativity* laws known from elementary algebra. The empty set $\varnothing$ plays the role of the *zero element*. We also have two *distributivity* properties: $A \cup (B \cap C) = (A \cup B) \cap (A \cup C)$ and $A \cap (B \cup C) = (A \cap B) \cup (A \cap C)$.

For any two sets A, B we can define the *relative difference* $A - B$ as the set consisting of those points of A which do not belong to B:

$$A - B = \{a : a \in A \quad \text{and} \quad a \notin B\}.$$

Thus $A - B$ is a subset of A and $A - B = A$ if and only if A and B are *disjoint*, i.e., $A \cap B = \varnothing$. Very often $A - B$ is called the *complement of B relative to A* and one writes $c_A B$ or $A \setminus B$ instead of $A - B$. If the set A is fixed throughout some discussion it is customary to use the simpler notation $-B$ or cB in place of $A - B$. This is particularly convenient when a set A is given and the reasoning involves only subsets of this universe A.

There are two important identities involving unions, intersections, and relative complements called *de Morgan's formulas*. The simplest case concerns two sets A, B and complementation with respect to a third fixed set X. In this special case de Morgan's laws are

$$c(A \cup B) = cA \cap cB \qquad \text{and} \qquad c(A \cap B) = cA \cup cB$$

or in full details

$$X - (A \cup B) = (X - A) \cap (X - B) \text{ and } X - (A \cap B) = (X - A) \cup (X - B).$$

Similarly, in the case of finitely many sets $A_1, ..., A_n$ we have

$$c(A_1 \cup ... \cup A_n) = cA_1 \cap ... \cap cA_n$$

and

$$c(A_1 \cap ... \cap A_n) = cA_1 \cup ... \cup cA_n .$$

The general laws concern an arbitrary set $\mathscr{A}$ of sets A and complementation with respect to a fixed set X:

$$c \bigcup \mathscr{A} = \bigcap \{cA : A \in \mathscr{A}\} \quad \text{and} \quad c \bigcap \mathscr{A} = \bigcup \{cA : A \in \mathscr{A}\}.$$

Thus the complement of the union of $\mathscr{A}$ is the intersection of the set consisting of the complements of the elements of $\mathscr{A}$ and a similar statement holds for the complement of the intersection of $\mathscr{A}$. The finite cases discussed earlier are obtained by taking as $\mathscr{A}$ the finite sets $\{A, B\}$ and $\{A_1, ..., A_n\}$.

The subsets of a given set X form a set $\mathscr{P}(X)$ which is called the *power set* of X. Since $\varnothing \subseteq X$ and $X \subseteq X$ we have $\varnothing$, $X \in \mathscr{P}(X)$ no matter what X is. If $X = \varnothing$, then of course $\mathscr{P}(X) = \{\varnothing\}$. The operations $\cup$ and $\cap$ induce an interesting algebraic structure on $\mathscr{P}(X)$. As we have seen already, both of these operations are *idempotent*, *commutative*, and *associative*. Furthermore, they jointly obey two *distributivity laws*. The algebraic structure can be further strengthened by considering also the unary operation derived from complementation relative to X and the partial ordering relation $\subseteq$ given by inclusion. Those who are familiar with the elements of abstract algebra recognize $\mathscr{P}(X)$ with this structure as a *Boolean algebra*.

The set $A \triangle B = (A - B) \cup (B - A)$ is called the *symmetric difference* of A and B. If $A, B \in \mathscr{P}(X)$, then $A \triangle B$ being a subset of $A \cup B$ we have also $A \triangle B \in \mathscr{P}(X)$. There are several identities involving the operations $\triangle$ and $\cap$. One finds that $\mathscr{P}(X)$ is a *commutative ring* with respect to these operations which has an identity, namely X, and in which every element is idempotent. In other words, $\mathscr{P}(X)$ is a *Boolean ring* under the *addition* $\triangle$ and *multiplication* $\cap$. The operations $\cup$ and $\cap$ are often called "cup" and "cap" or "join" and "meet." The same terminology occurs in lattice theory.

Let A and B be nonvoid sets and let $a \in A$, $b \in B$. The existence of the *ordered pair* (a, b) is intuitively obvious and we may also speak about the set of all these pairs (a, b). This set is called the *product* of the ordered pair (A, B) and it will be denoted by $(A \times B)$. Thus,

$$(A \times B) = \{(a, b) : a \in A \quad \text{and} \quad b \in B\}.$$

If $A \neq B$, then $(A \times B)$ and $(B \times A)$ are distinct sets, but if $A = B$, then these factors play symmetric roles and we have only one product which we denote by $A \times A$ or A^2. In axiomatic set theory the ordered pair (a, b) is constructed as follows: One of the axioms which was explicitly mentioned earlier states the existence of nonordered pairs. In particular, it implies the existence of the sets $\{a\}$, $\{b\}$, and $\{a, b\}$. By the same principle we may form the nonordered pair $\{\{a\}, \{a, b\}\}$ which we call the *ordered pair* and denote by (a, b). This set really has all the characteristic properties attributed to an ordered pair: It is asymmetric and the nonordered pair $\{a, b\}$ is determined by (a, b), namely, it is its sum. An alternative definition of an ordered pair could be $(a, b) = \{\{a\}, \{\{b\}\}\}$. *Ordered triples* can be easily defined in terms of ordered pairs: $(a, b, c) = (a, (b, c))$. More generally we can introduce *ordered n-tuples* by using ordered $(n - 1)$-tuples as follows:

$$(a_1, \ldots, a_n) = (a_1, (a_2, \ldots, a_n)).$$

The set

$$(A_1 \times \ldots \times A_n) = \{(a_1, \ldots, a_n) : a_1 \in A_1, \ldots, a_n \in A_n\}$$

is called the *product* of the ordered n-tuple $(A_1, \ldots, A_n)$ of sets $A_1, \ldots, A_n$. The sets $A_1, \ldots, A_n$ are the *factors* of the product $(A_1 \times \ldots \times A_n)$. It is important to realize that the product is defined only when the distinct sets of the finite family $A_1, \ldots, A_n$ have been arranged in a definite order. If all these sets coincide, say $A_1 = \ldots = A_n = A$, then the product $(A \times \ldots \times A) = A \times \ldots \times A = A^n$ is uniquely determined by A and the number of factors n.

Any subset R of the product set $(A \times B)$ is called a *binary relation* on the pair (A, B) or between the elements of the sets of the pair (A, B). If $A = B$ we speak about a binary relation on A or between the elements of A. If $(a, b) \in R$ we say that the relation R holds for the pair (a, b) and we express this fact by writing $a\,R\,b$. In practice, various other symbols might replace $a\,R\,b$, e.g., $a \perp b$, $a \parallel b$, $a \sim b$, or $a < b$, but even then it is worthwhile to interpret the relation as a particular subset R of $(A \times B)$.

The *inverse* of a binary relation R on an ordered pair (A, B) is defined as a binary relation on the pair (B, A):

$$R^{-1} = \{(b, a) : a \in A, \quad b \in B, \quad \text{and} \quad (a, b) \in R\}.$$

In the special case $A = B$ both R and R^{-1} are subsets of $A \times A$ and it might happen that $R = R^{-1}$ in which case R is called *symmetric*. The set

$$I = \{(a, a) : a \in A\}$$

is called the *diagonal* of the product $A \times A$. If R is such that $I \subseteq R$, then it is called a *reflexive* binary relation on the elements of A. *Antireflexivity* means that $a\ R\ a$ never holds and *antisymmetry* expresses the additional fact that at most one of the possibilities $a\ R\ b$ and $b\ R\ a$ can take place. For instance, if A is the set of all straight lines in the plane, then parallelism $\|$ is a symmetric and reflexive relation while orthogonality $\perp$ is symmetric and antireflexive. Set theoretical inclusion $\subset$ gives an example of an antisymmetric binary relation on the set of all subsets $\mathscr{P}(X)$ of a set X. If $a\ R\ c$ whenever $a\ R\ b$ and $b\ R\ c$, then R is called *transitive*. For instance, $\|$ and $\subset$ are transitive relations while $\perp$ is not.

Parallelism gives a simple example of one of the best-known types of binary relations: An *equivalence relation* is a reflexive, symmetric, and transitive relation on some set A. Another known type is *linear ordering*. This means an antisymmetric and transitive relation $<$ such that if $a \neq b$ then $a < b$ or $b < a$. If the last requirement is omitted we speak about an *antireflexive partial ordering*. A reflexive and transitive binary relation $\leqslant$ is called a *reflexive partial ordering*. Notice that $a \leqslant b$ and $b \leqslant a$ might hold simultaneously even if a and b are distinct elements of the set A. For instance, any equivalence relation is a reflexive partial ordering.

A *function* f on a set A with values in another set B can be most easily defined by its graph which is a subset of the product $(A \times B)$. A relation $F \subseteq A \times B$ will be called the *graph* of a *function* $f : A \to B$ if for any $a \in A$ there exists exactly one $b \in B$ such that $(a, b) \in F$. If the sets A and B are distinct no confusion can arise: The function $f : A \to B$ is determined by the ordering (A, B) which is now written as $(A \times B)$. It might happen that F and F^{-1} are both graphs in which case the associated functions are denoted by f and f^{-1}. If $A = B$, then f and f^{-1} are distinct or not accordingly as $F \neq F^{-1}$ or $F = F^{-1}$. If f^{-1} exists, then f is called *invertible* and f^{-1} is its *inverse*. By our definition F^{-1} is a graph only if for every $b \in B$ there is exactly one $a \in A$ such that $(a, b) \in F$. Thus an invertible function $f : A \to B$ maps A onto B and as such yields a *one-to-one correspondence* between the elements of A and B. Although a one-to-one correspondence could be viewed as a symmetric relation $A \rightleftarrows B$, it is preferable to keep the asymmetry so that a one-to-one correspondence is nothing but an invertible map $f : A \to B$. If distinct elements of A are mapped into distinct elements of B then $f : A \to B$ is called *injective* and if f maps A onto B then it is called *surjective*.

2. Set Theoretical Equivalence and Denumerability

In intuitive set theory the existence of infinite sets is taken for granted. For instance, one can be easily convinced that the natural numbers 1, 2, 3, ... may be collected in a single family $\{1, 2, 3, ...\}$ which is a set. As soon as an infinite set is given, others can be constructed by elementary set theoretic operations, e.g., by taking the set of all ordered pairs or the set of all subsets of the given set. For it is assumed that the ordered pairs (x, y) where $x \in X$ and $y \in Y$ can be considered as the elements of a single set $X \times Y$ and similarly there is a set $\mathscr{P}(X)$ called the *power set* of X which consists of the subsets of X:

$$\mathscr{P}(X) = \{A : A \subseteq X\}.$$

In axiomatic set theory the existence of the sets $\{1, 2, 3, ...\}$, $X \times Y$, and $\mathscr{P}(X)$ for any X, Y can be proved from the axioms. We can make an easy compromise by taking the existence of these sets as axioms.

The first significant set theoretic result of Cantor concerns a classification of infinite sets. Two sets A and B are called *equivalent*, or of the same *cardinality*, if there exists a one-to-one correspondence $f : A \to B$ between their elements. If such a one-to-one map exists between A and B we write $A \sim B$. It is clear that $A \sim A$ for any set A and also that $A \sim B$ implies $B \sim A$. A simple reasoning shows that if $A \sim B$ and $B \sim C$, then $A \sim C$. Two finite sets are equivalent if and only if they have the same number of elements. The concept of equivalence is of primary importance in the case of infinite sets when it can be used to distinguish between various infinite sets.

Following Cantor we prove:

No set X is equivalent to its power set $\mathscr{P}(X)$.

It will be sufficient to prove the following proposition: If $\mathscr{Q}$ is a subset of $\mathscr{P}(X)$ whose elements can be brought into a one-to-one correspondence $f : X \to \mathscr{Q}$ with the elements of X, then $\mathscr{Q}$ is a proper subset of $\mathscr{P}(X)$. In order to construct a subset A of X not belonging to $\mathscr{Q}$ we consider the image points $f(x)$ and distinguish between the possibilities $x \in f(x)$ and $x \notin f(x)$. Thus we define

$$A = \{x : x \in X \quad \text{and} \quad x \notin f(x)\}.$$

By the one-to-one correspondence $f : X \to \mathscr{Q}$ for every Q in $\mathscr{Q}$ there is a unique x in X such that $f(x) = Q$. If $x \in f(x) = Q$, then $x \notin A$, so $A \neq Q$, and if $x \notin f(x) = Q$, then $x \in A$, so again $A \neq Q$. Thus A is not an element of $\mathscr{Q}$ and consequently $\mathscr{Q}$ is a proper subset of $\mathscr{P}(X)$.

Cantor's theorem shows that there exist nonequivalent infinite sets. For example, $\{1, 2, 3, ...\}$ is not equivalent to the set of its subsets.

If a set X is equivalent to $\{1, 2, 3, ...\}$, then it is called *denumerable*. Thus if X is denumerable then its elements can be arranged in a sequence and named $x_1, x_2, x_3, \ldots$. Of course there is an infinity of ways in which the elements of X can be enumerated. Denumerability means the existence of such enumerations without the selection of any particular one of them. By choosing specific enumerations one can easily prove the following propositions:

If A and B are denumerable, then so are $A \cup B$ and $(A \times B)$.

If $\mathscr{A}$ is denumerable and if every element A of $\mathscr{A}$ is a denumerable set, then so is $\bigcup \mathscr{A}$.

For instance, if $a_1, a_2, a_3, \ldots$ and $b_1, b_2, \ldots$ are enumerations of A and B, then the map f given by the rule

$$f((a_k, a_l)) = \frac{(k+l-1)(k+l-2)}{2} + \frac{1+(-1)^{k+l}}{2}k + \frac{1+(-1)^{k+l-1}}{2}l$$

gives an enumeration of $(A \times B)$. Schematically we enumerate $(A \times B)$ in the following fashion:

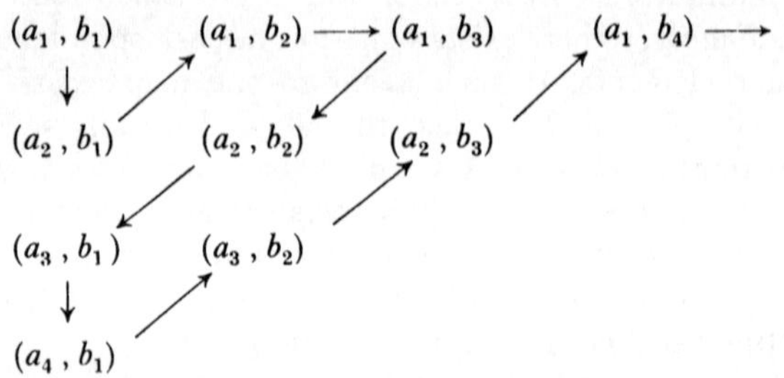

It is convenient to call a set *countable* if it is finite or denumerable. One readily proves the following results:

The subsets of a countable set are all countable sets.

If A and B are countable sets, then so is $(A \times B)$.

The union of countably many countable sets is countable.

The last shows that the set of rational numbers R is denumerable. Indeed R is the union of $\{0\}$, the denumerably many sets

$$R_m^+ = \left\{\frac{1}{m}, \frac{2}{m}, \frac{3}{m}, \ldots\right\} \qquad (m = 1, 2, 3, \ldots)$$

and

$$R_m^- = \left\{\frac{-1}{m}, \frac{-2}{m}, \frac{-3}{m}, \ldots\right\} \qquad (m = 1, 2, 3, \ldots)$$

each of which is obviously denumerable.

A complex number α is called *algebraic* over the field of rationals R if it is the root of an equation

$$r_0 + r_1 x + \ldots + r_{n-1} x^{n-1} + x^n = 0$$

where $r_0, \ldots, r_{n-1} \in R$. The lowest admissible degree n is called the degree of the algebraic number α. The set of ordered n-tuples $(r_0, \ldots, r_{n-1})$ is denumerable and each nth degree equation has at most n distinct roots. Hence the algebraic numbers of degree at most n form a denumerable set A_n. By considering the denumerable union $A_1 \cup A_2 \cup \ldots$ we obtain:

The set of all algebraic numbers is denumerable.

A nonalgebraic complex number is called *transcendental.* Since the algebraic numbers form a denumerable set the existence of such numbers can be proved by showing that the set of complex numbers contains a noncountable subset. For instance, it is sufficient to show that the set of reals x satisfying $0 < x \leqslant 1$ is not countable. Every such x has a unique representation of the form

$$x = \frac{\epsilon_1}{3} + \frac{\epsilon_2}{3^2} + \frac{\epsilon_3}{3^3} + \ldots$$

where $\epsilon_k = 0, 1$, or 2 and infinitely many ϵ_k's are different from 0. We are going to prove that every denumerable subset of $\{x : x \in \mathfrak{R}$ *and* $0 < x \leqslant 1\}$ is a proper subset. Since our set is not finite this will show that it is not countable and so the existence of transcendental numbers will be proved. Let a definite enumeration $x_1, x_2, \ldots$ of such a denumerable subset be given and let $\epsilon_{k1}, \epsilon_{k2}, \epsilon_{k3}, \ldots$ be the digits of x_k. We let $\epsilon_k = 1 + 2\epsilon_{kk} - \epsilon_{kk}^2$ so that $\epsilon_k = 1$ if $\epsilon_{kk} = 0$ or 2 and $\epsilon_k = 2$ if $\epsilon_{kk} = 1$. All the ϵ_k's are different from 0 and $\epsilon_k \neq \epsilon_{kk}$ for every $k = 1, 2, 3, \ldots$. Therefore the real number x $(0 < x \leqslant 1)$ defined by the infinite series $\Sigma\, \epsilon_k 3^{-k}$ is different from every x_k and so $\{x_1, x_2, \ldots\}$ is indeed a proper subset of $\{x : x \in \mathfrak{R}$ *and* $0 < x \leqslant 1\}$.

If $\mathscr{A}$ and $\mathscr{B}$ are countable sets of sets, then $\{A \cup B : A \in \mathscr{A}$ and $B \in \mathscr{B}\}$ is also countable.

Proof. Since $\mathscr{A}$ is countable its elements can be arranged in a finite or infinite sequence $(A_1, A_2, \ldots)$. For any fixed B consider the sequence $(A_1 \cup B, A_2 \cup B, \ldots)$. By omitting possible repetitions we are led to an enumeration of the set $\{A \cup B : A \in \mathscr{A}\}$ where B is a fixed element of $\mathscr{B}$. As B varies over the countable set $\mathscr{B}$ we obtain a countable family of these sets $\{A \cup B : A \in \mathscr{A}\}$ $(B \in \mathscr{B})$. Since each of them is countable, so is their union.

For any set A we let $A^{(n)}$ denote the set of those subsets of A which contain at most n elements. Thus $A^{(0)} = \{\emptyset\}$, $A^{(1)} = \{\{a\}: a \in A\}$ and

$$A^{(n+1)} = \{a^{(n)} \cup a^{(1)} : a^{(n)} \in A^{(n)} \quad \text{and} \quad a^{(1)} \in A^{(1)}\}$$

for $n \geqslant 0$. If A is countable, then by the foregoing result so is $A^{(2)}$. This in turn shows that $A^{(3)}$ is a countable set and by induction we see that $A^{(n)}$ is countable for every $n = 0, 1, 2, \ldots$. The union of these sets $A^{(n)}$ $(n = 0, 1, \ldots)$ is the *set of all finite subsets* of A. Hence we proved:

If A is countable, then so is the set of all finite subsets of A.

It is interesting to compare this with the earlier result which states that the set consisting of all subsets of a denumerable set is not countable.

The existence of nonequivalent infinite sets was proved from Cantor's theorem according to which no set is equivalent to its power set. However, this theorem alone is not sufficient to prove that there are many nonequivalent types among infinite sets, for one cannot be certain that X, $\mathscr{P}(X)$, $\mathscr{P}^2(X) = \mathscr{P}(\mathscr{P}(X))$, ... are nonequivalent. There are two ways by which this question can be settled, one of which consists of proving a stronger version of Cantor's theorem:

No set X is equivalent to $\mathscr{P}^k(X)$ for any $k = 1, 2, \ldots$.

Proof. We introduce the notation $a^0 = a$ and $a^k = \{a^{k-1}\}$ for $k > 1$ and $a \in X$ so that a^k consists of the single element $a^{k-1} \in \mathscr{P}^{k-1}(X)$. The following argument is a straightforward generalization of the one used in the special case $k = 1$: Given a one-to-one correspondence $f: X \to \mathscr{Q}^k$ between X and some subset $\mathscr{Q}^k$ of $\mathscr{P}^k(X)$ we define

$$\mathscr{A}^{k-1} = \{x^{k-1} : x^{k-1} \in \mathscr{P}^{k-1}(X) \quad \text{and} \quad x^{k-1} \notin f(x)\}.$$

Each $\mathscr{Q}^{k-1} \in \mathscr{Q}^k$ is the image of some element of X under the map f, say $\mathscr{Q}^{k-1} = f(x)$, for some $x \in X$. If $x^{k-1} \in \mathscr{Q}^{k-1} = f(x)$, then $x^{k-1} \notin \mathscr{A}^{k-1}$ and so $\mathscr{A}^{k-1} \neq \mathscr{Q}^{k-1}$. Conversely, if $x^{k-1} \notin \mathscr{Q}^{k-1} = f(x)$, then $x^{k-1} \in \mathscr{A}^{k-1}$ and again $\mathscr{A}^{k-1} \neq \mathscr{Q}^{k-1}$. Therefore $\mathscr{A}^{k-1}$ is not an element of $\mathscr{Q}^k$ and so $\mathscr{Q}^k$ is a proper subset of $\mathscr{P}^k(X)$. Hence there can be no one-to-one correspondence between X and $\mathscr{P}^k(X)$.

The foregoing result implies immediately that X; $\mathscr{P}(X)$, $\mathscr{P}^2(X)$, ... are always nonequivalent sets. By choosing $X = \{1, 2, 3, \ldots\}$ we obtain an infinite sequence of sets, no two of which are equivalent. The same can be proved from the special case $k = 1$ with the help of the following important theorem due to Cantor and Bernstein:

If X and X' are arbitrary sets such that X is equivalent to a subset of X' and also X' is equivalent to a subset of X, then $X \sim X'$.

Proof. By hypothesis there is a one-to-one mapping f from X onto a subset of X' and there is another one-to-one mapping g from X'

onto a subset of X. Using f and g we divide the sets X and X' into mutually disjoint subsets such that there is a natural one-to-one correspondence between the subsets of X and the subsets of X': Given any element $a \in X$ we define the elements $\dots, a_{-2n}, \dots, a_{-2}, a_0, a_2, \dots, a_{2n}, \dots$ belonging to X and $\dots, a_{-2n+1}, \dots, a_{-1}, a_1, \dots, a_{2n-1}, \dots$ belonging to X' by induction as follows: We put $a_0 = a$ and a_1 is defined to be the image of a_0 under the mapping f; furthermore, a_2 is the image of a_1 under g. In general we let a_{2n-1} $(n \geqslant 1)$ be the image of a_{2n-2} under f and a_{2n} be the image of a_{2n-1} under g. This construction is valid for every $n = 1, 2, \dots$ but it is possible that some of the a_n's coincide. For negative indices we define the elements as follows: a_{-1} is defined to be the element of X' (if there is any such element) whose image under g is a_0. If a_{-1} exists, then a_{-2} is defined to mean the element belonging to X (if there is such an element) whose image under f is a_{-1}. In general we define a_{-2n+1} such that its image under g is a_{-2n+2} provided such an element exists in X'. Similarly, a_{-2n} is defined to mean the unique element in X (if it exists) whose image under f is a_{-2n+1}. The set of even elements a_{2n} we denote by $[a]$ and the set of odd elements will be noted by $[a]'$, so that $[a] \subseteq X$ and $[a]' \subseteq X'$.

Since f and g are one-to-one maps, by the definition of $[a]$ any two sets $[a]$ and $[b]$ are either disjoint or identical. A similar statement holds for the sets $[a]'$ and $[b]'$. Hence every element of X belongs to exactly one set $[a]$ and every element of X' belongs to exactly one set $[a]'$, namely, a of X belongs to $[a]$ and a' of X' belongs to $[a]'$ where a is the image of a' under the mapping g. Therefore it is sufficient to construct a one-to-one map between the elements of each set $[a]$ and the corresponding $[a]'$ as this will define a one-to-one correspondence between the sets X and X'.

There are two different types of $[a]$ sets: First, it is possible that no repetition occurs in the sequence (a_{2n}) and consequently no repetition takes place in (a_{2n-1}). Then both $[a]$ and $[a]'$ consist of denumerably many distinct elements and so they can be brought into a one-to-one correspondence. The other possibility is that there is a repetition in $[a]$, say $a_{2m} = a_{2n}$, for some $m \neq n$. Then $a_{2m+1} = a_{2n+1}$ and hence $a_{2m+2} = a_{2n+2}$ and in general $a_{2m+k} = a_{2n+k}$ for every $k = 0, 1, \dots$. Similarly, $a_{2m-1} = a_{2n-1}$ and $a_{2m-2} = a_{2n-2}$ and so $a_{2m-k} = a_{2n-k}$ for every $k = 0, 1, \dots$. Given m, choose n such that the elements $a_{2m}, a_{2m+2}, \dots, a_{2n-2}$ are all distinct. Then $a_{2m+1}, a_{2m+3}, \dots, a_{2n-1}$ are distinct elements of X' and

$$a_{2m} \to a_{2m+1}, \quad a_{2m+2} \to a_{2m+3}, \dots, a_{2n-2} \to a_{2n-1}$$

determines a one-to-one correspondence between the elements of $[a]$

and $[a]'$. Consequently one can define a one-to-one map between the elements of $[a]$ and $[a]'$ for every a in X and these maps automatically yield a one-to-one map between the elements for X and X'.

We can now easily give a second proof for the existence of infinitely many nonequivalent infinite sets. It is sufficient to show that a set X is not equivalent to any of its power sets $\mathscr{P}^k(X)$ where $k = 1, 2, \ldots$. For $k = 1$ this proposition holds by Cantor's theorem. We suppose that none of the sets $\mathscr{P}(X), \ldots, \mathscr{P}^k(X)$ are equivalent to X and prove that X and $\mathscr{P}^{k+1}(X)$ are inecuivalent sets. For on the one hand there is a natural one-to-one correspondence between X and a subset of $\mathscr{P}^k(X)$ mapping x of X into $f(x) = x^k$. A similar one-to-one map, called the *injection map*, exists between $\mathscr{P}^k(X)$ and a subset of $\mathscr{P}^{k+1}(X)$, namely, each element $\mathscr{Q}^{k1-}$ of $\mathscr{P}^k(X)$ corresponds to the set $\{\mathscr{Q}^{k-1}\}$ lying in $\mathscr{P}^{k+1}(X)$. If there were a one-to-one correspondence between $\mathscr{P}^{k+1}(X)$ and X, then it could be combined with the injection map to give a one-to-one map g between $\mathscr{P}^k(X)$ and a subset of X. The existence of the maps $f: X \to \mathscr{P}^k(X)$ and $g: \mathscr{P}^k(X) \to X$ would imply the equivalence of X and $\mathscr{P}^k(X)$ which is contradictory to our hypothesis. Hence X and $\mathscr{P}^{k+1}(X)$ are not equivalent sets.

If A and B are sets and if there exists a one-to-one map between A and some subset of B, then we write $A \leqslant B$ and say that the *cardinality* of A is at most as large as that of B. Clearly $A \sim B$ implies $A \leqslant B$ and the last theorem states that $A \leqslant B$ and $B \leqslant A$ together imply that $A \sim B$. If $A \leqslant B$ but A and B are not equivalent, we write $A < B$ and say that A is of smaller *cardinality* than B. Hence $A < B$ takes place if and only if A can be mapped in a one-to-one fashion onto some subset of B, but not conversely. For instance, the cardinality of the rational numbers is smaller than that of the reals.

It follows immediately from the definition of the relation $\leqslant$ that $A \leqslant B$ and $B \leqslant C$ imply $A \leqslant C$. Moreover, the last theorem shows that $A < B$ and $B \leqslant C$ or $A \leqslant B$ and $B < C$ imply $A < C$. It is natural to expect that for any pair of sets A, B exactly one of the three possibilities $A < B$, $A \sim B$, and $B < A$ takes place. Since, as we have already seen, at most one of these relations can hold, the real problem is to show that at least one of the two possibilities $A \leqslant B$ and $B \leqslant A$ will take place for any two sets A and B. The proof of this simple sounding fact is considerably harder than anything else we have seen so far. For the proof essentially depends on an important and deep axiom of set theory called the *axiom of choice*. Once we are familiar with the various equivalent formulations of this axiom the *trichotomy* problem "$A < B$ or $A \sim B$ or $B < A$" will easily be settled.

3. The Axiom of Choice and Its Equivalents

The most famous and important axiom of set theory concerns nonempty sets $\mathscr{A}$ whose elements A are themselves nonvoid sets. A function f defined on $\mathscr{A}$ with values in $\bigcup \mathscr{A}$ is called a *choice function* for $\mathscr{A}$ if $f(A) \in A$ for every $A \in \mathscr{A}$. Indeed such a function $f : \mathscr{A} \to \bigcup \mathscr{A}$ "chooses" an element $f(A)$ from each of the sets A belonging to $\mathscr{A}$, and conversely if we claim that there is some process by which an element can be selected simultaneously from each of the sets A of the family $\mathscr{A}$, then we merely say that some kind of choice function f is given. The *axiom of choice* states:

If neither $\mathscr{A} = \{A\}$ nor any of its elements is void, then there exists at least one choice function for $\mathscr{A}$.

This is a very simple sounding requirement whose validity had never been questioned until 1904 when Zermelo showed that it has far reaching consequences, in particular, the so called *well-ordering principle* can be derived from it. A linearly ordered set X is called *well ordered* if every nonvoid subset A of X has a first element, i.e., an element a such that $a \leqslant x$ for every $x \in A$ or equivalently $a < x$ for every $x \neq a$ and lying in A. For instance, the set of natural numbers is well ordered, but the integers do not form a well-ordered set under their natural ordering. Cantor firmly believed in the following principle:

Every nonvoid set admits at least one linear ordering which well orders it.

Subsequent studies have shown that the axiom of choice and the well-ordering principle are actually equivalent propositions, each one can be derived from the other one by using, also, the more elementary axioms of set theory. Today, already many equivalent formulations are known and there are also several important theorems which we cannot prove without using the axiom of choice or one of its known equivalents. Some of these theorems, like the Stone representation theorem, might actually turn out to be equivalents of the axiom of choice.

The closest equivalent of the axiom of choice is the *product axiom:* Let I be a nonvoid set and let a nonvoid set A_i be associated with each element i of I. This means that a set $\mathscr{A}$ of nonvoid sets A and a function $\varphi : I \to \mathscr{A}$ are given, the function values being denoted by A_i instead of $\varphi(i)$. We may suppose that every A in $\mathscr{A}$ is attained at some element i of I so that $\mathscr{A}$ is the range of $\varphi : I \to \mathscr{A}$. The *product axiom* or *multiplicative axiom* states:

If I and the associated sets A_i are not void, then there exists at least one function $f : I \to \bigcup \mathscr{A} = \bigcup A_i$ such that $f(i) \in A_i$ for every $i \in I$.

The equivalence of the two axioms is clear: On the one hand, the axiom of choice concerns the special situation when the associated sets A_i are all distinct. On the other hand, if $\mathscr{A}$ denotes the set of A_i's then every choice function $f : \mathscr{A} \to \bigcup \mathscr{A}$ gives rise to a function $f : I \to \bigcup A_i$, the values of which are given by the rule $f(i) = f(A_i)$.

The family of functions $f : I \to \bigcup \mathscr{A}$ such that $f(i) \in A_i$ is a set which is called the *product* or *Cartesian product* of the sets A_i and is denoted by $\prod A_i$ or sometimes by $\times A_i$. The product axiom states that if none of the sets A_i is void, then $\prod A_i$ is not the empty set. The product $\prod A_i$ is defined also in the degenerate case when at least one of the sets A_i is void: Then we let $\prod A_i$ be the empty set $\emptyset$. It is important to realize that the family $\mathscr{A}$ alone does not yet determine the product. For instance, if $I = \{1, 2\}$ and if $\mathscr{A}$ consists of two elements, then there are two distinct products $\prod A_i$. In the case of a small finite index set it is convenient to write $A_1 \times \dots \times A_n$ instead of $\prod A_i$. Notice that $A_1 \times \dots \times A_n$ *is not the set of ordered n-tuples* $(A_1 \times \dots \times A_n)$. For $A_1 \times \dots \times A_n$ depends on the particular index set I used while $(A_1 \times \dots \times A_n)$ is uniquely determined by the ordering $(A_1, \dots, A_n)$. Moreover, the Cartesian product is defined also if the number of factors is infinite. In practice, one does not distinguish between the set of ordered n-tuples $(A_1 \times \dots \times A_n)$ and the Cartesian product $A_1 \times \dots \times A_n$ associated with the standard index set $\{1, \dots, n\}$.

The sets A_i are called the *factors* of the product $\prod A_i$. The typical element of $\prod A_i$ is usually denoted by the symbol a, its value at i is called the ith *coordinate* of a and is denoted by a_i instead of $a(i)$. A particularly simple situation occurs when all the factors are identical, say $A_i = A$ for every index i. Then the product is uniquely determined by I and A and it is denoted by A^I. For example, $\mathfrak{R}^{\mathfrak{C}}$ denotes the set of all real-valued functions defined over the domain of complex numbers.

One of the most useful equivalents of the axiom of choice is *Zorn's lemma.* If the axiom of choice is needed in the course of a proof, then the proof can generally be reformulated more elegantly by using Zorn's lemma instead. A similar pattern was followed by researchers working before the discovery of Zorn's lemma in 1922 and again in 1935, except they had to be satisfied by using the well-ordering principle in their reasonings. Such proofs were said to depend on *transfinite induction.*

Zorn's lemma concerns partially ordered sets $\mathscr{A}$ and it is best to suppose that the ordering $\leqslant$ is reflexive. A linearly ordered subset of such a set $\mathscr{A}$ is usually called a *chain* of $\mathscr{A}$. By an *upper bound* of a chain $\mathscr{C}$ we mean any element U in $\mathscr{A}$ such that $C \leqslant U$ for every C in $\mathscr{C}$. Thus an upper bound is comparable with each element of $\mathscr{C}$ and it need not belong to the chain $\mathscr{C}$. An element M of $\mathscr{A}$ is called a *maximal element* of $\mathscr{A}$ if $A \leqslant M$ for all those elements A of $\mathscr{A}$ which are comparable with M. Zorn's lemma states:

If the partially ordered set $\mathscr{A}$ *is such that every chain has an upper bound in* $\mathscr{A}$, *then* $\mathscr{A}$ *has at least one maximal element.*

In order to show how Zorn's lemma is used in practice, it is best to discuss a few applications. A simple illustration is obtained by considering a nonvoid family D of disks d lying in the plane and asking for a subfamily M such that

(1) *the disks* d *of* M *are disjoint, and*

(2) M *is maximal with respect to this property.*

After realizing that the solution of the problem requires more than trying to pick disks one by one until one hits on M, the *first* thing to do is to collect all subfamilies A of D satisfying requirement (1) into one set $\mathscr{A}$. The *second* step consists of checking that $\mathscr{A}$ is not void. Here this is trivial. *Third*, it is necessary to introduce a reflexive partial ordering $\leqslant$ in $\mathscr{A}$ which is suitable for the purpose. In our example the right partial ordering is given by inclusion, so $A_1 \leqslant A_2$ if A_1 is a subfamily of A_2. The *fourth* step consists of proving that every chain $\mathscr{C}$ of $\mathscr{A}$ has an upper bound U in $\mathscr{A}$. In our example it is sufficient to choose as U the union of all families C of disks belonging to the chain $\mathscr{C}$. In more involved applications it might be harder to find an upper bound U and one should carefully check that U indeed lies in $\mathscr{A}$ and not only in some longer partially ordered set enveloping $\mathscr{A}$. By Zorn's lemma $\mathscr{A}$ contains a maximal element M. Since $\mathscr{A}$ is ordered by inclusion and it consists of families satisfying (1), M will be a solution of our problem.

The next illustrative example is an important application as it shows, among other things, the existence of *maximal orthonormal systems.* Let X be a nonvoid set and let R be a reflexive and symmetric binary relation on X, i.e., let $R \subseteq X \times X$ be such that $I \subseteq R$ and $R^{-1} = R$. The object is to determine a subset M of X such that

(1) $M \times M \subseteq R$, i.e., $x_1 \, R \, x_2$ *holds for any* $x_1, x_2 \in M$, *and*

(2) M *is maximal with respect to this property.*

In order to prove the existence of such maximal subsets, we apply Zorn's lemma to the set $\mathscr{A}$ of all those subsets A of X for which $A \times A \subseteq R$. The set A is not void as it contains every $A = \{a\}$ consisting of a single element a of X. The proper partial ordering of $\mathscr{A}$ is set theoretic inclusion so that $A_1 \leqslant A_2$ means $A_1 \subseteq A_2$. If $\mathscr{C}$ is a chain in $\mathscr{A}$, then any possible upper bound of $\mathscr{C}$ will necessarily contain $U = \bigcup\{C : C \in \mathscr{C}\}$. We prove that the set U lies in $\mathscr{A}$: If $x_1, x_2 \in U$, then $x_1 \in C_1$ and $x_2 \in C_2$ for some sets $C_1, C_2 \in \mathscr{C}$ and $\mathscr{C}$ being a chain either $x_1, x_2 \in C_2 \supseteq C_1$ or $x_1, x_2 \in C_1 \supseteq C_2$. Hence by $C_1, C_2 \in \mathscr{A}$ we have in either case $x_1 \, R \, x_2$ and this shows that $x_1 \, R \, x_2$ holds for any two elements x_1, x_2 of U. Since U lies in $\mathscr{A}$ and is an upper bound of $\mathscr{C}$, Zorn's lemma applies and the existence of a maximal set M follows. Actually we proved more than is necessary to apply Zorn's lemma: We proved that every chain $\mathscr{C}$ has a least upper bound M in $\mathscr{A}$.

A very simple application of the last result is the following: Let X be the set of all straight lines of the ordinary three-dimensional space and let R be the perpendicularity-identity relation: $x_1 \, R \, x_2$ if, and only if, $x_1 = x_2$ or if x_1 and x_2 are perpendicular. Our result states the existence of a set of lines $M = \{x\}$ such that any two lines of the family M are perpendicular, and if $x \notin M$ then x is not perpendicular to at least one line of the family M. The same type of reasoning shows the existence of maximal orthonormal systems in inner product spaces.

Let V be a vector space over some field F. For instance, V can be the set of real numbers, F the field of rationals. For vector addition we can choose ordinary addition of reals and as multiplication of vectors by elements of $F = R$ we can use ordinary multiplication of reals by rationals. A *linearly independent set* means any subset L of V such that no nontrivial linear combination of elements of L gives the zero element of V. In other words, L is linearly independent over F if $v_1, ..., v_n \in L$ and $\lambda_1 v_1 + ... + \lambda_n v_n = 0$ with $\lambda_1, ..., \lambda_n$ in F imply that $\lambda_1 = ... = \lambda_n = 0$. There exist linearly independent sets; for instance, every set consisting of one nonzero vector is linearly independent over the scalar field F. A *maximal linearly independent set* means a linearly independent set M which is not contained in any other linearly independent set, i.e., M is such that $M \subseteq L$ implies $L = M$. The existence of such maximal sets M follows from Zorn's lemma: We let $\mathscr{A}$ be the set of all linearly ordered subsets of V and partially order it by inclusion. Each chain $\mathscr{C}$ of $\mathscr{A}$ has an upper bound in $\mathscr{A}$, for $U = \bigcup\{L : L \in \mathscr{C}\}$ belongs to $\mathscr{A}$ and is the least upper bound of $\mathscr{C}$: If $v_1, ..., v_n \in U$, then $v_i \in L_i$ for some $L_i \in \mathscr{C}$ $(i = 1, ..., n)$ and $\mathscr{C}$ being linearly ordered there is one among the L_i's which contains all the others. Since $v_1, ..., v_n$ belongs to this largest L_i a linear combination $\lambda_1 v_1 + ... + \lambda_n v_n$ can vanish only if $\lambda_1 = ... = \lambda_n = 0$.

A *base* for a vector space V over a field F means a subset B of V such that every element v of V can be expressed in the form

$$v = \lambda_1 v_1 + ... + \lambda_n v_n$$

where $v_1, ..., v_n \in B$ and $\lambda_1, ..., \lambda_n \in F$. Usually one is interested in bases B such that the representation of every vector v is unique. The uniqueness requirement holds if, and only if, B is a linearly independent set in which case one speaks about a *linearly independent base.* Every such base is a maximal linearly independent set. Conversely, if M is a maximal linearly independent set and v is an arbitrary vector outside of M, then

$$\lambda v + \lambda_1 v_1 + \dots + \lambda_n v_n = 0$$

for suitable $v_1, \dots, v_n \in M$ and scalars $\lambda, \lambda_1, \dots, \lambda_n$ not all of which are 0. Since M itself is independent, $\lambda \neq 0$ and so v is expressible as a linear combination of $v_1, \dots, v_n$. Thus for vector spaces maximal linearly independent sets and linearly independent bases mean the same thing. Our earlier reasoning shows the existence of linearly independent bases for arbitrary vector spaces V over some field F. The special case when V is the additive group of the reals and F is the field of rationals is of special interest. The corresponding linearly independent bases are called *Hamel bases,* after their discoverer. They can be used to find discontinuous solutions for the functional equation $f(x + y) = f(x) + f(y)$ or to construct sets which are not measurable in the Lebesgue sense.

It was mentioned already that the *trichotomy law of cardinals* follows from the axiom of choice. As a last illustration of the power and versatility of this axiom, we derive the trichotomy property from Zorn's lemma: Given two nonvoid sets S_1 and S_2 we wish to prove the existence of a one-to-one map of S_1 into S_2 or of S_2 into S_1. If both types of maps exist, then by the Cantor-Bernstein theorem $S_1 \sim S_2$, otherwise we shall have $S_1 < S_2$ or $S_2 < S_1$. We start with the set $\mathscr{A}$ of graphs F of invertible functions from S_1 into S_2. By this we mean functions whose domain of definition is part of S_1 and whose range lies in S_2. Since S_1 and S_2 are nonvoid, such graphs F exist. We order $\mathscr{A}$ by letting $F_1 \leqslant F_2$ if $F_1 \subseteq F_2$; that is, if the second function is an extension of the first. If $\mathscr{C}$ is a chain in $\mathscr{A}$, then $\bigcup \{F : F \in \mathscr{C}\}$ is an upper bound of every F in $\mathscr{C}$ and it lies in $\mathscr{A}$. The hypothesis of Zorn's lemma being satisfied we can find a maximal element M in $\mathscr{A}$. Let

$$M_1 = \{s_1 : (s_1, s_2) \in M\} \qquad \text{and} \qquad M_2 = \{s_2 : (s_1, s_2) \in M\}.$$

Since $M_1 \subset S_1$ and $M_2 \subset S_2$ together contradict the maximality of M, we have $M_1 = S_1$ or $M_2 = S_2$, possibly both. Thus M is the graph of a map of S_1 into S_2 or M^{-1} is the graph of a map of S_2 into S_1. We proved that $S_1 \leqslant S_2$ or $S_2 \leqslant S_1$.

NOTES

The two papers by Cantor mentioned in the beginning are both entitled "Beiträge zur Begründung der transfiniten Mengenlehre" [1]. They were translated into English and provided with a very interesting historical introduction and remarks by Philip E. B. Jourdain [2]. The membership relation which we denote by $\in$ was originally introduced by Peano who used ϵ as an abbreviation for the Greek word $\epsilon\sigma\tau\iota$.

The informal set theory developed here should not be confused with strictly axiomatic set theories. The axioms mentioned here belong to the formal systems developed by a number of outstanding mathematicians. At this stage the reader is urged to continue the present text and start the study of general topology. Later, however, he might want to get acquainted with some of the formal axiom systems for set theory. He might then consult the series of articles by Bernays [3] and the beginning of Gödel's book [4].

Zermelo wrote two papers on well ordering. His first proof of the well-ordering theorem [5] is clearer to follow than the second [6]. Although the first proof was correct it had been criticized by several people and in order to answer these objections Zermelo published the second version which avoids these critical steps in the reasoning.

There are several closely related variants of Zorn's lemma known as Kneser's lemma, Kuratowski's lemma, Tukey's lemma, and Hausdorff's maximal principle. It is fair to point out that Kuratowski's lemma [7] is not only simple in form ("each chain of a partially ordered set is contained in some maximal chain") but was also published considerably earlier than the version due to Zorn [8]. Kneser's formulation is analogous to Zorn's but his hypothesis is considerably weaker: It is required only that well-ordered chains should have an upper bound in the partially ordered set [9]. An even weaker hypothesis appears in Bourbaki's version [10].

The axiom of choice is needed in the derivation of many interesting results, or at least it makes the proofs easier. In addition to the applications already discussed in the text we add the following: The existence of algebraic closure; Banach limits; Tychonoff's compactness theorem; the existence of invariant measure; the existence of nonmeasurable sets; the Hahn-Banach theorem; the existence of noncontinuous solutions of the functional equation $f(x + y) = f(x) + f(y)$; the Banach-Tarski paradox and the existence of ultrafilters. Tychonoff's theorem is known to be equivalent to the axiom of choice while the existence of invariant measures on topological groups can be proved also without using the axiom of choice. It was anounced recently that the existence of ultrafilters can also be proved without applying this axiom.

References

1. G. Cantor, *Math. Ann.* **46**, 481–512 (1895); **49**, 207–246 (1897).
2. G. Cantor, "Contributions to the Founding of the Theory of Transfinite Numbers." Dover, New York.
3. P. Bernays, A system of axiomatic set theory, Parts I–VII. *J. Symb. Logic* **2**, 65–77 (1937); **6**, 1–17 (1941); **7**, 65–89, 133–145 (1942); **8**, 89–106 (1943); **13**, 65–79 (1948); **19**, 81–96 (1954).
4. K. Gödel, "The Consistency of the Axiom of Choice and of the Generalized Continuum-Hypothesis with the Axioms of Set Theory" (Ann. Math. Studies, No. 3) Princeton Univ. Press, Princeton, New Jersey, 1940.
5. E. Zermelo, Beweis dass jede Menge wohlgeordnet werden kann. *Math. Ann.* **59**, 514–516 (1904).
6. E. Zermelo, Neuer Beweis für die Möglichkeit einer Wohlordnung. *Math. Ann.* **65**, 107–128 (1908).
7. C. Kuratowski, Une méthode d'élimination des nombres transfinis des raisonnements mathématiques. *Fund. Math.* **3**, 76–108 (1922).
8. M. Zorn, A remark on method in transfinite algebra. *Bull. Amer. Math. Soc.* **41**, 667–670 (1935).
9. H. Kneser, Eine direkte Ableitung des Zornschen Lemmas aus dem Auswahlaxiom. *Math. Z.* **53**, 110–113 (1950).
10. N. Bourbaki, "Éléments de mathématique." Part I. Livre I. Théorie des ensembles (Fascicule de résultats). (Actual. Sci. Ind., no. 846.) Hermann, Paris, 1939.

CHAPTER I

Topological Spaces

1. Open Sets and Closed Sets

We shall deal with mathematical systems which consist of a set X and a family of subsets of X which are subject to a few simple axioms. These systems are called topological spaces and the family $\mathcal{O}$ of subsets is the family of open sets.

Definition 1. *A topological space is a set X and a family of subsets O called the open sets of the space such that the following axioms are satisfied:*

(O.1) $$\emptyset \in \mathcal{O} \quad \textit{and} \quad X \in \mathcal{O}.$$

(O.2) $$\textit{If} \quad O_1 \in \mathcal{O} \quad \textit{and} \quad O_2 \in \mathcal{O}, \quad \textit{then} \quad O_1 \cap O_2 \in \mathcal{O}.$$

(O.3) $$\textit{If} \quad O_i \in \mathcal{O} \quad \textit{for every} \quad i \in I, \quad \textit{then} \quad \bigcup \{O_i : i \in I\} \in \mathcal{O}.$$

The second axiom implies that $\mathcal{O}$ contains with every finite collection $O_1, \ldots, O_n$ also the intersection $O_1 \cap \ldots \cap O_n$. Infinite intersections need not belong to $\mathcal{O}$ even if each factor O_i is an element of $\mathcal{O}$. The third axiom states that $\mathcal{O}$ contains all finite and infinite unions of sets $O_i \in \mathcal{O}$. If a family $\mathcal{O}$ subject to these axioms is given we say that a *topology* $\mathcal{T}$ is defined on X. As far as topology is concerned the particular method used to describe the family $\mathcal{O}$ is of no importance. We shall often use the expression *X is a topological space*. This means that X is a nonvoid set and a topology $\mathcal{T}$ is given on X.

It is possible to define a topological space on any nonvoid set X: One trivial way of defining open sets is by choosing $\mathcal{O} = \{\emptyset, X\}$. The space so obtained is called a *nondiscrete topological space*. Another trivial choice is $\mathcal{O} = \mathcal{P}(X)$, the set of all subsets of X, in which case every subset is open. Then we speak about a *discrete topological space*. If X contains more than one element, then the discrete and nondiscrete topological spaces formed on X will be distinct. If X is infinite, then we can define another simple topological space: We put in $\mathcal{O}$ the set $\emptyset$ and all those sets O whose complement $X - O$ is finite. In this case we say that X is topologized by the *topology of finite complements*.

The family of all possible topologies for a fixed set X can be ordered as follows: $\mathcal{T}_1$ is called *coarser* or *weaker* than $\mathcal{T}_2$ and $\mathcal{T}_2$ is called *finer* or *stronger* than $\mathcal{T}_1$ if the family $\mathcal{O}_1$ of open sets associated with $\mathcal{T}_1$ is a subfamily of the family $\mathcal{O}_2$ belonging to $\mathcal{T}_2$. In symbols, $\mathcal{T}_1 \leqslant \mathcal{T}_2$ and $\mathcal{T}_2 \geqslant \mathcal{T}_1$ if and only if $\mathcal{O}_1 \subseteq \mathcal{O}_2$. It is clear that $\leqslant$ is a reflexive ordering relation for the set of all topologies for X. If $\mathcal{T}_1 \leqslant \mathcal{T}_2$ and $\mathcal{T}_2 \leqslant \mathcal{T}_1$, then $\mathcal{T}_1$ and $\mathcal{T}_2$ are identical, that is to say, $\mathcal{O}_1 = \mathcal{O}_2$. If $\mathcal{T}_1 \leqslant \mathcal{T}_2$ but $\mathcal{T}_1$ and $\mathcal{T}_2$ are not identical, we say that $\mathcal{T}_1$ is *strictly coarser* than $\mathcal{T}_2$ and $\mathcal{T}_2$ is *strictly finer* than $\mathcal{T}_1$. We write $\mathcal{T}_1 < \mathcal{T}_2$ and $\mathcal{T}_2 > \mathcal{T}_1$. For instance, in case of an infinite set the topology of finite complements is strictly finer then the nondiscrete topology and it is strictly coarser then the discrete topology.

Definition 2. *The topologies $\mathcal{T}_x$ and $\mathcal{T}_y$ formed on the sets X and Y are called homeomorphic if there is a one-to-one correspondence between the elements of X and Y such that the open sets of $\mathcal{T}_x$ correspond to those of $\mathcal{T}_y$ and conversely. If $\mathcal{T}_x$ and $\mathcal{T}_y$ are homeomorphic we write $\mathcal{T}_x \sim \mathcal{T}_y$.*

The existence of a homeomorphism between two topologies formed on the same set X does not necessarily imply their identity. For example, let X be a finite set, say $\{x_1, \ldots, x_n\}$, and let the topologies $\mathcal{T}_i$ $(i = 1, \ldots, n)$ be defined as follows: A nonvoid set O is open if it contains the point x_i. It is clear that these topologies are homeomorphic, but not identical. The discrete and nondiscrete topologies of a set consisting of more than one element are not homeomorphic. If the set is infinite, there is a third nonhomeomorphic topology, namely, the topology of finite complements. The relation $\sim$ is reflexive, symmetric, and transitive; that is, $\mathcal{T}_x \sim \mathcal{T}_x$; $\mathcal{T}_x \sim \mathcal{T}_y$ implies $\mathcal{T}_y \sim \mathcal{T}_x$; and if $\mathcal{T}_x \sim \mathcal{T}_y$ and $\mathcal{T}_y \sim \mathcal{T}_z$, then $\mathcal{T}_x \sim \mathcal{T}_z$.

Definition 3. *A set C of a topological space X is called closed if its complement is open. The family of closed sets is denoted by $\mathcal{C}$.*

It is obvious that ø and X are open and also closed and there might be other such sets. For example, if the topology is discrete, then every subset is both open and closed. Sets having this twofold property will be called *open-closed* or *closed-open* sets. If the topology is nondiscrete or if it is the topology of finite complements on an infinite X, then the only open-closed sets are ø and X. If only the improper subsets ø and X are open-closed, then the topological space is called *connected*.

Theorem 1. *The union of finitely many closed sets is closed. The intersection of an arbitrary finite or infinite family of closed sets is closed.*

Proof. Let $\{C_i\}$ $(i \in I)$ be a family of closed sets. Then $c \bigcap C_i = \bigcup cC_i$ and $c \bigcup C_i = \bigcap cC_i$ where cC_i is open for every $i \in I$. Using axioms

(O.2) and (O.3) we see that $c \bigcap C_i$ is open and if I is finite then so is $c \bigcup C_i$.

Theorem 2. *A family $\mathscr{C}$ of subsets of a set X is the family of closed sets of some topology on X if and only if the following axioms are satisfied:*

(C.1) $\varnothing \in \mathscr{C}$ *and* $X \in \mathscr{C}$.

(C.2) *If* $C_1 \in \mathscr{C}$ *and* $C_2 \in \mathscr{C}$, *then* $C_1 \cup C_2 \in \mathscr{C}$.

(C.3) *If* $C_i \in \mathscr{C}$ *for every* $i \in I$, *then* $\bigcap \{C_i : i \in I\} \in \mathscr{C}$.

Proof. Axioms (O.1), (O.2), and (O.3) for the family $\mathscr{O}$ of sets $O = cC$ $(C \in \mathscr{C})$ follow from the above axioms by complementation. The necessity is a consequence of Theorem 1.

Axioms (C.1), (C.2), and (C.3) are often used as a starting point in place of axioms (O.1), (O.2), and (O.3).

We end this section with a very useful lemma:

Lemma 1. *A set A of a topological space X is open if and only if it contains with each point x an open set O_x containing x.*

Proof. If A is open, then we can choose $O_x = A$ for every $x \in A$. If on the other hand A is such that for every $x \in A$ there is an open set O_x satisfying $x \in O_x \subseteq A$, then by $A = \bigcup \{O_x : x \in A\}$ the set A is the union of open sets and so it is open.

EXERCISES

1. Let X be a nonvoid set and let A be a subset of X. Denote by $\mathscr{T}(A)$ the topology on X whose open sets are $\varnothing$ and the sets O which contain A. Show that:

(a) We have $\mathscr{T}(A_1) \geqslant \mathscr{T}(A_2)$ if and only if $A_1 \subseteq A_2$ and $\mathscr{T}(A_1) > \mathscr{T}(A_2)$ if and only if $A_1 \subset A_2$.

(b) If $\mathscr{T}$ is a topology on X such that $\mathscr{T}(A_i) \leqslant \mathscr{T}$ for $i = 1, \ldots, n$, then $\mathscr{T}(\bigcap A_i) \leqslant \mathscr{T}$.

(c) If $\mathscr{T}$ is a topology on X such that $\mathscr{T}(A_i) \geqslant \mathscr{T}$ for every $i \in I$, then $\mathscr{T}(\bigcup A_i) \geqslant \mathscr{T}$.

2. Let X be a set and let m be an infinite cardinal. Show that $\varnothing$ and the subsets O of X which satisfy card $cO \leqslant m$ form the open sets of a topology. Determine the closed sets.

3. Let X be a linearly ordered set having the least upper bound property and let $\mathscr{C}$ be the family consisting of $\varnothing$ and X and of all sets $[x, +\infty) = \{\xi : x \leqslant \xi\}$ where $x \in X$. Show that $\mathscr{C}$ is the family of closed sets for a topology on X.

4. Find a topology $\mathscr{T}$ different from the discrete and nondiscrete topologies and such that $\mathscr{O} = \mathscr{C}$.

(Let X be the set of reals and let O be open if $x \in O$ implies $-x \in O$.)

2. Interior, Exterior, Boundary, and Closure

In the special case when X is the real line or the plane with its usual topology these notions reduce to the well-known concepts which they generalize.

Definition 1. *Let X be a topological space and let $A \subseteq X$. The interior of A is the union of all open sets contained in A:*

$$A^i = \bigcup\{O : O \subseteq A\}.$$

The exterior of A is the union of all open sets not intersecting A:

$$A^e = \bigcup\{O : O \subseteq cA\}.$$

Therefore the interior of A is the largest open set contained in A and the exterior of A is the largest open set not intersecting A. It is clear that $A^e = (cA)^i$ and $A^i = (cA)^e$. Both A^i and A^e can be expressed in terms of closed sets. It is easy to see that A is open if and only if $A^i = A$ and A is closed if and only if $A^e = cA$. The exterior and interior never intersect.

Theorem 1. *The interior and exterior operators satisfy the following rules:*

$$\varnothing^i = \varnothing; \qquad X^i = X; \qquad A^i \subseteq A; \qquad A^{ii} = A^i; \qquad (A \cap B)^i = A^i \cap B^i.$$

$$\varnothing^e = X; \qquad X^e = \varnothing; \qquad A^e \subseteq cA; \qquad A^{ee} \supseteq A^i; \qquad (A \cup B)^e = A^e \cap B^e.$$

If $A \subseteq B$, then $A^i \subseteq B^i$ and $A^e \supseteq B^e$.

Proof. Most of these rules are obvious from the definitions. Since A^i is open $A^{ii} = (A^i)^i = A^i$. The inclusion $(A \cap B)^i \subseteq A^i \cap B^i$ follows from $A \cap B \subseteq A$ and $A \cap B \subseteq B$. The opposite inclusion is also obvious because $A^i \subseteq A$ and $B^i \subseteq B$ imply that $A^i \cap B^i$ is an open set contained in $A \cap B$. Finally, using the rule on the interior of an intersection we see that

$$(A \cup B)^e = (c(A \cup B))^i = (cA \cap cB)^i = (cA)^i \cap (cB)^i = A^e \cap B^e.$$

For infinite unions the equality need not hold and we can assert only the inclusion $(\bigcup A_\gamma)^e \subseteq \bigcap A_\gamma^e$. Similarly, for an infinite intersection only $(\bigcap A_\gamma)^i \subseteq \bigcap A_\gamma^i$ holds in general.

Definition 2. *A point $x \in X$ is a boundary point of the set A if every open set O_x containing x intersects both A and cA. The set of boundary points is called the boundary of A and is denoted by A^b.*

It is obvious that $A^b = (cA)^b$. Moreover, x is a boundary point of a set A if and only if $x \notin A^i \cup A^e$. Therefore the boundary A^b is the complement of $A^i \cup A^e$. Since A^i and A^e are disjoint we obtain:

Theorem 2. *For every set A of a topological space X its interior A^i, exterior A^e, and boundary A^b are disjoint sets whose union is X. The sets A^i and A^e are open and A^b is closed.*

A simple consequence is:

Theorem 3. *A set A is closed if and only if $A^b \subseteq A$ and it is open if and only if $A^b \subseteq cA$.*

Proof. If A is closed, then cA is open and so $cA = A^e = c(A^i \cup A^b) \subseteq cA^b$ which shows that $A^b \subseteq A$. Conversely, if $A^b \subseteq A$, then $A^i \cup A^b \subseteq A$ and using $X = A^i \cup A^b \cup A^e$ together with $A^e \subseteq cA$ we see that $A \subseteq A^i \cup A^b$. Hence $A = A^i \cup A^b$ and so $A = cA^e$ is closed.

Definition 3. *The closure $\bar{A}$ of a set A of a topological space X is the intersection of all closed sets containing $A : \bar{A} = \bigcap \{C : A \subseteq C\}$.*

Since the intersection of closed sets is closed, $\bar{A}$ is a closed set. For closed A we have $\bar{A} = A$ and conversely. Hence this is a characteristic property:

Theorem 4. *A set A is closed if and only if $\bar{A} = A$.*

We could have obtained this also from the following:

Theorem 5. *For every set A we have $\bar{A} = A^i \cup A^b$.*

Proof. Since $\bar{A} = \bigcap \{C : A \subseteq C\}$ we see that

$$c\bar{A} = \bigcup \{cC : cC \subseteq cA\} = \bigcup \{O : O \subseteq cA\} = A^e.$$

Therefore by Theorem 2 $\bar{A} = cA^e = A^i \cup A^b$.

The basic properties of the closure can be summarized as follows:

Theorem 6. *The closure operator satisfies the rules:*

$$\bar{\varnothing} = \varnothing; \qquad \bar{X} = X; \qquad A \subseteq \bar{A}; \qquad \bar{\bar{A}} = \bar{A}; \qquad \overline{A \cup B} = \bar{A} \cup \bar{B}.$$

If $A \subseteq B$, then $\bar{A} \subseteq \bar{B}$.

Note. We have $\overline{A_1 \cup \dots \cup A_n} = \bar{A}_1 \cup \dots \cup \bar{A}_n$ for every finite family of sets, but for an infinite family only $\bigcup \bar{A}_i \subseteq \overline{\bigcup A_i}$ holds in general. By $\bigcap A_i \subseteq A_i \subseteq \bar{A}_i$ $(i \in I)$ we have $\overline{\bigcap A_i} \subseteq \bigcap \bar{A}_i$ for any family but equality need not hold even if the number of sets is finite.

Proof. Rules $A \subseteq \bar{A}$ and $A \subseteq B \Rightarrow \bar{A} \subseteq \bar{B}$ are clear from the definition of the closure. Since ø, X, and $\bar{A}$ for any A are closed sets we have $\bar{ø} = ø$, $\bar{X} = X$, and $\bar{\bar{A}} = \bar{A}$. Finally, $\overline{A \cup B} = \bar{A} \cup \bar{B}$ can be proved by showing that $\overline{A \cup B} \subseteq \bar{A} \cup \bar{B}$ and $\overline{A \cup B} \supseteq \bar{A} \cup \bar{B}$: The first inclusion follows from the fact that $\bar{A} \cup \bar{B}$ is a closed set which contains $A \cup B$. To see the second we use $A \subseteq A \cup B$ and $B \subseteq A \cup B$ to obtain $\bar{A} \subseteq \overline{A \cup B}$ and $\bar{B} \subseteq \overline{A \cup B}$.

Definition 4. *A point x is an accumulation point of a set A if every open set containing x contains at least one point of A which is different from x.*

The set of all accumulation points of a set A is called the derived set of A and is denoted by A'.

We have the following:

Theorem 7. *A set A is closed if and only if $A' \subseteq A$.*

Proof. If A is closed, then cA is open and so no point of cA is an accumulation point of A. Therefore $A' \subseteq A$. Conversely, if $A' \subseteq A$, then for every $x \in cA$ there is an open set O_x such that $x \in O_x \subseteq cA$ and so by Lemma 1.1 cA is open.

This last result can also be derived from:

Theorem 8. *For every A we have $\bar{A} = A \cup A'$.*

Proof. It is sufficient to show that a point $x \notin A$ is an accumulation point of A if and only if $x \in A^b$. In fact, if $x \in A'$ and O_x is an open set containing x, then there is an $a \neq x$ in A such that $a \in O_x$. Hence if $x \notin A$, then O_x intersects both A and cA and so $x \in A^b$. Conversely, if $x \in A^b$ but $x \notin A$, then $x \in A'$.

The derived set of X is not necessarily X itself. For instance, if the topology is discrete, then $X' = ø$. A point x does not belong to X' if and only if $\{x\}$ is an open set. These are the *isolated points* of the space X. The set X' is sometimes called *dense part* of the space X.

In some cases $\bar{O}^i = O$ but not always. For example, let $X = \{a, b, c\}$ and let the open sets be ø, $\{a\}$, $\{b\}$, $\{a, b\}$, and $\{a, b, c\}$. Then the closed sets are ø, $\{c\}$, $\{a, c\}$, $\{b, c\}$, and $\{a, b, c\}$. The smallest closed set containing the open set $\{a, b\}$ is $\{a, b, c\}$ and so $\overline{\{a, b\}}^i = \{a, b, c\}$. If O is such that $\bar{O}^i = O$, then O is called a *regular open set*. The union of regular open sets need not be regular, e.g., in the preceding example $\{a\}$ and $\{b\}$

are regular but $\{a, b\}$ is not. Occasionally one uses a modified union operation called *regular union* which preserves regularity: We define $A \underset{r}{\cup} B = (\overline{A \cup B})^i$ for any pair of sets A, B of a topological space. The set $A \underset{r}{\cup} B$ is always regular open. More generally we have:

Theorem 9. *For every A the set $\bar{A}^i$ is regular open.*

Proof. We wish to show that $\overline{\bar{A}^i}^i = \bar{A}^i$. Since $\overline{\bar{A}^i} \supseteq \bar{A}^i$, on the one hand we have $\overline{\bar{A}^i}^i \supseteq \bar{A}^{ii} = \bar{A}^i$. On the other hand, $\bar{A} \supseteq \bar{A}^i$ and so $\bar{A} = \bar{\bar{A}} \supseteq \overline{\bar{A}^i}$ and $\bar{A}^i \supseteq \overline{\bar{A}^i}^i$.

In a similar fashion we can speak about *regular intersections:* $A \underset{r}{\cap} B = (\overline{A \cap B})^i$. One can also define *regular closed sets*; these are sets such that $\overline{C^i} = C$. Later we shall need the following:

Lemma 1. *For any pair of sets A, B we have $\bar{A} \cap B^i = \overline{A \cap B} \cap B^i$.*

Proof. It is sufficient to show that $\overline{A \cap B} \cap B^i \supseteq \bar{A} \cap B^i$ because by $\overline{A \cap B} \subseteq \bar{A}$ the other inclusion is obvious. If $x \in \bar{A} \cap B^i$ and O_x is an open set containing x, then $O_x \cap B^i$ is open and contains x, so by $x \in \bar{A}$ it intersects A. Therefore $O_x \cap B \cap A$ is not void and so $x \in \overline{A \cap B}$. Hence if $x \in \bar{A} \cap B^i$, then $x \in \overline{A \cap B} \cap B^i$.

EXERCISES

1. Show that O is open if and only if $A \cap O = ø$ implies $\bar{A} \cap O = ø$ for every A in X.

2. Show the following: If $A \cup B = X$ then $\bar{A} \cup B^i = X$, and if $A \cap B = ø$ then $\bar{A} \cap B^i = ø$.

(First, if $x \notin \bar{A}$, then there is an open set O_x containing x such $A \cap O_x = ø$. Hence $O_x \subseteq B$ and $x \in B^i$. Next, if $x \in \bar{A}$, then for every O_x we have $O_x \cap A \neq ø$ and so $O_x \nsubseteq B$.)

3. Determine A^i, A^e, A^b, $\bar{A}$, $\bar{A}^i$, and $\overline{A^i}$ for every set A when X is topologized by the topology of finite complements.

(If cA is finite then $A^i = A$, and if cA is infinite then $A^i = ø$. If A is finite then $A^e = cA$, and if A is infinite then $A^e = ø$. If A is finite then $A^b = A$, if both A and cA are infinite then $A^b = X$, and if cA is finite then $A^b = cA$. If A is finite then $\bar{A} = A$, and if A is infinite then $\bar{A} = X$. If A is finite then $\bar{A}^i = ø$, and if A is infinite then $\bar{A}^i = X$. If cA is finite then $\overline{A^i} = \bar{A}$, and if cA is infinite then $A^i = ø$.)

4. Prove that in every topological space $(A \cup B)' = A' \cup B'$.

[Clearly $(A \cup B)' \supseteq A' \cup B'$. If $x \in (A \cup B)'$, then for every open set O_x the intersection $O_x \cap (A \cup B)$ contains a point which is different from x. If there were open sets O_x and Q_x such that $O_x \cap A \subseteq \{x\}$ and $Q_x \cap B \subseteq \{x\}$, then $O_x \cap Q_x$ would be an open set containing x and such that $(O_x \cap Q_x) \cap (A \cup B) \subseteq \{x\}$. Since this is not possible either every O_x intersects A or every O_x intersects B in some point different from x. Hence either $x \in A'$ or $x \in B'$.]

5. Prove that if points form closed sets then every derived set is closed.

(We show that cA' is open. If $x \in cA'$, then there is an open set O_x such that $A \cap O_x \subseteq \{x\}$. If $y \in O_x$ and O_y is another open set containing y, then $O_x \cap O_y$ is open, it contains y, and $A \cap O_x \cap O_y \cap c\{x\} = \emptyset$. Hence $y \in cA'$ and so $O_x \subseteq cA'$. By Lemma 1.1 the set cA' is open.)

6. Let X be the set of reals and let $\mathcal{O}$ consist of $\emptyset$, X, and the intervals $(-\infty, x)$ where $x \in X$. Find the derived set of $\{0\}$ and show that it is not closed.

7. Give an example of a space and of a family of sets A_i $(i \in I)$ such that $\overline{\bigcup A_i}$ and $\bigcup \bar{A}_i$ are different.

(Let X be an infinite set and let it be topologized by the topology of finite complements. Let $A_1, A_2, \ldots$ be finite sets whose union is not X.)

8. Verify the following rules: $A^{bb} \subseteq A^b$; $(A \cap B)^b \subseteq A^b \cup B^b$; $(A \cup B)^b \subseteq A^b \cup B^b$; $\bar{A}^b \subseteq A^b$; and $A^{ib} \subseteq A^b$.

[The boundary A^b is closed and so it contains its own boundary: $A^{bb} \subseteq A^b$.

$$(A \cap B)^b = \overline{A \cap B} \cap \overline{c(A \cap B)} \subseteq \bar{A} \cap \bar{B} \cap (\overline{cA \cup cB}) \subseteq (\bar{A} \cap \overline{cA}) \cup (\bar{B} \cap \overline{cB})$$
$$= A^b \cup B^b.$$

The other inclusion follows by taking complements. Next

$$\bar{A}^b = (A \cup A^b)^b \subseteq A^b \cup A^{bb} = A^b$$

and so

$$A^{ib} = (cA^i)^b = (\overline{cA})^b \subseteq (cA)^b = A^b.]$$

9. Prove: C is regular closed if and only if cC is regular open.

[If cC is regular open, then $cC = (\overline{cC})^i$ and so $C = c\overline{c\overline{\overline{C}}} = \overline{C^i}$. If C is regular closed, then $C = \overline{C^i} = \overline{c\overline{c\overline{C}}}$ and $cC = c\overline{c\overline{c\overline{C}}} = (\overline{cC})^i$.]

10. The intersection of finitely many regular open sets is regular open and the union of finitely many regular closed sets is regular closed.

[First $(\overline{O_1 \cap O_2})^i \subseteq (\bar{O}_1 \cap \bar{O}_2)^i = \bar{O}_1{}^i \cap \bar{O}_2{}^i = O_1 \cap O_2$. Next $(\overline{O_1 \cap O_2})^i \supseteq (O_1 \cap O_2)^i = O_1 \cap O_2$ The other proposition can be proved by a similar argument or by taking complements and using the preceding exercise.]

11. Show that an open set O is regular open if and only if $O^b = \bar{O}^b$. (For $O^b = \bar{O} \cap cO = \bar{O} \cap c\bar{O}^i = \bar{O} \cap c\bar{O} = \bar{O}^b$.)

12. Show that a closed set C is regular closed if and only if $C^b \subseteq \overline{C^i}$. [If C is regular closed, then $C^b = (\overline{C^i})^b \subseteq \overline{C^i}$, and if $C^b \subseteq \overline{C^i}$, then $\bar{C} = C^i \cup C^b \subseteq C^i \cup \overline{C^i} = \overline{C^i} \subseteq \bar{C}$ and so $\bar{C} = C = \overline{C^i}$.]

3. Closure Operators

There are many practical methods which can be used to introduce topologies on sets. As Kuratowski first proved, the closure operator leads to such a method. One can also use the interior, exterior, boundary, and derivation operators since any of these completely determines the topology of the set. These will be discussed in the exercises at the end of this section.

Theorem 1. *Consider an operator associating with each subset A of a set X another subset $\bar{A}$ and subject to the axioms:*

(K.1) $$\bar{\emptyset} = \emptyset.$$

(K.2) $$A \subseteq \bar{A} \quad \textit{for every} \quad A \subseteq X.$$

(K.3) $$\bar{\bar{A}} = \bar{A} \quad \textit{for every} \quad A \subseteq X.$$

(K.4) $$\overline{A \cup B} = \bar{A} \cup \bar{B} \quad \textit{for every} \quad A, B \subseteq X.$$

Then there exists exactly one topology $\mathscr{T}$ on X such that the closure of every set A relative to $\mathscr{T}$ is $\bar{A}$.

Note. Sometimes it is more convenient to use instead of (K.4) the following pair of axioms:

(K.4a) $$\overline{A \cup B} = A \cup B \quad \textit{whenever} \quad \bar{A} = A \textit{ and } \bar{B} = B.$$

(K.4b) $$\bar{A} \subseteq \bar{B} \quad \textit{whenever} \quad A \subseteq B \subseteq X.$$

The equivalence will follow easily.

Proof. We define the topology $\mathscr{T}$ by specifying the family of closed sets: Let $\mathscr{C}$ be the family of those subsets C of X for which $\bar{C} = C$. Then

$\mathscr{C}$ satisfies axioms (C.1), (C.2), and (C.3) in Theorem 1.2: Indeed, by (K.1) we have $\emptyset \in \mathscr{C}$ and by (K.2) we obtain $X \subseteq \bar{X} \subseteq X$, i.e., $\bar{X} = X$ and $X \in \mathscr{C}$. Axiom (C.2) is a direct consequence of (K.4). To prove (C.3) we first show that $\bar{A} \subseteq \bar{B}$ whenever $A \subseteq B$. In fact, if $A \subseteq B$, then $B = A \cup B$ and so by (K.4) we get $\bar{B} = \overline{A \cup B} = \bar{A} \cup \bar{B} \supseteq \bar{A}$. Now let $\{C_i\}$ $(i \in I)$ be an arbitrary family of sets $C_i \in \mathscr{C}$. Then $\bigcap C_i \subseteq C_i$ for every $i \in I$ and so $\overline{\bigcap C_i} \subseteq \bar{C}_i = C_i$ for every $i \in I$. Therefore $\overline{\bigcap C_i} \subseteq \bigcap C_i$. Since by (K.2) we have $\bigcap C_i \subseteq \overline{\bigcap C_i}$ we see that $\overline{\bigcap C_i} = \bigcap C_i$, that is to say, $\bigcap C_i \in \mathscr{C}$. This shows axiom (C.3).

The reader will notice that axiom (K.3) was not used when we verified axioms (C.1), (C.2), and (C.3). Hence any operator satisfying (K.1), (K.2), and (K.4) defines some topology $\mathscr{T}$ on the set X. Axiom (K.3) assures that the closure operator of the topology so obtained is exactly the operator $\bar{A}$: Let $\tilde{A}$ denote the closure of A under the topology $\mathscr{T}$ having closed sets $C \in \mathscr{C}$. Since $\bar{\bar{A}} = \bar{A}$ by (K.3), the set $\bar{A}$ belongs to $\mathscr{C}$ and $A \subseteq \bar{A}$ by (K.2). Thus $\tilde{A} = \bigcap\{C : A \subseteq C\} \subseteq \bar{A}$. On the other hand, if $C \in \mathscr{C}$ and if $A \subseteq C$, then $\bar{A} \subseteq \bar{C} = C$ and so $\bar{A} \subseteq \bigcap\{C : A \subseteq C\} = \tilde{A}$. Therefore $\tilde{A} = \bar{A}$ for every subset A of X.

Now we show that the axiom systems (K.1)–(K.4) and (K.1)–(K.4a), (K.4b) are equivalent. First, (K.4a) is a special case of (K.4). Next if $A \subseteq B$, then $B = A \cup B$ and so by (K.4) we obtain $\bar{B} = \overline{A \cup B} = \bar{A} \cup \bar{B} \supseteq \bar{A}$. Thus (K.4) implies (K.4b). Let us suppose now that axioms (K.1)–(K.4a) and (K.4b) are satisfied. Then by (K.4b) from $A \subseteq A \cup B$ and $B \subseteq A \cup B$ we get $\bar{A} \subseteq \overline{A \cup B}$ and $\bar{B} \subseteq \overline{A \cup B}$. Therefore $\bar{A} \cup \bar{B} \subseteq \overline{A \cup B}$. By (K.2) we have $A \subseteq \bar{A}$ and $B \subseteq \bar{B}$ so that $A \cup B \subseteq \bar{A} \cup \bar{B}$. Hence by (K.2), (K.3), and (K.4a) $\overline{A \cup B} \subseteq \overline{\bar{A} \cup \bar{B}} = \bar{\bar{A}} \cup \bar{\bar{B}} = \bar{A} \cup \bar{B}$. This shows that $\overline{A \cup B} = \bar{A} \cup \bar{B}$ for any A, B in X.

As an example let X be a linearly ordered set. For every nonvoid set A in X define $\bar{A} = \bigcap\{(-\infty, x] : a \leqslant x \textit{ for all } a \in A\}$. If $A = \emptyset$, then let $\bar{\emptyset} = \emptyset$. It is easy to show that the operator $\bar{A}$ follows axioms (K.1)–(K.4) and so it is the closure operator of some topology on X. The topology $\mathscr{T}$ associated with this closure operator will be called the *order topology* of the partially ordered set X. (Some authors use the expression "order topology" for another topology on X. This second topology is the *interval topology* of X.)

The use of the Kuratowski axiom system (K.1)–(K.4) is often inconvenient and sometimes would cause difficulties. However, there are a number of variants which can often be used with success. For instance, we can consider the closure operator only on a subset $\mathscr{A}$ of the power set of X and construct a topology on X by using this skeleton of the closure operator.

Theorem 2. *Suppose that an operator $\bar{A}$ is defined on a subset $\mathscr{A}$ of the power set of X such that the following axioms hold:*

(Kb.1) $$\text{If} \quad \emptyset \in \mathscr{A}, \quad \text{then} \quad \bar{\emptyset} = \emptyset.$$

(Kb.2) $$\text{If} \quad A \in \mathscr{A}, \quad \text{then} \quad A \subseteq \bar{A}.$$

(Kb.3) $$\text{If} \quad A \quad \text{and} \quad \bar{A} \quad \text{belong to} \quad \mathscr{A}, \text{then} \quad \bar{\bar{A}} = \bar{A}.$$

(Kb.4) $$\text{If} \quad A, B \in \mathscr{A}, \quad \text{then} \quad A \cup B \in \mathscr{A} \quad \text{and} \quad \overline{A \cup B} = \bar{A} \cup \bar{B}.$$

Then the operator $\bar{A}$ $(A \in \mathscr{A})$ defines a topology $\mathscr{T}$ on X such that the closure of every $A \in \mathscr{A}$ is $\bar{A}$.

Proof. We define $\bar{\emptyset} = \emptyset$ and $\bar{S} = \bigcap \{\bar{A} : S \subseteq \bar{A} \textit{ and } A \in \mathscr{A}\}$. If there is no $A \in \mathscr{A}$ with $S \subseteq \bar{A}$ then $\bar{S} = X$, and if $S \in \mathscr{A}$ then the new and old definitions of $\bar{S}$ coincide. We show that $\bar{S}$ satisfies axioms (K.1)–(K.4a) and (K.4b) and so a topology can be defined on X such that the operator $\bar{S}$ is the closure operator of this topology. Only axiom (K.4a) needs proving. Suppose that $\bar{S}_1 = S_1$ and $\bar{S}_2 = S_2$. If $x \notin S_1 \cup S_2$, then there exist A_1 and A_2 in $\mathscr{A}$ such that $x \notin \bar{A}_1 \supseteq S_1$ and $x \notin \bar{A}_2 \supseteq S_2$. Hence $x \notin \overline{A_1 \cup A_2}$ and so $x \notin \overline{S_1 \cup S_2}$. Therefore $\overline{S_1 \cup S_2} \subseteq S_1 \cup S_2$.

A simple special case of the preceding theorem is the one in which $\mathscr{A}$ is the family of closed sets of a topological space. Then $A = \bar{A}$ for every A in $\mathscr{A}$.

EXERCISES

1. Let an operator $\bar{A}$ be defined on the set of all subsets of a set X such that the following axioms are satisfied:

(I) $$\overline{A \cup B} = \bar{A} \cup \bar{B}.$$

(II) $$A = \bar{A} \quad \text{for every set} \quad A \quad \text{consisting of a single point.}$$

(III) $$\bar{\bar{A}} = \bar{A} \quad \text{for every } A \quad \text{in} \quad X.$$

Show that if X contains at least two elements then the operator $\bar{A}$ satisfies axioms (K.1)–(K.4) and points are closed sets in the topology generated by the operator $\bar{A}$.

[To show (K.2) let $a \in A$ and put $A = (A - \{a\}) \cup \{a\}$. Then by (I) and (II) we have $\bar{A} = \overline{A - \{a\}} \cup \{a\}$. Hence $a \in \bar{A}$ and $A \subseteq \bar{A}$. To prove (K.1) let $x \in X$ and let $y \neq x$ in X. Take $\emptyset \cup \{y\} = \{y\}$ and apply

(I) and (II): $\{y\} = \{y\} \cup \bar{ø}$. This shows that $x \notin \bar{ø}$. Thus $\bar{ø} = ø$. If X consists of a single point, then one can define $\bar{ø} = \bar{X} = X$ and axioms (I), (II), and (III) will hold but (K.1) obviously fails.]

2. Let $\gamma : A \to \gamma(A)$ be an operator on the power set of X subject to the following axioms:

(1) $$\gamma(ø) = ø.$$

(2) $$\textit{If} \quad x \in \gamma(A), \quad \textit{then} \quad x \in \gamma(A - \{x\}).$$

(3) $$\gamma(\gamma(A)) \subseteq \gamma(A) \quad \textit{for every} \quad A.$$

(4) $$\gamma(A \cup B) = \gamma(A) \cup \gamma(B) \quad \textit{for every} \quad A, B.$$

Show that there is a unique topology on X such that $\gamma(A) = A'$ for every set A in X.

[Let $\bar{A} = A \cup \gamma(A)$ and verify axioms (K.1)–(K.4) for the operation $\bar{A}$:

$$\begin{aligned}\bar{\bar{A}} = \overline{A \cup \gamma(A)} &= A \cup \gamma(A) \cup \gamma(A \cup \gamma(A)) = A \cup \gamma(A) \cup \gamma(\gamma(A)) \\ &= A \cup \gamma(A) = \bar{A}\end{aligned}$$

and

$$\overline{A \cup B} = A \cup B \cup \gamma(A \cup B) = A \cup \gamma(A) \cup B \cup \gamma(B) = \bar{A} \cup \bar{B}.$$

Next show that $A' \subseteq \gamma(A)$: The set

$$O = c((A - \{x\}) \cup \gamma(A - \{x\})) = c(A - \{x\}) \cap c\gamma(A - \{x\})$$

is open and $O \cap A \subseteq \{x\}$. Hence if $x \in A'$, then $x \notin O$ and so $x \in \gamma(A - \{x\})$. Thus $x \in \gamma(A - \{x\}) \subseteq \gamma(A)$ and $A' \subseteq \gamma(A)$. To show that $\gamma(A) \subseteq A'$ suppose that $x \notin A'$. Then there is an O_x, say $O_x = c(B \cup \gamma(B))$, such that $O_x \cap A \subseteq \{x\}$. Then $A - \{x\} \subseteq B \cup \gamma(B)$ and $\gamma(A - \{x\}) \subseteq \gamma(B \cup \gamma(B)) = \gamma(B)$. Since $x \in O_x = c(B \cup \gamma(B))$ we have $x \notin \gamma(B)$ and so $x \notin \gamma(A - \{x\})$. Therefore by (2) $x \notin \gamma(A)$ and $\gamma(A) \subseteq A'$.]

3. Consider an operator $\gamma(A)$ for the set of all subsets of a set X subject to the axioms:

(1) $$\gamma(X) = X.$$

(2) $$\gamma(A) \subseteq A \quad \textit{for every} \quad A.$$

(3) $$\gamma(\gamma(A)) = \gamma(A) \quad \textit{for every} \quad A.$$

(4) $$\gamma(A \cap B) = \gamma(A) \cap \gamma(B).$$

Find a topology $\mathscr{T}$ on X such that $\gamma(A) = A^i$ where the interior is taken with respect to $\mathscr{T}$.

[The family of open sets is $\mathscr{O} = \{\gamma(A) : A \subseteq X\}$. It is the simplest to notice that $\hat{\gamma}(A) = c\gamma(cA)$ satisfies (K.1)–(K.4).]

4. Bases and Subbases

Here we describe the most common method of introducing topologies on sets. The basic idea is to define a family of some of the open sets from which the remaining open sets can be easily constructed. There are two types of families which are important in practice; the more primitive one is called a subbase and the other is a base. One can also introduce bases and subbases for closed sets.

Definition 1. *A family $\mathscr{B}$ of subsets B of a set X is a base for the open sets of a topology $\mathscr{T}$ on X if it has the following properties:*

(B.1) *For each $x \in X$ there is a $B \in \mathscr{B}$ such that $x \in B$.*

(B.2) *If $B_1, B_2 \in \mathscr{B}$ and if $x \in B_1 \cap B_2$, then there is a $B \in \mathscr{B}$ such that $x \in B \subseteq B_1 \cap B_2$.*

The family of open sets of $\mathscr{T}$ is defined as follows:

A nonvoid set O is open if and only if it is the union of some sets B belonging to $\mathscr{B}$. The topology $\mathscr{T}$ is called the topology generated by the base $\mathscr{B}$.

For example, let X be the set of real numbers and let $\mathscr{B}$ be the set of all open intervals $(a, b) = \{x : a < x < b\}$. Axioms (B.1) and (B.2) are clearly satisfied. A set $O \subseteq X$ is open in the topology generated by $\mathscr{B}$ if it is the union of an arbitrary family of open intervals (a, b). This is called the *topology of the reals* or the "usual topology of the reals."

Theorem 1. *A necessary and sufficient condition that the topology generated by the base $\mathscr{B}_1$ be finer than the topology generated by the base $\mathscr{B}_2$ is the following: Given any $B_2 \in \mathscr{B}_2$ and any $x \in B_2$, there is a $B_1 \in \mathscr{B}_1$ such that $x \in B_1 \subseteq B_2$.*

Corollary. *Two bases $\mathscr{B}_1$ and $\mathscr{B}_2$ generate the same topology if and only if given any $B_1 \in \mathscr{B}_1$ and any $x \in B_1$ there is a $B_2 \in \mathscr{B}_2$ such that $x \in B_2 \subseteq B_1$ and also given any $B_2 \in \mathscr{B}_2$ and any $x \in B_2$ there is a $B_1 \in \mathscr{B}_1$ such that $x \in B_1 \subseteq B_2$.*

Proof. Let O be open with respect to the topology generated by $\mathscr{B}_2$. Then for each $x \in O$ there is a $B_2 \in \mathscr{B}_2$ such that $x \in B_2 \subseteq O$. Hence if the condition holds there is a $B_1 \in \mathscr{B}_1$ such that $x \in B_1 \subseteq O$. Consequently O is the union of sets $B_1 \in \mathscr{B}_1$ and so O is open also with respect to the topology generated by $\mathscr{B}_1$. The necessity is clear.

Theorem 2. *A necessary and sufficient condition for $\mathscr{B}$ to be a base for a given topology is: Given $O \in \mathcal{O}$ and $x \in O$ there is $B \in \mathscr{B}$ such that $x \in B \subseteq O$ and if $B \in \mathscr{B}$, then $B \in \mathcal{O}$.*

Proof. If $\mathscr{B}$ is a base for the topology $\mathscr{T}$, then every $O \in \mathcal{O}$ is the union of sets $B \in \mathscr{B}$ and so the condition is satisfied. Conversely, if the condition is fulfilled, then on the one hand every $O \in \mathcal{O}$ is the union of sets B and so every $O \in \mathcal{O}$ is open with respect to the topology generated by $\mathscr{B}$. On the other hand, by $\mathscr{B} \subseteq \mathcal{O}$ and by axiom (O.3) every set which is open with respect to the topology generated by $\mathscr{B}$ is also open with respect to the given topology.

The family $\mathcal{O}$ of open sets of a topological space is always a base for the topology of the space and this is the largest possible base. It is often important to go toward the other extreme, namely, to find bases with as few elements as possible.

Definition 2. *A family $\mathscr{S}$ of subsets S of a set X is called a subbase for a topology if for each $x \in X$ there is an $S \in \mathscr{S}$ such that $x \in S$.*

By the topology generated by $\mathscr{S}$ is meant the topology generated by the base $\mathscr{B}$ which consists of all finite intersections of sets $S \in \mathscr{S}$.

It is easy to see that the family $\mathscr{B}$ satisfies axioms (B.1) and (B.2) of a base. Hence indeed $\mathscr{S}$ determines a topology $\mathscr{T}$ on the set X and every $S \in \mathscr{S}$ is open relative to $\mathscr{T}$.

There is no difficulty in introducing *bases* and *subbases* for *closed sets.* The corresponding axioms and the analogs of Theorems 1 and 2 can be obtained by using complementation.

Consider the set of all topologies on a fixed set X. This set is partially ordered by inclusion: $\mathscr{T}_1 \leqslant \mathscr{T}_2$ if $\mathcal{O}_1 \subseteq \mathcal{O}_2$. Using the concept of a subbase we can easily show that every family $\{\mathscr{T}_i\}$ $(i \in I)$ of topologies has a least upper bound and a greatest lower bound. For let $\mathcal{O}_i$ $(i \in I)$ denote the family of open sets of the topology $\mathscr{T}_i$ and let $\mathscr{S} = \bigcup \mathcal{O}_i$. Denote by $\operatorname{lub}\{\mathscr{T}_i\}$ the topology generated on X by the subbase $\mathscr{S}$. Since $\mathcal{O}_i \subseteq \mathscr{S}$ this topology is finer than any of the topologies $\mathscr{T}_i$. On the other hand, if a topology is finer than every $\mathscr{T}_i$, then $\mathcal{O}_i \subseteq \mathcal{O}$ for every $i \in I$ and so $\mathscr{S} \subseteq \mathcal{O}$. Therefore $\operatorname{lub}\{\mathscr{T}_i\}$ is the least fine topology on X which is at least as fine as the topologies $\mathscr{T}_i$ $(i \in I)$. This is the *least upper bound of the topologies $\mathscr{T}_i$ $(i \in I)$.* Now consider the family $\bigcap \mathcal{O}_i$. Axioms (O.1), (O.2), and (O.3) are satisfied, so $\bigcap \mathcal{O}_i$ is the family of open sets of a topology $\operatorname{glb}\{\mathscr{T}_i\}$. This is the finest topology on X which is at least as coarse as any of the topologies $\mathscr{T}_i$. It is called the *greatest lower bound of the topologies $\mathscr{T}_i$ $(i \in I)$.*

EXERCISES

1. Let $\mathscr{T}_x$ be the topology on X whose open sets are X and all sets not containing the point $x \in X$. Determine $\operatorname{lub}\{\mathscr{T}_x : x \in X\}$ and $\operatorname{glb}\{\mathscr{T}_x : x \in X\}$.

2. Find an axiom system for bases consisting of closed sets and state the analogs of Theorems 1 and 2.

3. Let $[a, b]$ denote the closed interval with end points a and b in a linearly ordered set X. Show that the family of all closed intervals $[a, b]$ $(a, b \in X)$ is a subbase for closed sets. Show also that if X is the set of real numbers, then this is not the usual topology.

(Relative to the usual topology of the reals there exist disjoint nonvoid open sets.)

4. Which is the weakest topology in which points form closed sets?

5. Topologies on Linearly Ordered Sets

A simple but important case where bases are used to introduce a topology on a set occurs in the definition of topologies on linearly ordered sets. If X is a linearly ordered set, then for any $a < b$ in X we denote by (a, b), $(a, b]$, $[a, b)$, and $[a, b]$ the intervals

$$\begin{aligned}(a, b) &= \{x : a < x < b\}\\ (a, b] &= \{x : a < x \leqslant b\}\\ [a, b) &= \{x : a \leqslant x < b\}\\ [a, b] &= \{x : a \leqslant x \leqslant b\}.\end{aligned}$$

(a, b) is the open and $[a, b]$ is the closed interval with end points a and b. The others are the half-open intervals. As usual we let

$$\begin{aligned}(-\infty, b) &= \{x : x < b\}\\ (a, +\infty) &= \{x : a < x\}.\end{aligned}$$

The sets $(-\infty, b]$ and $[a, +\infty)$ are defined similarly. Occasionally we write $(-\infty, +\infty)$ instead of X. These sets are the *improper intervals* of X.

In the sequel we define what is meant by the *interval topology* of a linearly ordered set and prove an important theorem on the characterization of the open sets of this topology when the linearly ordered set has the least upper bound property.

The family $\mathscr{S}$ of improper open intervals of a linearly ordered set X is a subbase for a topology which is called the interval topology of X.

In fact, if X consists of at least two elements, then every $x \in X$ belongs to at least one interval of the form $(-\infty, b)$ or $(a, +\infty)$ and so $\mathscr{S}$ is a subbase for a topology on X. If X has no largest nor smallest element then the topology can be generated by a base which consists of all proper open intervals (a, b) in X. This is the origin of the name interval topology. For example, if X is the set of real numbers and if $\mathscr{B}$ is the family of all proper open intervals, then a topology is generated on X by $\mathscr{B}$ which is called the ordinary or *usual topology of the reals*. The same topology can be obtained by different methods, e.g., by introducing a metric. Similarly we can consider as X subsets of the set of real numbers and form the interval topologies on these sets. Some of these will be trivial, for instance, if X is the set of all integers, then the interval topology is the discrete topology.

A linearly ordered set X is said to have the *least upper bound property* if every subset bounded from above has a least upper bound. X itself need not have a maximal element. The greatest lower bound property is defined similarly. Either one implies the other. The set of real numbers and the set of integers for example have the least upper bound property but the set of rationals does not.

Theorem 1. *Let X be linearly ordered such that the least upper bound property holds. Then a set O is open relative to the interval topology if and only if it is the union of a family of disjoint open intervals.*

Proof. Let $x \in O$ and let $S \subseteq O$ be the set of those points $\xi \in O$ for which $[x, \xi] \subseteq O$. Then S is not void and so it has a proper or improper least upper bound u. We have $S = [x, u]$ or $S = [x, u)$ according as u does or does not belong to S. If $u \notin S$, we let $b_x = u$ so that $b_x \in X$ or b_x denotes the improper element $+\infty$ and $S = [x, b_x)$. If $u \in S$, then $u \in O$ and so there is an interval (a, b) such that $u \in (a, b) \subseteq O$. It follows that there exists no $\xi \in X$ satisfying $u < \xi < b$: For by $(a, b) \subseteq O$ such a ξ would belong to S and so it would be an element larger than the least upper bound. Therefore choosing $b_x = b$ again we have $S = [x, b_x)$.

Now we consider the set $T \subseteq O$ of those points $\xi \in O$ for which $[\xi, x] \subseteq O$. In the same way as before we can find a proper or an improper element a_x such that $T = (a_x, x]$. Consequently, for any $x \in O$ we found extremal elements a_x and b_x such that $x \in S \cup T = (a_x, b_x) \subseteq O$. If $x \neq y$, then either $a_x = a_y$ and $b_x = b_y$ or the open intervals (a_x, b_x) and (a_y, b_y) are disjoint. Therefore $\{(a_x, b_x)\}$ $(x \in O)$ is a family of disjoint open intervals whose union is the open set O.

The theorem has an important special case:

Theorem 2. *Let X be the space of the real numbers with its usual topology. Then a set O is open if and only if it is the union of countably many disjoint open intervals.*

Proof. Using Theorem 1 we see that any open set O consists of disjoint open intervals, say (a_i, b_i), where $i \in I$. For every positive integer n there are at most denumerably many among these intervals such that

$b_i - a_i \geqslant 1/n$. Since every interval satisfies this inequality for a sufficiently large value of n we see that $\{(a_i, b_i)\}$ is a countable family.

One can introduce topologies on a linearly ordered set X also by using the half-open intervals: The family of all intervals $[a, b)$ $(a, b \in X)$ and $[a, +\infty)$ $(a \in X)$ is a base for the open sets of a topology called the *right half-open interval topology* $\mathscr{T}_+$ of the ordered set X. The *left half-open interval topology* $\mathscr{T}_-$ is generated by the base consisting of the intervals $(a, b]$ $(a, b \in X)$ and $(-\infty, b]$ $(b \in X)$. These topologies are finer than the interval topology because the open intervals (a, b) can be expressed in the form

$$(a, b) = \bigcup \{[\alpha, b) : a < \alpha < b\} = \bigcup \{(a, \beta] : a < \beta < b\}.$$

If X is the set of reals, then the half-open interval topologies are strictly finer than the usual topology. They have many unexpected properties and so can serve as examples.

EXERCISES

1. Let $\mathscr{O}$ be the family of those subsets of a linearly ordered set X which are unions of disjoint open intervals. Prove Theorem 1 by showing that if X has the least upper bound property, then $\mathscr{O}$ is the family of open sets of some topology on X.

2. Let $\mathscr{T}_+$ denote the right half-open interval topology of a linearly ordered set X having the least upper bound property. Prove the following representation theorem: A set O is open relative to $\mathscr{T}_+$ if and only if it is the union of disjoint open and right half-open intervals.

3. A subset O of a partially ordered set X is called a *segment* if $x \leqslant y$ and $y \in O$ imply $x \in O$. Show that the family $\mathscr{O}$ of all segments satisfies the axioms for open sets. Determine the least upper bound and the greatest lower bound of the topologies associated with the ordering and the inverse ordering of X.

4. Let $\mathscr{T}$ be the least upper bound of the usual topology and of the topology of countable complements on the set of reals. Show that O is open relative to $\mathscr{T}$ if and only if $O = Q - A$ where Q is open in the usual sense and A is countable.

(The open sets of the topology of countable complements are ø and those sets A whose complement cA is countable.)

5. Let X be a linearly ordered set and let $\mathscr{B}$ consist of X and of all intervals $(-\infty, x)$ $(x \in X)$. Show that $\mathscr{B}$ is a base for a topology. If X has no maximal element, then X can be dropped from $\mathscr{B}$.

6. Metric Spaces

These spaces are the classical types of topological spaces. Most of the fundamental concepts and results of modern analytical topology were first developed for metric spaces and many important instances of topological spaces belong to this group.

Definition 1. *A metric space is a set X together with a real-valued function d given on $X \times X$ which satisfies the following axioms:*

(M.1a) $d(x, y) \geqslant 0$ *for every pair* $x, y \in X$ *and* $d(x, x) = 0$.

(M.1b) $d(x, y) = 0$ *implies* $x = y$.

(M.2) $d(x, y) = d(y, x)$ *for every* $x, y \in X$.

(M.3) $d(x, y) + d(y, z) \geqslant d(x, z)$ *for any* $x, y, z \in X$.

The first axiom states that the distance from x to y is nonnegative and it is zero only if the two points coincide. If this second requirement, i.e., (M.1b), is omitted from the axioms we speak about a *pseudometric space*. Axiom (M.2) states that d is symmetric so that we may speak about the distance between the point pair $\{x, y\}$ without specifying their ordering. The inequality which occurs in (M.3) is the so-called *triangle inequality*. It is a well-known property of the Euclidean distance in the plane or in the space.

The set of reals X furnishes some simple examples of metric spaces. For example, we may define d to be $d(x, y) = | x - y |$ or $d(x, y) = \sqrt{| x - y |}$. The function given by $d(x, y) = (x - y)^2$ is not admissible because it violates axiom (M.3). The set X of all ordered n-tuples of real numbers becomes a metric space if we set $d(x, y) = | x_1 - y_1 | + \dots + | x_n - y_n |$. The same definition holds good if X is the set of all ordered n-tuples of complex numbers.

Theorem 1. *Every pseudometric space can be considered as a topological space by choosing as a base $\mathscr{B}$ the set of all sets*

$$S_\epsilon[x] = \{y : d(x, y) < \epsilon\}$$

where $\epsilon > 0$ and $x \in X$. $S_\epsilon[x]$ is called the ϵ-neighborhood of x or the ϵ-ball with center at x.

Note. When we are speaking about a metric space X it is assumed that X is a topological space with the topology which we just described. The expression "*ϵ-sphere*" is usually reserved for the set $\{y : d(x, y) = \epsilon\}$.

Proof. We must verify axioms (B.1) and (B.2) of a base. The first of these is clearly satisfied because, for instance, $x \in S_1[x]$. To show the second let us suppose that $x \in S_{\epsilon_1}[y_1]$ and $x \in S_{\epsilon_2}[y_2]$. Then by the definition of these balls, $\epsilon_1 - d(x, y_1) > 0$ and $\epsilon_2 - d(x, y_2) > 0$. Let $\epsilon > 0$ be the smaller one of these positive differences. Now if $d(x, y) < \epsilon$ for some point y, then by (M.3),

$$d(y, y_i) \leqslant d(y, x) + d(x, y_i) < \epsilon + d(x, y_i) \leqslant \epsilon_i .$$

Hence, if $y \in S_\epsilon[x]$, then $y \in S_{\epsilon_i}[y_i]$ and so $S_\epsilon[x] \subseteq S_{\epsilon_1}[y_1] \cap S_{\epsilon_2}[y_2]$.

The open sets O of any topological space which is defined by a base $\mathscr{B}$ are characterized by the property that O contains with each point x some set $B \in \mathscr{B}$ such that $x \in B$. For a metric space this criterion states that if $x \in O$, then there is a ball $S_\epsilon[y]$ about some center $y \in O$ such that $x \in S_\epsilon[y] \subseteq O$. At present, however, the criterion can be further strengthened:

Lemma 1. *A set $O \subseteq X$ is open relative to a pseudometric topology given on X if and only if O contains with each point x an ϵ-ball $S_\epsilon[x]$ about x as its center.*

Proof. We know that the condition is sufficient. To show its necessity suppose O is open and $x \in O$. Then there is some ball $S_\epsilon[y]$ such that $x \in S_\epsilon[y] \subseteq O$. By $x \in S_\epsilon[y]$ we have $\epsilon - d(x, y) > 0$. Consider the ball with radius $r = \epsilon - d(x, y)$ and center x. If $z \in S_r[x]$, then $d(x, z) < r = \epsilon - d(x, y)$ and so

$$d(y, z) \leqslant d(y, x) + d(x, z) < \epsilon,$$

that is to say, $z \in S_\epsilon[y]$. Hence with each $x \in O$ a whole ball $S_r[x]$ belongs to O.

Different metrics may lead to the same topological space; for instance, a distance function d and kd for any positive constant k determine the same topology. There is a simple necessary and sufficient condition under which two metrics d_1 and d_2 give the same topology.

Lemma 2. *The topology $\mathscr{T}_1$ defined by the pseudometric d_1 on a set X is finer than the topology $\mathscr{T}_2$ defined by the pseudometric d_2 if and only if for each $x \in X$ and $\epsilon_2 > 0$ there is an $\epsilon_1 > 0$ such that $d_2(x, y) < \epsilon_2$ whenever $d_1(x, y) < \epsilon_1$.*

Proof. A set $O \subseteq X$ is open with respect to $\mathscr{T}_i$ if it contains with each point x an entire ball $S^i_{\epsilon_i}[x]$. Hence if O is open relative to $\mathscr{T}_2$, then for each $x \in O$ there is a ball $S^2_{\epsilon_2}[x]$ such that $S^2_{\epsilon_2}[x] \subseteq O$. If we choose $\epsilon_1 > 0$ such that $d_2(x, y) < \epsilon_2$ whenever $d_1(x, y) < \epsilon_1$, then we get $S^1_{\epsilon_1}[x] \subseteq S^2_{\epsilon_2}[x] \subseteq O$. Therefore O is open relative to the topology $\mathscr{T}_1$. Since $S^2_{\epsilon_2}[x]$ is open relative to $\mathscr{T}_2$ the necessity of the condition follows from Lemma 1. The finite-dimensional *Euclidean spaces* E_n are important instances of metric spaces. E_n is defined to be the metric space formed by the ordered n-tuples $x = (x_1, ..., x_n)$ of arbitrary real numbers $x_1, ..., x_n$ when the distance function is

$$d(x, y) = \sqrt{|x_1 - y_1|^2 + ... + |x_n - y_n|^2}.$$

Axioms (M.1a), (M.1b), and (M.2) are obviously satisfied and (M.3) can be verified, for instance, by using *Minkowski's inequality:* If $p > 1$ and if $a_1, ..., a_n \geqslant 0$ and $b_1, ..., b_n \geqslant O$, then

$$((a_1 + b_1)^p + ... + (a_n + b_n)^p)^{1/p} \leqslant (a_1{}^p + ... + a_n{}^p)^{1/p} + (b_1{}^p + ... + b_n{}^p)^{1/p}.$$

Equality takes place only if the sets of numbers $a_1, ..., a_n$ and $b_1, ..., b_n$ are proportional. For $p = 1$ the equality is obvious.

The finite-dimensional Euclidean spaces are special cases of several other types of metric spaces all having the same topology but different metrics. For example, if the distance of the ordered n-tuples x and y is defined to be

$$d(x, y) = (|x_1 - y_1|^p + ... + |x_n - y_n|^p)^{1/p},$$

then by Minkowski's inequality the axioms of a metric space are satisfied for every $p \geqslant 1$. Let the spaces so obtained be denoted by $E_n{}^p$ so that $E_n{}^2 = E_n$ is the n-dimensional Euclidean space. Using Lemma 2 we can easily see that as far as the topology of $E_n{}^p$ is concerned, the specific value of $p \geqslant 1$ is irrelevant.

Usually it makes little difference which particular metric is used to define the topology on a set X. If we wish to emphasize that the topology was introduced by a certain metric d, we speak about a *topological space with metric d*. If the topology $\mathcal{T}$ on a set X can be introduced by some metric and we wish to stress that we have no particular metric in mind, we say that X is a *metrizable space* or $\mathcal{T}$ is a *metrizable topology* on X. A metric d is said to be *compatible* with a topology $\mathcal{T}$ on a set X if the topology generated by d is $\mathcal{T}$. Similarly, we speak about a *topological space with pseudometric d*, about *pseudometrizable spaces*, and *pseudometrizable topologies*.

Not every topological space is metrizable. For instance, if X is an infinite set, then the topology of finite complements is not metrizable. Indeed, in order that a metric d be compatible with the topology of finite complements it is necessary that $d(x, y)$ be positive for any pair of distinct points x, y. This, however, would imply the existence of disjoint open sets containing x and y, respectively. For instance, the ϵ-balls $S_\epsilon[x]$ and $S_\epsilon[y]$ would satisfy this requirement provided $2\epsilon < d(x, y)$.

EXERCISES

1. Show that the metric spaces X and Y formed over the integers x and the ordered pairs of integers $y = (y_1, y_2)$ by using the metrics $d(x', x'') = |x' - x''|$ and $d(y', y'') = |y_1' - y_1''| + |y_2' - y_2''|$ are homeomorphic.

2. Let X be the set of integers and let Y be the set of rationals. Consider in both cases the same metric d defined as $d(x', x'') = |x' - x''|$ and $d(y', y'') = |y' - y''|$. Show that the metric spaces X and Y are not homeomorphic.

3. Let X be a metric space with metric d. Prove that the functions defined by

$$\delta(x, y) = \frac{d(x, y)}{1 + d(x, y)} \quad \text{and} \quad \delta(x, y) = \begin{cases} d(x, y) & \text{if } d(x, y) \leqslant 1 \\ 1 & \text{if } d(x, y) \geqslant 1 \end{cases}$$

satisfy the axioms of a metric. Show that these new metrics generate the same topology on X as the original d.

4. Let f be a real-valued function defined for $d \geqslant 0$ and such that (i) $f(0) = 0$; (ii) $f(d_1) < f(d_2)$ *for every* $d_1 < d_2$; (iii) $f(d_1+d_2) \leqslant f(d_1)+f(d_2)$ *for every* $d_1, d_2 \geqslant 0$. Show that if the function d is a metric on the set X, then $f(d)$ is also a metric for X.

Find a necessary and sufficient condition under which the metrics d and $f(d)$ generate the same topology.

(The functions f and f^{-1} must be continuous at the origin.)

5. Let S be a nonvoid set in a metric space with metric d and let x_1, x_2 be points in X. Prove that if $d(x_i, S) = \text{glb}\{d(x_i, x) : x \in S\}$ is the distance of x_i $(i = 1, 2)$ from the set S, then

$$|d(x_1, S) - d(x_2, S)| \leqslant d(x_1, x_2).$$

[We have $d(x_1, S) \leqslant d(x_1, x)$ for every $x \in S$ and given $\epsilon > 0$ there is an $x \in S$ such that $d(x_2, S) \geqslant d(x_2, x) - \epsilon$. Hence

$$d(x_1, S) - d(x_2, S) \leqslant d(x_1, x) - d(x_2, x) + \epsilon \leqslant d(x_1, x_2) + \epsilon.]$$

6. Consider the set X of all disks $S_r[x]$ $(r > 0, x \in E_2)$ lying in the Euclidean plane. Show that relative to the distance

$$d(S_{r_1}[x_1], S_{r_2}[x_2]) = |r_1 - r_2| + |x_1 - x_2|$$

the set X is a metric space which is homeomorphic to the upper half $\{x : x_3 > 0\}$ of the Euclidean space E_3.

7. Show that if $d_1, \ldots, d_n$ are metrics on X which generate the topologies $\mathscr{T}_1, \ldots, \mathscr{T}_n$, then $d_1 + \ldots + d_n$ and $\max\{d_1, \ldots, d_n\}$ generate $\text{lub}\{\mathscr{T}_i\}$.

(Let d denote either one of the metrics $d_1 + \ldots + d_n$ and $\max\{d_1, \ldots, d_n\}$ and let $\mathscr{T}$ be the topology generated by d. Since $d_i \leqslant d$ for every $i = 1, \ldots, n$ we obtain $\mathscr{T}_i \leqslant \mathscr{T}$ and $\text{lub}\{\mathscr{T}_i\} \leqslant \mathscr{T}$. If O is open relative to $\mathscr{T}$ and $x \in O$, then for some $\epsilon > 0$, $S_\epsilon[x] \subseteq O$. The ball $S^i_{\epsilon/n}[x]$ is an open set in the topology $\mathscr{T}_i$; therefore $\bigcap S^i_{\epsilon/n}[x]$ is an open set of $\text{lub}\{\mathscr{T}_i\}$. Since $x \in \bigcap S^i_{\epsilon/n}[x] \subseteq S_\epsilon[x] \subseteq O$, O is also open relative to $\text{lub}\{\mathscr{T}_i\}$ so that $\mathscr{T} \leqslant \text{lub}\{\mathscr{T}_i\}$.)

8. Let $\{d_i\}$ $(i \in I)$ be a family of metrics for the set X having the property that for every $x, y \in X$ there is a positive constant M_{xy} such that $d_i(x, y) < M_{xy}$ for every $i \in I$. Show that the function d defined by $d(x, y) = \text{lub}\{d_i(x, y) : i \in I\}$ is a metric on X.

Find an example where every d_i $(i \in I)$ generates the same topology $\mathscr{T}$ but the topology generated by d is different.

[Only the triangle inequality needs proving. The inequalities $d_i(x, z) \leqslant d_i(x, y) + d_i(y, z)$ imply $d_i(x, z) \leqslant d(x, y) + d(y, z)$ and so $d(x, z) \leqslant d(x, y) + d(y, z)$. Let X be the set of real numbers and let for every $u \neq v$ the metric d_{uv} be defined as follows:

$$d_{uv}(x, y) = \begin{cases} \left| \dfrac{x - y}{u - v} \right| & \text{if } |x - y| \leqslant |u - v| \\ 1 & \text{if } |x - y| \geqslant |u - v| \end{cases}$$

Then $d_{uv}(x, y) \leqslant 1$ for every $x, y \in X$ and $d_{uv}(u, v) = 1$. Hence $d(u, v) = 1$ for every $u \neq v$ and $\mathscr{T}$ is the discrete topology. The topology generated by d_{uv} is the usual topology of the reals.]

9. Find an example which shows that in general the closure of the ϵ-ball $S_\epsilon[x] = \{\xi : d(x, \xi) < \epsilon\}$ is a proper subset of $\{\xi : d(x, \xi) \leqslant \epsilon\}$.

7. Neighborhood Filters

We shall often deal with sets N_x such that a given point x is contained in the interior of N_x. These sets are called the *neighborhoods* of the point $x \in S$. If N_x is an open set, then it is called an *open neighborhood* of x. Similarly, if N_x is closed and $x \in N_x^i$, then N_x is a *closed neighborhood* of x. It is possible to give a more primitive definition of neighborhoods and as a matter of fact it is this alternative definition which one uses in practice:

Lemma 1. *A set N_x is a neighborhood of a point x if and only if there is an open set O_x such that $x \in O_x \subseteq N_x$.*

The equivalence of the two definitions is obvious from the definition of the interior of the set N_x.

Definition 1. *The family $\mathcal{N}(x)$ of all neighborhoods N_x of a point x is called the neighborhood filter of x.*

Let $\mathcal{F}$ be a family of subsets F of a set X subject to the following requirements:

$$\varnothing \notin \mathcal{F}. \tag{1}$$

$$\textit{If}\quad F_1 \in \mathcal{F} \quad \textit{and} \quad F \supseteq F_1, \quad \textit{then} \quad F \in \mathcal{F}. \tag{2}$$

$$\textit{If}\quad F_1, F_2 \in \mathcal{F}, \quad \textit{then} \quad F_1 \cap F_2 \in \mathcal{F}. \tag{3}$$

If $\mathcal{F}$ has these properties, then it is called a *filter* over X.

It is easy to check that $\mathcal{N}(x)$ is a filter and actually (1) holds in the following strengthened form: $x \in N_x$ for every $N_x \in \mathcal{N}(x)$.

The family of open neighborhoods O_x of a point $x \in X$ is not a filter in X because rule (2) does not hold for $\mathcal{O}(x)$. Similarly, the family $\mathcal{C}(x)$ of closed neighborhoods does not follow rule (2) and so $\mathcal{C}(x)$ is not a filter. A family $\mathcal{B}$ of sets $B \subseteq X$ is called a *filter base* over the set X if $\varnothing \notin \mathcal{B}$ and if $B^1, B^2 \in \mathcal{B}$ implies the existence of a set $B \in \mathcal{B}$ such that $B \subseteq B^1 \cap B^2$. Every filter base $\mathcal{B}$ determines a filter $\mathcal{F}$, called the filter generated by $\mathcal{B}$; namely, $\mathcal{F}$ is defined to be the family of those sets $F \subseteq X$ which contain some set $B \in \mathcal{B}$:

$$\mathcal{F} = \{F : F \supseteq B \quad \textit{for some} \quad B \in \mathcal{B}\}.$$

Rules (1), (2), and (3) can be easily verified.

It is clear that the family $\mathcal{O}(x)$ of open neighborhoods O_x of a point x is a filter base and the filter generated by $\mathcal{O}(x)$ is $\mathcal{N}(x)$. The closed neighborhoods of x form a family $\mathcal{C}(x)$ which is a filter base but the filter generated by $\mathcal{C}(x)$ is not necessarily $\mathcal{N}(x)$. This is the case for instance if X is infinite and its topology is the topology of finite complements. Then the only closed neighborhood of any point x is the entire space X.

The problem which we discuss here is twofold: First suppose that the neighborhood filter $\mathcal{N}(x)$ of each point x of a topological space X is known. The object is to describe the family of open sets in terms of these filters. Next, suppose that no topology is given but to each point x in X there corresponds a filter $\mathcal{N}(x)$ consisting of subsets of X. The problem is to decide whether or not a topology can be determined by means of these families such that $\mathcal{N}(x)$ is the neighborhood filter of each $x \in X$. The answers are contained in the following:

Theorem 1. *Suppose that to every point x of a set X there corresponds a nonvoid family $\mathcal{N}(x)$ of subsets N_x. These families $\mathcal{N}(x)$ are the neighborhood filters of a topology on X if and only if the following axioms are satisfied:*

(Nb.1) $x \in N_x$ *for every* $N_x \in \mathcal{N}(x)$.

(Nb.2) *If* $N_x \in \mathcal{N}(x)$ *and* $N_x \subseteq N$, *then* $N \in \mathcal{N}(x)$.

(Nb.3) *If* $N_x^1 \in \mathcal{N}(x)$ *and* $N_x^2 \in \mathcal{N}(x)$ *then* $N_x^1 \cap N_x^2 \in \mathcal{N}(x)$.

(Nb.4) *If* $N_x \in \mathcal{N}(x)$ *then there is a subset* $O_x \subseteq N_x$ *such that* $x \in O_x \in \mathcal{N}(x)$ *and* $N_x \in \mathcal{N}(y)$ *for every* $y \in O_x$.

If all these axioms are satisfied, then there exists exactly one topology $\mathcal{T}$ *on* X *whose neighborhood filters are the families* $\mathcal{N}(x)$. *A set* O *is open with respect to* $\mathcal{T}$ *if and only if it contains with each* $x \in O$ *a whole set* $N_x \in \mathcal{N}(x)$.

Proof. If X is a topological space and $\mathcal{N}(x)$ is the neighborhood filter of x, then the axioms are satisfied: In (Nb.4) it is sufficient to choose O_x to be the interior of N_x. Hence the axioms are necessary.

Now suppose that $\mathcal{N}(x)$ is subject to axioms (Nb.1)–(Nb.4). We prove that the family $\mathcal{O} = \{O\}$ which is given in the second half of the theorem satisfies the axioms of a topological space: The sets $\emptyset$ and X clearly belong to $\mathcal{O}$. If O_i for each $i \in I$ belongs to $\mathcal{O}$, then so does $\bigcup O_i$ because with every $x \in O_i$ also some $N_x \in \mathcal{N}(x)$ belongs to O_i and hence to $\bigcup O_i$. Finally let $O_1, O_2 \in \mathcal{O}$ and let $x \in O_1 \cap O_2$. Then there exist N_x^1 and N_x in $\mathcal{N}(x)$ such that $N_x^1 \subseteq O_1$ and $N_x \subseteq O_2$. Hence $N_x = N_x^1 \cap N_x^2 \subseteq O_1 \cap O_2$ and $x \in N_x$. By axiom (Nb.3) we have $N_x \in \mathcal{N}(x)$ and so $O_1 \cap O_2$ contains with each point x an entire set N_x of $\mathcal{N}(x)$. Therefore if O_1 and O_2 are in $\mathcal{O}$, then $O_1 \cap O_2$ is also an element of $\mathcal{O}$. Thus $\mathcal{O}$ is the family of open sets of a topology $\mathcal{T}$.

We must prove that the neighborhood filter of the point x relative to $\mathcal{T}$ is exactly $\mathcal{N}(x)$. Let N be a neighborhood of x with respect to $\mathcal{T}$. Then there is an $O \in \mathcal{O}$ such that $x \in O \subseteq N$ and so by the definition of the family $\mathcal{O}$ there is an N_x in $\mathcal{N}(x)$ such that $N_x \subseteq O$. By $N_x \subseteq O \subseteq N$ and by (Nb.2) we have $N \in \mathcal{N}(x)$. This shows that x has no other neighborhoods than elements of the family $\mathcal{N}(x)$. Conversely, we show that every $N_x \in \mathcal{N}(x)$ is a neighborhood of x with respect to $\mathcal{T}$; namely, we prove the existence of an $O \in \mathcal{O}$ such that $x \in O \subseteq N_x$: In fact, let

$$O = \{y : N_x \in \mathcal{N}(y)\}.$$

Then $x \in O$ because by our choice $N_x \in \mathcal{N}(x)$. Moreover, $O \subseteq N_x$ because if $N_x \in \mathcal{N}(y)$, then by (Nb.1) $y \in N_x$. To prove $O \in \mathcal{O}$ let $y \in O$ be given. By the definition of O we have $N_x \in \mathcal{N}(y)$ and so by (Nb.4) there exists an O_y in $\mathcal{N}(y)$ such that $N_x \in \mathcal{N}(z)$ for every $z \in O_y$. However, if $N_x \in \mathcal{N}(z)$, then $z \in O$ and so if $z \in O_y$, then $z \in O$, that is, $O_y \subseteq O$. Therefore O contains with each y a set O_y of $\mathcal{N}(y)$ whence $O \in \mathcal{O}$.

The uniqueness of the topology $\mathcal{T}$ is obvious from the above reasoning because the open sets of $\mathscr{T}$ can be characterized by the families $\mathscr{N}(x)$ $(x \in X)$: O is open if and only if it contains with each point x a neighborhood N_x.

EXERCISES

1. Usually it is more convenient to describe instead of $\mathscr{N}(x)$ $(x \in X)$ a suitable base $\mathscr{B}(x)$ $(x \in X)$ for $\mathscr{N}(x)$ $(x \in X)$. Find a necessary and sufficient condition that $\mathscr{B}(x)$ $(x \in X)$ be a base for some $\mathscr{N}(x)$ $(x \in X)$.

2. Give a simple description of the neighborhood filters of a point in the following cases: A metric space, the interval topology, the half-open interval topologies.

3. Show that $\mathscr{B}_1(x)$ and $\mathscr{B}_2(x)$ generate the same neighborhood filter $\mathscr{N}(x)$ if and only if every $B_x{}^1$ contains some $B_x{}^2$ and conversely.

4. Prove that in any metric space the closed balls

$$C_\epsilon[x] = \{y : d(x, y) \leqslant \epsilon\} \qquad (\epsilon > 0)$$

form a base for the neighborhood filter of x.

5. Show that the filter $\mathscr{F}(x)$ generated by $\mathscr{C}(x)$ is related to $\mathscr{N}(x)$ by the inclusion $\mathscr{F}(x) \subseteq \mathscr{N}(x)$.

8. Uniform Structures

These are natural generalizations of metric spaces. In a metric space a distance function d is defined on the set $X \times X$ and this function determines a topology on X. In uniform structures a family $\mathscr{U}$ of subsets U of $X \times X$ is given and a topology, called the uniform topology associated with $\mathscr{U}$, is described in terms of these sets U. The family is subject to a set of axioms which are analogous to those of a metric. The importance of these structures lies in the fact that uniform spaces preserve the main features of metric spaces. For instance, in terms of these structures one can speak about uniform convergence. Cauchy sequences, and completeness. If X and Y are sets with uniform structures, then one can define the uniform continuity of a function f from X into Y.

Definition 1. *A uniform structure for a set X is a nonvoid family $\mathscr{U}$ of subsets U of $X \times X$ such that the following axioms are satisfied:*

(U.1) *If* $U \in \mathcal{U}$, *then* U *is not void.*

(U.2) *If* $U_1 \in \mathcal{U}$ *and if* $U_1 \subseteq U$, *then* $U \in \mathcal{U}$.

(U.3) *If* $U_1 \in \mathcal{U}$ *and* $U_2 \in \mathcal{U}$, *then* $U_1 \cap U_2 \in \mathcal{U}$.

(U.4) *If* $U \in \mathcal{U}$, *then* $I \subseteq U$.

(U.5) *If* $U \in \mathcal{U}$, *then* $U^{-1} \in \mathcal{U}$.

(U.6) *If* $U \in \mathcal{U}$, *then there is a* $V \in \mathcal{U}$ *such that* $V \circ V \subseteq U$.

The sets U are called *uniformities* or *surroundings*.

As usual, I denotes the *diagonal*: $I = \{(x, x) : x \in X\}$. The first axiom (U.1) is an immediate consequence of (U.4). The reason for including it in the axiom system is that axioms (U.1), (U.2), and (U.3) define $\mathcal{U}$ to be a filter. As we shall see, the remaining axioms (U.4), (U.5), and (U.6) correspond to axioms (M.1), (M.2), and (M.3) of a metric space. The composition operator $\circ$ is defined by

$$U \circ V = \{(x, z) : (x, y) \in U \quad \textit{and} \quad (y, z) \in V \quad \textit{for some} \quad y.\}.$$

A simple consequence of (U.4) is: $U \subseteq U \circ U$ for every uniformity $U \in \mathcal{U}$. Indeed if $(x, y) \in U$, then by $(y, y) \in U$ we have $(x, y) \in U \circ U$. As usual $U^{-1} = \{(y, x) : (x, y) \in U\}$ and a uniformity U is called symmetric if $U = U^{-1}$. The axioms imply a sharper form of axiom (U.6):

Lemma 1. *Given any* $U \in \mathcal{U}$ *there is a symmetric* $W \in \mathcal{U}$ *such that* $W \circ W \subseteq U$.

Proof. By (U.6) there is a $V \in \mathcal{U}$ such that $V \circ V \subseteq U$. By (U.5) and (U.3) $W = V \cap V^{-1}$ is symmetric and $W \circ W \subseteq V \circ V \subseteq U$. Hence W has the required properties.

In practice, a uniform structure $\mathcal{U}$ is often described by specifying the elements V of a filter base $\mathcal{U}_B$ instead of giving all the elements of $\mathcal{U}$. It is easy to see that every filter base $\mathcal{U}_B$ of a uniform structure $\mathcal{U}$ satisfies the following axioms, and conversely if $\mathcal{U}_B$ is such that axioms (Ub.3)–(Ub.6) hold, then the filter generated by $\mathcal{U}_B$ is a uniform structure. Thus we have:

Theorem 1. *A nonvoid family* $\mathcal{U}_B$ *of subsets* V *of* $X \times X$ *is a base for a uniform structure* $\mathcal{U}$ *if and only if the following axioms hold:*

(Ub.3) *If* $V_1 \in \mathcal{U}_B$ *and* $V_2 \in \mathcal{U}_B$, *then there is a* $V \in \mathcal{U}_B$ *such that* $V \subseteq V_1 \cap V_2$.

(Ub.4) *If* $V \in \mathcal{U}_B$, *then* $I \subseteq V$.

(Ub.5) *If* $V \in \mathcal{U}_B$, *then there exists a* $W \in \mathcal{U}_B$ *such that* $W \subseteq V^{-1}$.

(Ub.6) *If* $V \in \mathcal{U}_B$, *there is a* $W \in \mathcal{U}_B$ *such that* $W \circ W \subseteq V$.

If $\mathscr{U}_B$ satisfies these axioms, $\mathscr{U}$, the uniform structure associated with $\mathscr{U}_B$, is the family of those subsets U of $X \times X$ for which $V \subseteq U$ for some $V \in \mathscr{U}_B$.

Every uniform structure $\mathscr{U}$ has a symmetric structure base, i.e., a base $\mathscr{U}_B$ consisting of symmetric uniformities. In fact, one can choose for $\mathscr{U}_B$ the set of all symmetric uniformities of $\mathscr{U}$.

It is possible to define uniform structures also by subbases. The notion of a subbase is meaningful for every filter but here we use it only for uniform structures:

Theorem 2. *A family $\mathscr{U}_S$ of nonvoid subsets S of $X \times X$ is a subbase for a uniform structure if it satisfies the following axioms:*

(Us.4) $\quad I \subseteq S$ *for every* $S \in \mathscr{U}_S$.

(Us.5) *If* $S \in \mathscr{U}_S$, *then there exist sets* $T_1, \ldots, T_m \in \mathscr{U}_S$ *such that* $T_1 \cap \ldots \cap T_m \subseteq S^{-1}$.

(Us.6) *If* $S \in \mathscr{U}_S$, *then there exist sets* $T_1, \ldots, T_m \in \mathscr{U}_S$ *such that* $(T_1 \cap \ldots \cap T_m) \circ (T_1 \cap \ldots \cap T_m) \subseteq S$.

The uniform structure generated by $\mathscr{U}_S$ is determined by the base $\mathscr{U}_B$ consisting of all finite intersections $V = S_1 \cap \ldots \cap S_n$ where $S_1, \ldots, S_n \in \mathscr{U}_S$.

Note. Axioms (Us.5) and (Us.6) can be replaced by a single axiom: *If $S \in \mathscr{U}_S$, then there exist sets $T_1, \ldots, T_m \in \mathscr{U}_S$ such that*

$$(T_1 \cap \ldots \cap T_m) \circ (T_1 \cap \ldots \cap T_m) \subseteq S \cap S^{-1}.$$

Proof. It is simple to show that $\mathscr{U}_B$ satisfies the axioms of a structure base. In fact, (Ub.3) is a consequence of the definition of $\mathscr{U}_B$ and (Ub.4) is implied by (Us.4). Similarly, (Ub.5) can be derived from (Us.5). To prove (Ub.6) one needs the relation

$$(T^1 \cap \ldots \cap T^n) \circ (T^1 \cap \ldots \cap T^n) \subseteq (T^1 \circ T^1) \cap \ldots \cap (T^n \circ T^n).$$

This follows from the definition of the composition operator: If $(x, z) \in (T^1 \cap \ldots \cap T^n) \circ (T^1 \cap \ldots \cap T^n)$, then there is a y such that (x, y) and (y, z) belong to $T^1 \cap \ldots \cap T^n$. Hence $(x, z) \in T^k \circ T^k$ for every $k = 1, \ldots, n$, and so $(x, z) \in (T^1 \circ T^1) \cap \ldots \cap (T^n \circ T^n)$. Now let $V \in \mathscr{U}_B$ be given. Then $V = S_1 \cap \ldots \cap S_n$ for some $S_1, \ldots, S_n \in \mathscr{U}_S$. By axiom (Us.6) there exist $T_1^k, \ldots, T_{m_k}^k \in \mathscr{U}_S$ for every $k = 1, \ldots, n$ such that $T^k = T_1^k \cap \ldots \cap T_{m_k}^k$ satisfies $T^k \circ T^k \subseteq S_k$. Hence if we put $T = T^1 \cap \ldots \cap T^n$, then by the preceding relation

$$T \circ T \subseteq (T^1 \circ T^1) \cap \ldots \cap (T^n \circ T^n) \subseteq S_1 \cap \ldots \cap S_n = V.$$

This shows that axiom (Ub.6) holds.

Next we associate a topology $\mathcal{T}$ of X with the uniform structure $\mathcal{U}$:

Theorem 3. *Given a uniform structure $\mathcal{U}$ on the set $X \times X$ consider for each $x \in X$ the family $\mathcal{N}(x)$ of all sets*

$$U[x] = \{y : (x, y) \in U\} \quad \textit{where} \quad U \in \mathcal{U}.$$

Then each $\mathcal{N}(x)$ satisfies axioms (Nb.1)–(Nb.4). *The topology generated by these neighborhood filters $\mathcal{N}(x)$ is called the uniform topology associated with $\mathcal{U}$ or the topology derived from $\mathcal{U}$.*

Note. Sometimes $U[x]$ is referred to as "*U evaluated at x.*" One defines $U[A]$ for every $A \subseteq X$ as

$$U[A] = \bigcup \{U[x] : x \in A\} = \{y : (x, y) \in U \quad \textit{for some} \quad x \in A\}.$$

This set is "*U evaluated on A.*" One easily verifies that

$$(U \circ V)[A] = V[U[A]].$$

Proof. Axiom (Nb.1) is a consequence of (U.4). To show (Nb.2) let $U[x]$ and $N \supseteq U[x]$ be given. We consider the set

$$V = U \cup \{(x, y) : y \in N\}.$$

Since $U \subseteq V$ we have $V \in \mathcal{U}$ and by our construction $V[x] = N$. Hence if $N \supseteq U[x]$, then $N \in \mathcal{N}(x)$. Next, in order to prove (Nb.3) we notice that $U \cap V \in \mathcal{U}$ for every pair $U, V \in \mathcal{U}$ and

$$U[x] \cap V[x] = (U \cap V)[x].$$

Hence if $U[x], V[x] \in \mathcal{N}(x)$, then also $U[x] \cap V[x] \in \mathcal{N}(x)$. Finally we must prove (Nb.4): Given $U[x] \in \mathcal{N}(x)$ consider $V \in \mathcal{U}$ such that $V \circ V \subseteq U$. By axiom (U.6) such a uniformity V exists. The set $V[x]$ can be chosen as O_x of (Nb.4): In fact, if $y \in V[x]$, then $V[y] \subseteq U[x]$ because $y \in V[x]$ implies $(x, y) \in V$ and so if $z \in V[y]$, i.e., if $(y, z) \in V$, then $(x, z) \in V \circ V \subseteq U$, i.e., $z \in U[x]$. Since $V[y] \in \mathcal{N}(y)$ we have: If $y \in V[x]$, then $U[x] \in \mathcal{N}(y)$. Therefore the families $\mathcal{N}(x)$ $(x \in X)$ satisfy axioms (Nb.1)–(Nb.4).

Earlier it was mentioned that not every topology can be defined by a metric. Similarly, there are topologies which are not associated with any uniform structure. For instance, the topology of finite complements of the set of integers gives such a space. If a topology can be derived from some uniform structure $\mathcal{U}$, then it is called a *uniformizable topology* and one speaks about a *uniformizable space*. The expression *uniform space* is also used.

Theorem 4. *Every pseudometric space is uniformizable.*

Proof. Given a pseudometric d we consider the sets

$$V_\epsilon = \{(x, y) : d(x, y) < \epsilon\}$$

where $\epsilon > 0$. The family $\mathscr{U}_B$ of all these sets V_ϵ is a base for a uniform structure $\mathscr{U}$: We have $V_{\epsilon_1} \cap V_{\epsilon_2} = V_\epsilon$ where $\epsilon = \min(\epsilon_1, \epsilon_2)$ and so (Ub.3) holds. Axioms (Ub.4) and (Ub.5) are immediate consequences of (M.1a) and of (M.2), respectively. To see (Ub.6) let $W = V_{\epsilon/2}$. The relation $W \circ W \subseteq V$ is assured by the triangle inequality. Hence by Theorem 3 there is a topology $\mathscr{T}$ associated with the uniform structure $\mathscr{U}$. A set $O \subseteq X$ is open relative to $\mathscr{T}$ if and only if it contains with each point x an entire $U[x]$ where $U \in \mathscr{U}$. Since $U \in \mathscr{U}$ if and only if $U \supseteq V_\epsilon$ for some $\epsilon > 0$ the set O is open relative to $\mathscr{T}$ if and only if it contains with each x a set $V_\epsilon[x]$. Thus O is open relative to $\mathscr{T}$ if and only if it is open relative to the pseudometric topology derived from d.

The uniform structure $\mathscr{U}_d$ generated by the structure base $\mathscr{U}_{Bd}$ consisting of the sets $V_\epsilon = \{(x, y) : d(x, y) < \epsilon\}$ $(\epsilon > 0)$ is called the *uniform structure associated with the pseudometric d.* If $\mathscr{U}$ is a uniform structure such that $\mathscr{U} = \mathscr{U}_d$ for a suitable pseudometric d, then $\mathscr{U}$ is called a *pseudometrizable structure.*

EXERCISES

1. Let X be a nonvoid set and let $\mathscr{U}$ consist of the single set $X \times X$. Show that $\mathscr{U}$ is a uniform structure for X. It is called the *nondiscrete structure.*

[Usually only the last axiom (U.6) needs proving. In the present case even this is trivial.]

2. Let $\mathscr{U}$ be the set of those $U \subseteq X \times X$ which contain the diagonal I. Show that $\mathscr{U}$ is a uniform structure for X. This is the *discrete structure.*

3. Let X be the set of reals. Consider for every $\epsilon > 0$ the set $V_\epsilon = \{(x, y) : |x - y| < \epsilon\}$ and verify that $\mathscr{U}_B = \{U_\epsilon\}$ $(\epsilon > 0)$ satisfies the axioms of a structure base. The structure generated by $\mathscr{U}_B$ is *the uniform structure of the additive group of the reals* or the "*usual*" *structure of the reals.*

4. Let $\mathscr{U}_B$ be a structure base for the uniform structure $\mathscr{U}$. Prove that the sets $V[x]$ $(V \in \mathscr{U}_B)$ form a base for the neighborhood filter $\{U[x]\}$ $(U \in \mathscr{U})$ of the point x relative to the uniform topology generated by $\mathscr{U}$.

(Given $U \in \mathscr{U}$ there is a $V \in \mathscr{U}_B$ satisfying $V \subseteq U$ and so $V[x] \subseteq U[x]$.)

5. Show that the uniform topology associated with the "usual" uniform structure of the reals is the usual topology of the reals.

(A base for the complete neighborhood filter of x relative to the uniform topology consists of the sets

$$V_\epsilon[x] = \{y : x - \epsilon < y < x + \epsilon\} = (x - \epsilon, x + \epsilon).$$

These are open neighborhoods of x relative to the interval topology. Conversely, if N_x is a neighborhood of x in the interval topology, then there is an $\epsilon > 0$ such that $V_\epsilon[x] = (x - \epsilon, x + \epsilon) \subseteq N_x$. Hence the base $\{V_\epsilon[x]\}$ $(\epsilon > 0)$ generates exactly the neighborhoods of x relative to the interval topology.)

6. Let X be the set of the reals and consider for every $a < b$ the sets

$$S_{ab} = \{(x, y) : \textit{both} \quad x, y < b \quad \textit{or} \quad \textit{both} \quad x, y > a\}.$$

Show that $\{S_{ab}\}$ $(a < b$ and $a, b \in X)$ is a subbase for a uniform structure.

[Only the last axiom (Us.6) needs proving. Given $a < b$ choose reals c and d such that $a < c < d < b$. Then $T = S_{ac} \cap S_{cd} \cap S_{db}$ has the property $T \circ T \subseteq S_{ab}$.]

7. Show that the uniform topology associated with the uniform structure $\mathscr{U}$ constructed in the preceding exercise is the usual topology of the reals.

(If $x \leqslant a$, then $S_{ab}[x] = \{y : y < b\}$; if $b \leqslant x$, then

$$S_{ab}[x] = \{y : a < y\};$$

and if $a < x < b$, then $S_{ab}[x] = X$. Hence every $S_{ab}[x]$ is open relative to the interval topology and so $\{S_{ab}[x]\} \subseteq \mathscr{N}(x)$. Now given $x \in X$ and an open interval (a, b) containing x we can find real numbers c, d such that $a < c \leqslant x \leqslant d < b$. Then $S_{ac}[x] = (a, +\infty)$ and $S_{db}[x] = (-\infty, b)$ so that $(S_{ad} \cap S_{cb})[x] = (a, b)$. The sets $T[x]$ where T is a finite intersection of sets S_{ab} $(a < b$ and $a, b \in X)$ form a base for the neighborhood filter of x relative the interval topology of the reals.)

8. Show that the uniform structure constructed in Exercise 6 is not the usual uniform structure of the reals.

(The following difference between the two structures is very important: In the case of the structure given in Exercise 6 for every uniformity U there are finitely many points $x_1, \ldots, x_m$ such that

$$U[x_1] \cup \ldots \cup U[x_m] = X;$$

namely, if $U \supseteq S_{a_1 b_1} \cap \ldots \cap S_{a_m b_m}$, then it is sufficient to choose $x_i \in (a_i, b_i)$ for each $i = 1, \ldots, m$. In the case of the usual structure there are uniformities such that $U[x_1] \cup \ldots \cup U[x_m] \subset X$ for all choices of finitely many points.)

9. Show that the metrics d, $\min(1, d)$, and $d/(1 + d)$ generate the same uniform structure.

[Let V_ϵ^1, V_ϵ^2, and V_ϵ^3 denote the ϵ-uniformities generated by the metrics d, $\min(1, d)$, and $d/(1 + d)$, respectively. If $\epsilon < \frac{1}{2}$, then $V_\epsilon^1 \subseteq V_\epsilon^3 \subseteq V_{2\epsilon}^1$ and so $\{V_\epsilon^1\}$ $(\epsilon > 0)$ and $\{V_\epsilon^3\}$ $(\epsilon > 0)$ are equivalent filter bases. If $\epsilon \leqslant 1$, then $V_\epsilon^1 = V_\epsilon^2$ and so $\{V_\epsilon^1\}$ $(\epsilon > 0)$ and $\{V_\epsilon^2\}$ $(\epsilon > 0)$ are also equivalent.]

10. Let X be the set of the reals and let for $n = 0, 1, \ldots$ the metric d_n be defined by $d_n(x, y) = | x^{2n+1} - y^{2n+1} |$. Show that the structure generated by d_0 is distinct from every structure generated by d_n $(n > 0)$. Using a similar argument extend this result to show that the structures generated by these metrics are all distinct.

[Let V_ϵ^n denote the ϵ-uniformity generated by the metric d_n. We show that no V_ϵ^0 $(\epsilon > 0)$ is contained entirely in any V_1^n for $n > 0$. For let x be a large positive number and let y be such that $x - y = \delta$ where $0 < \delta < \epsilon$. Then $(x, y) \in V_\epsilon^0$ but $d_n(x, y) > 1$ if x is sufficiently large because

$$d_n(x, y) = x^{2n+1} - y^{2n+1} = x^{2n}(\delta + y) - y^{2n+1} = \delta x^{2n} + y\,(x^{2n} - y^{2n}) > \delta x^{2n}.$$

In general, when $m < n$ are arbitrary positive integers we choose for a given $x > 0$ the real number y such that $x^{2m+1} - y^{2m+1} = \delta$ and use the inequality

$$d_n(x, y) = x^{2n+1} - y^{2n+1} > \delta x^{2(n-m)}$$

to show that $(x, y) \notin V_1^n$ for all sufficiently large vales of $x > 0$.]

11. Show that the topologies generated by the metrics d_n of the preceding exercise are all the same.

(Compare every topology $\mathscr{T}_n$ with $\mathscr{T}_0$.)

12. Let X be a linearly ordered set and for every $a \in X$ let

$$S_a = \{(x, y) : \textit{both} \quad x, y < a \quad \textit{or both} \quad x, y \geqslant a\}.$$

Show that $\mathscr{U}_S = \{S_a\}$ $(a \in X)$ is a subbase for a uniform structure $\mathscr{U}$ and the uniform topology associated with $\mathscr{U}$ is the right half-open interval topology.

[Axioms (Us.4)–(Us.6) hold trivially because $S_a \circ S_a = S_a$ for every $a \in X$. Given $a < b$ in X we have $U_a[a] = \{y : y \geqslant a\}$ and $U_b[a] = \{y : y < b\}$. Hence $(U_a \cap U_b)\,[a] = [a, b)$ and so every element $[a, b)$ of a base of the right half-open interval topology can be represented in the form $T[a]$ where T belongs to the structure base generated by $\mathscr{U}_S$. Since the intersection of finitely many right half-open intervals is itself a right half-open interval, only such intervals are representable in the form $T[a]$ $(a \in X$ and $T \in \mathscr{U}_B)$.]

13. Prove that if X is the set of the real numbers, then every structure

base of the uniform structure $\mathscr{U}$ constructed in the preceding exercise contains more than countably many elements and so it is not pseudometrizable.

[First we prove that no countable base can be selected from the structure base $\mathscr{U}_B$ generated by the subbase $\mathscr{U}_S$. If $V \in \mathscr{U}_B$, then $V = S_{a_1} \cap \dots \cap S_{a_n}$ for some $a_1 < \dots < a_n$. Suppose that $a_i < a < a_{i+1}$. Choose x and y such that $a_i < x < a < y < a_{i+1}$. Then $(x, y) \in V$ but $(x, y) \notin S_a$. The situation is similar if $a < a_1$ or if $a_n < a$ and so $S_a \subseteq V$ only if $a = a_i$ for some index i. Since the set of real numbers is not countable $\mathscr{U}_B$ has no countable subset which is a structure base. If some other countable structure base existed we could select a countable base also from $\mathscr{U}_B$.]

9. Simple Results on Uniform Structures and Uniform Spaces

Here we collect a number of elementary lemmas. We introduce also an order classification for uniform structures of a given set and define the least upper bound and the greatest lower bound of a family of structures formed on a fixed set.

Lemma 1. *For every $U \in \mathscr{U}$ the point x lies in the interior of $U[x]$.*

Proof. By definition the neighborhood filter of x consists of the sets $U[x]$ $(U \in \mathscr{U})$. Hence every $U[x]$ contains an open set O_x such that $x \in O_x \subseteq U[x]$.

Lemma 2. *A point x belongs to the interior of a set A if and only if there is a $U \in \mathscr{U}$ such that $U[x] \subseteq A$.*

Corollary. *A point x lies in the exterior of A if and only if there is a U such that $U[x] \subseteq cA$, and x is a boundary point of A if and only if $U[x]$ intersects both A and cA for every $U \in \mathscr{U}$.*

Proof. If there is a U such that $U[x] \subseteq A$, then by Lemma 1 we have $x \in U[x]^i \subseteq A^i$. Conversely, if $x \in A^i$, then A^i being open there is a U in $\mathscr{U}$ such that $U[x] \subseteq A^i$ and so $U[x] \subseteq A$.

Lemma 3. *For every nonvoid open set O in a uniform space X and for every $x \in O$ there is a U such that $O = U[x]$.*

Proof. Let $x \in O$ and let $V \in \mathscr{U}$ be such that $V[x] \subseteq O$. We define $U = \{(x, y) : y \in O\} \cup V$. Since $V \subseteq U$ and $V \in \mathscr{U}$ by axiom (U.2) we have $U \in \mathscr{U}$. Also $U[x] = V[x] \cup O = O$ because $V[x] \subseteq O$.

Lemma 4. *If $U[A] \subseteq B$ for some U, then $A \subseteq B^i$.*

Proof. By Lemma 1 we have $A \subseteq U[A]^i$ and so by the hypothesis $A \subseteq B^i$.

Lemma 5. *If $U[A] \subseteq B$ for some U, then $\bar{A} \subseteq B$.*

Proof. We may suppose that U is symmetric since there is a symmetric V such that $V[A] \subseteq (V \circ V)[A] \subseteq U[A]$. If $x \in \bar{A}$, then $a \in U[x]$ for some $a \in A$. Then $(x, a) \in U$ and so by the symmetry $(a, x) \in U$ and $x \in U[a] \subseteq U[A] \subseteq B$.

The preceding two lemmas can be united in the following stronger result:

Lemma 6. *If $U[A] \subseteq B$ for some U, then $\bar{A} \subseteq B^i$.*

Proof. By axiom (U.6) there is a $V \in \mathscr{U}$ such that $V \circ V \subseteq U$. Therefore $V[V[A]] = (V \circ V)[A] \subseteq U[A] \subseteq B$ and so by Lemma 4 we have $V[A] \subseteq B^i$. Next we apply Lemma 5 and obtain $\bar{A} \subseteq B^i$.

A useful special case of the last lemma is:

Lemma 7. *If $U, V \in \mathscr{U}$ and if there is a $W \in \mathscr{U}$ such that $U \circ W \subseteq V$, then for every A we have $\overline{U[A]} \subseteq V[A]^i$.*

Proof. We have $W[U[A]] = (U \circ W)[A] \subseteq V[A]$ and hence by the above lemma $\overline{U[A]} \subseteq V[A]^i$.

Let $\mathscr{U}_1$ and $\mathscr{U}_2$ be uniform structures for the same set X. We define $\mathscr{U}_1$ to be *coarser* than $\mathscr{U}_2$ and $\mathscr{U}_2$ to be *finer* than $\mathscr{U}_1$ if $\mathscr{U}_1 \subseteq \mathscr{U}_2$, and we write $\mathscr{U}_1 \leqslant \mathscr{U}_2$ or $\mathscr{U}_2 \geqslant \mathscr{U}_1$. If $\mathscr{U}_1 \subset \mathscr{U}_2$, then $\mathscr{U}_1$ is called *strictly coarser* than $\mathscr{U}_2$ and $\mathscr{U}_2$ is *strictly finer* than $\mathscr{U}_1$; in symbols, $\mathscr{U}_1 < \mathscr{U}_2$ and $\mathscr{U}_2 > \mathscr{U}_1$. This leads to an order classification of the family of all uniform structures for X. The coarsest structure is the nondiscrete structure and its finest structure is the discrete one. If $\mathscr{U}_1 \leqslant \mathscr{U}_2$, then the uniform topologies $\mathscr{T}_1$ and $\mathscr{T}_2$ are related by $\mathscr{T}_1 \leqslant \mathscr{T}_2$. For O is open relative to $\mathscr{T}_i$ if and only if it contains with each point x a whole set $U_i[x]$ where $U_i \in \mathscr{U}_i$, and so if O is open relative to $\mathscr{U}_1$, then it is also open relative to $\mathscr{U}_2$. It is possible that $\mathscr{U}_1 < \mathscr{U}_2$ and $\mathscr{T}_1 = \mathscr{T}_2$. Moreover, it can happen that $\mathscr{U}_1$ and $\mathscr{U}_2$ are not comparable; nevertheless, $\mathscr{T}_1 = \mathscr{T}_2$.

Let $\mathscr{U}_i$ for every index $i \in I$ be a uniform structure for the fixed set X. A structure $\mathscr{U}$ is an upper bound of the family if $\mathscr{U}_i \leqslant \mathscr{U}$ for every $i \in I$. We prove that there is a *least upper bound.* In fact, $\mathscr{U}_S = \bigcup\{\mathscr{U}_i : i \in I\}$ satisfies axioms (Us.4)–(Us.6) of a subbase and the uniform structure $\text{lub}\{\mathscr{U}_i\}$ generated by $\mathscr{U}_S$ satisfies the requirements: On the one hand,

$\text{lub}\{\mathscr{U}_i\}$ is finer than every $\mathscr{U}_i$ $(i \in I)$, and, on the other hand, if a structure $\mathscr{U}$ is finer than every $\mathscr{U}_i$ $(i \in I)$, then $\bigcup \{\mathscr{U}_i : i \in I\} \subseteq \mathscr{U}$ and so $\text{lub}\{\mathscr{U}_i\} \leqslant \mathscr{U}$. There are families which do not have a greatest lower bound. As a matter of fact, there are families such that the uniform topology associated with $\mathscr{U}_i$ is the same for every i; nevertheless, no greatest lower bound exists for $\{\mathscr{U}_i\}$ $(i \in I)$.

In Lemma 6.2 we gave a simple criterion for the equivalence of metric topologies formed on the same set X. This result can also be extended to uniform spaces:

Lemma 8. *The topology $\mathscr{T}_1$ generated on the set X by the uniform structure $\mathscr{U}_1$ is finer than the topology $\mathscr{T}_2$ generated by $\mathscr{U}_2$ if and only if for each $x \in X$ and for every $U_2 \in \mathscr{U}_2$ there is a $U_1 \in \mathscr{U}_1$ such that $U_1[x] \subseteq U_2[x]$.*

Note. It is understood that the uniformity $U_1 \in \mathscr{U}_1$ may depend on $x \in X$ and the lemma can be used only to show that $\mathscr{T}_2 \leqslant \mathscr{T}_1$. It tells nothing about the relationship between the uniform structures $\mathscr{U}_1$ and $\mathscr{U}_2$.

Proof. If the condition is satisfied, then every set O which is open relative to $\mathscr{T}_2$ is also open relative to $\mathscr{T}_1$. Hence $\mathscr{T}_1$ is finer than $\mathscr{T}_2$ and the condition is sufficient. Next we suppose that $\mathscr{T}_1$ is finer than $\mathscr{T}_2$ and show that the condition is satisfied: Given x and U_2, the point x belongs to the interior of the set $U_2[x]$. Let $U_2[x]^i$ denote the interior relative to $\mathscr{T}_2$. Since $U_2[x]^i$ is open relative to $\mathscr{T}_2$ and $\mathscr{T}_1$ is finer than $\mathscr{T}_2$, it is open also relative to $\mathscr{T}_1$. Hence there is a uniformity $U_1 \in \mathscr{U}_1$ such that $x \in U_1[x] \subseteq U_2[x]^i$ and so $U_1[x] \subseteq U_2[x]$. This proves the necessity of the condition.

EXERCISES

1. Suppose that the uniform topology associated with $\mathscr{U}_i$ is the same topology $\mathscr{T}$ for every $i \in I$ where I is an arbitrary index set. Show that the topology associated with $\text{lub}\{\mathscr{U}_i\}$ is also $\mathscr{T}$.

(Let $\mathscr{T}_1$ be the topology associated with the least upper bound. Since $\mathscr{U}_i \leqslant \text{lub}\{\mathscr{U}_i\}$ we have $\mathscr{T} \leqslant \mathscr{T}_1$. Let $U[x]$ be a neighborhood of x relative to $\mathscr{T}_1$. Then $U \in \text{lub}\{\mathscr{U}_i\}$ and so $\bigcup \{\mathscr{U}_i : i \in I\}$ being a subbase we can find $U_{i_1}, \ldots, U_{i_n}$ such that $U \supseteq U_{i_1} \cap \ldots \cap U_{i_n}$. Hence

$$U[x] \supseteq (U_{i_1} \cap \ldots \cap U_{i_n})[x] = U_{i_1}[x] \cap \ldots \cap U_{i_n}[x]$$

where $U_{i_1}[x], \ldots, U_{i_n}[x]$ are neighborhoods of x relative to $\mathscr{T}$. Therefore the neighborhood $U[x]$ relative to $\mathscr{T}_1$ is a neighborhood also relative to $\mathscr{T}$ and $\mathscr{T}_1 \leqslant \mathscr{T}$.)

2. Give an example to show that lub$\{\mathscr{T}_1, \mathscr{T}_2\}$ may be uniformizable while $\mathscr{T}_1$ and $\mathscr{T}_2$ are not.

[Let the open sets of $\mathscr{T}_i$ $(i = 1, 2)$ be ø and all sets containing a fixed point a_i $(a_1 \neq a_2)$. Then $\mathscr{T}_i$ is not uniformizable because if $x \neq a_i$ there is a neighborhood of a_i not containing x but there is no neighborhood of x not containing a_i. In a uniform space this situation is not possible. Since lub$\{\mathscr{T}_1, \mathscr{T}_2\}$ is the discrete topology it is uniformizable.]

3. Let the uniform topology associated with the uniform structure $\mathscr{U}_i$ be $\mathscr{T}_i$. Show that the topology associated with lub$\{\mathscr{U}_i\}$ is lub$\{\mathscr{T}_i\}$.

10. Subspaces

Let X be a topological space with topology $\mathscr{T}$ and let Y be a nonvoid subset of X. The family $\mathscr{O}_Y = \{O \cap Y\} (O \in \mathscr{O})$ where $\mathscr{O}$ is the family of open sets of X satisfies the axioms for the open sets of a topological space on Y. This topology is called the *relativization* of $\mathscr{T}$ to the set Y or the *relative topology* or the *induced topology* and Y under this topology is called a subspace of X.

Let X be a topological space and let Y and Z be subsets of X such that $Z \subseteq Y \subseteq X$. Then Z can be considered as a subspace of X or as a subspace of the subspace Y. These two topologies on Z are identical. In fact, if O is open in X, then $O \cap Z = (O \cap Y) \cap Z$ where by definition $O \cap Y$ is an open set of the subspace Y. A simple consequence of this rule is the following: If X is a topological space and A and B are subsets of X, then $A \cap B$ as a subspace of the subspace A and $A \cap B$ as a subspace of the subspace B are identical and as a matter of fact, their topology is the relativization of the topology of X to $A \cap B$.

Lemma 1. *Let X be a topological space, let Y be a subspace of X and let S be a set in Y. Then*

(i) *the set S is an open set of Y if and only if there is a set O open in X and such that $S = O \cap Y$;*

(ii) *the set S is a closed set of Y if and only if there is a set C closed in X and such that $S = C \cap Y$;*

(iii) *the closure of S relative to the topology of Y is $\bar{S} \cap Y$ where $\bar{S}$ is the closure of S in X;*

(iv) *the interior of S relative to the topology of Y contains the interior of S in X;*

(v) *the boundary of S relative to the topology of Y is contained in the boundary of S in X.*

Proof. Statement (i) is the definition of a subspace. The set S is closed in Y if and only if the complement of S relative to Y is open in Y, i.e., if and only if $c_Y(S) = O \cap Y$ where O is open in X. Hence S is closed if and only if $S = c(O \cap Y) \cap Y = (cO \cup cY) \cap Y = cO \cap Y$, i.e., when $S = C \cap Y$ where C is closed in X. This proves (ii). Let $\bar{S}_Y$ denote the closure of S with respect to Y. Then by the definition of the closure $\bar{S}_Y = \bigcap (C \cap Y)$ where the intersection is taken for every closed set C in X satisfying $S \subseteq C \cap Y$. Therefore by $S \subseteq Y$ we have

$$\bar{S}_Y = \bigcap \{C \cap Y : S \subseteq C\} = \bigcap \{C : S \subseteq C\} \cap Y = \bar{S} \cap Y.$$

This proves (iii). Statement (iv) follows similarly:

$$S^i_Y = \bigcup \{O \cap Y : O \cap Y \subseteq S\} \supseteq \bigcup \{O : O \subseteq S\} \cap Y = S^i \cap Y = S^i.$$

To see (v) let $y \in Y$ be a boundary point of S relative to the topology on Y. Then for every open set O in X containing the point y the set $O \cap Y$ intersects both S and $c_Y(S)$. Hence every open O containing y intersects both S and cS and so y is a boundary point of S in X.

If the set on which the subspace is formed is an open or closed subset of the space X, then the results given in the last lemma become much simpler:

Lemma 2. *Let X be a topological space, let Y be a closed subset of X, and let S be a subset of Y. Then S is a closed set of the subspace Y if and only if it is closed in X. The closure of S in Y is the same as its closure in X.*

Proof. If the set S is closed in Y by part (ii) of Lemma 1 we have $S = C \cap Y$ where C is closed in X. Since Y is closed in X the set S is also closed in X. Conversely, if S is closed in X, then by $S = S \cap Y$ it is closed in the subspace Y. If Y is closed and $S \subseteq Y$, then $\bar{S} \subseteq Y$ and so by (iii) we have $\bar{S}_Y = \bar{S} \cap Y = \bar{S}$.

Lemma 3. *Let X be a topological space, let Y be an open subset of X and let S be a subset of Y. Then the set S is open in the subspace Y if and only if S is open in X. The boundary of S in Y is $S^b{}_Y = S^b \cap Y$. The interior of S in Y is the same as in X.*

Proof. If S is open in Y, then $S = O \cap Y$ where O is open in X and so S is open in X. Conversely, if S is open in X, then it is an open set of the subspace Y. By part (v) of Lemma 1 every boundary point of S in Y is a boundary point of S in X and so it is sufficient to show that $S_Y{}^b \supseteq S^b \cap Y$. If $y \in S^b \cap Y$, then every open set O in X which contains y intersects both S and cS. However, every $O \subseteq Y$ which is open in Y is also open in X and so O intersects both S and cS. Since $O \subseteq Y$ the set O intersects $c_Y(S)$ and this shows that $y \in S_Y{}^b$. The interiors of S in Y and in X are identical because O is open in Y if and only if it is open in X.

Lemma 4. *A set N is a neighborhood of a point $y \in Y$ in the subspace Y if and only if $N = N_y \cap Y$ where N_y is a neighborhood of y in X.*

Proof. If N is a neighborhood of y relative to Y, then there is an open set in X such that $y \in O \cap Y \subseteq N$. Therefore $N = N_y \cap Y$ where $N_y = N \cup O$. Conversely, if $N = N_y \cap Y$ where N_y is a neighborhood of y in X, then there is an open set O in X such that $y \in O \subseteq N_y$ and so $y \in O \cap Y \subseteq N_y \cap Y = N$. Since $O \cap Y$ is open in Y the set N is a neighborhood of y in Y.

Suppose that X is a metric space with metric d and $Y \subseteq X$. Then the restriction of the distance function d to points of Y gives a metric on Y. It is easy to see that the topology generated by this metric is the induced topology of Y. More generally, let X be a uniform space with uniform structure $\mathscr{U}$ and let $Y \subseteq X$. Then a uniform structure $\mathscr{U}_Y$ can be defined for Y, namely, $\mathscr{U}_Y$ consists of the sets $U \cap (Y \times Y)$ where $U \in \mathscr{U}$. This $\mathscr{U}_Y$ is called the *relativization* of $\mathscr{U}$ to Y. We can look at Y as a subspace of the uniform space X and we can also consider Y as a uniform space relative to the uniform topology associated with $\mathscr{U}_Y$. We prove that these two topologies are identical and so we obtain:

Theorem 1. *Every subspace Y of a uniform space X is uniformizable and if $\mathscr{U}$ is a structure for X, then $\mathscr{U}_Y = \{U \cap (Y \times Y)\}\ (U \in \mathscr{U})$ is a structure for the subspace Y.*

Note. $U_Y = U \cap (Y \times Y)$ is called the *trace* of U on $Y \times Y$.

Proof. Let $y \in Y$ be fixed. By Lemma 4, the neighborhoods of y in the subspace Y are the neighborhoods of y in X intersected by Y, i.e., they are the sets $U[y] \cap Y$ where $U \in \mathscr{U}$. The neighborhoods of Y in the uniform topology associated with $\mathscr{U}_Y$ are the sets $U_Y[y]$ where $U_Y \in \mathscr{U}_Y$. However $U_Y = U \cap (Y \times Y)$ and so

$$U_Y[y] = \{\eta : \eta \in Y \quad \textit{and} \quad (y, \eta) \in U\} = U[y] \cap Y.$$

This shows that y has the same neighborhoods in both topologies.

EXERCISES

1. Let X be the real line and let $Y = (a, b)$ where $a < b$. Show that Y as a subspace of X is homeomorphic to X.

2. Determine the topology induced on the set of integers by the usual topology of the reals.

3. We define the "circle" or "1-sphere" to be the topological space formed on the plane set $S^1 = \{x : x_1^2 + x_2^2 = 1\}$ by the induced topology. Any topological space homeomorphic to S^1 is called a *simple closed path*. Show that the circle is not homeomorphic to the real line. (The complement of a point of S^1 relative to S^1 is homeomorphic to the real line. The complement of a point of the real line is not homeomorphic to the real line because it is the union of two nonvoid open sets.)

4. Let X be a linearly ordered set topologized by the interval topology and let Y be a subset of X such that if $x \in X$ and $y \in Y$ where $x \neq y$ then the open interval determined by x and y contains points of Y. Show that the topology induced on Y is the interval topology of Y. [The family $\{(a, b) \cap Y\}$ $(a, b \in X \cup \{-\infty, +\infty\})$ is a base for the induced topology of Y. If the hypothesis is satisfied, every $(a, b) \cap Y$ can be expressed as a union of open intervals in Y and the two topologies are identical.]

5. Let X be a topological space and let Y be a nonvoid subset of X. Define the closure of every set $A \subseteq Y$ as $\bar{A}_Y = \bar{A} \cap Y$ where $\bar{A}$ denotes the closure of A in X. Show that the axioms of a closure operator are satisfied and the topology defined in this way is the induced topology on Y.

11. Product Spaces

Let us consider the Cartesian product $X = \prod X_s$ of an arbitrary family of nonvoid sets X_s $(s \in S)$. We assume that a topology $\mathscr{T}_s$ is given on each of the sets X_s and using these we wish to introduce a topology $\mathscr{T}$ on the set X. Of course, in general a great number of topologies can be defined in terms of the topologies $\mathscr{T}_s$ and many of these will be related to the individual $\mathscr{T}_s$'s in a simple way. We wish to select one among these which will be called the product of the topologies $\mathscr{T}_s$ or concisely the *product topology*. This topology is distinguished by the simplicity of its definition and its relation to the individual topologies $\mathscr{T}_s$, but beyond that the final deciding factor which determines the product topology is its usefulness.

Definition 1. *Let* $X = \prod X_s$ *where* S *is a finite or infinite index set and each* X_s *is a topological space. Denote by* $\mathscr{S}$ *the family of all sets*

$$Z(s, O_s) = \{x : x \in X \quad \text{and} \quad x_s \in O_s\}$$

where $s \in S$ *and* O_s *is open in* X_s. *Then* $\mathscr{S}$ *is a subbase for a topology on* X *called the product topology.*

A base for the product topology consists of all finite intersections

$$Z(s_1, O_{s_1}) \cap \dots \cap Z(s_n, O_{s_n}) = \{x : x \in X \quad \text{and} \quad x_{s_i} \in O_{s_i} \quad \text{for} \quad i = 1, \dots, n\}$$

where n is arbitrary. These sets are called cylinders with base $O_{s_1} \times \dots \times O_{s_n}$. In particular, $Z(s, O_s)$ is a cylinder with base O_s. The product topology is the weakest topology on X relative to which the cylinders $Z(s, O_s)$ are open sets. It is natural to ask whether a base for the product topology should consist of all finite intersections of cylinders $Z(s, O_s)$ or whether, if it should also include all sets of the form $\bigcap \{Z(s, O_s) : s \in S\}$. There are several results which indicate that the first choice is preferable.

Let A be a set in the Cartesian product $X = \prod X_s$. We denote by $\pi_s(A)$ or occasionally by A_s the projection of A in X_s, i.e., $\pi_s(A)$ is the set of those points $\xi_s \in X_s$ for which there is a point x in A whose sth coordinate is ξ_s:

$$\pi_s(A) = \{\xi_s : x_s = \xi_s \quad \text{for some} \quad x \in A\}.$$

Lemma 1. *If* O *is open in* X, *then its projection* $\pi_s(O)$ *is open in* X_s *for every* $s \in S$.

Proof. If $x_s \in \pi_s(O)$, then there is an x in X such that its sth coordinate is x_s. Since O is open we have $x \in \bigcap Z(s, O_s) \subseteq O$ for suitable open sets $O_s \subseteq X_s$ such that $O_s = X_s$ for all but finitely many indices $s \in S$. Let all coordinates of y be the same as those of x except for the sth coordinate y_s. If $y_s \in O_s$, then $y \in \bigcap Z(s, O_s) \subseteq O$ and so $y_s \in \pi_s(O)$. Therefore $x_s \in O_s \subseteq \pi_s(O)$ which shows that $\pi_s(O)$ is open.

Lemma 2. *A set of the form* $\prod \{O_s : s \in S\}$ *is open in the product topology if and only if every* O_s *is open in* X_s *and if in addition* $O_s = X_s$ *for all but finitely many indices* $s \in S$.

Proof. The condition is sufficient by the definition of the product topology. Moreover, if $\prod \{O_s : s \in S\} = \bigcap \{Z(s, O_s) : s \in S\}$ is open and x is contained in it, then there are finitely many indices $s_1, \dots, s_n$ such that

$$x \in Z(s_1, O_{s_1}) \cap \dots \cap Z(s_n, O_{s_n}) \subseteq \bigcap \{Z(s, O_s) : s \in S\}.$$

Hence $O_s = X_s$ for every $s \neq s_1, ..., s_n$ and by the preceding lemma O_s is open for all indices including $s_1, ..., s_n$.

Lemma 3. *If C_s is closed for every $s \in S$, then $\Pi\{C_s : s \in S\}$ is a closed set in X.*

Proof. Since $\Pi C_s = \bigcap Z(s, C_s)$ it is sufficient to show that $Z(s, C_s)$ is closed for every $s \in S$. We have $cZ(s, C_s) = \{x : x \in X \text{ and } x_s \in cC_s\}$ and this is an open set in X. Hence $Z(s, C_s)$ is closed.

It is easy to find closed sets C in a product space such that $\pi_s(C)$ is not closed. For instance, if X is the Euclidean plane, then $C = \{x : x = (x_1, x_2) \text{ and } x_1x_2 \geqslant 1\}$ is such a set. This shows that there is no counterpart to Lemma 1 for closed sets.

Lemma 4. *If A_s is a subset of X_s, then $\overline{\Pi A_s} = \Pi \bar{A}_s$.*

Corollary. *The set ΠA_s is closed in X if and only if A_s is closed in X_s for each index $s \in S$.*

Proof. Since $A_s \subseteq \bar{A}_s$ we see that $\Pi A_s \subseteq \Pi \bar{A}_s$ and so by Lemma 3 $\overline{\Pi A_s} \subseteq \Pi \bar{A}_s$. On the other hand, if $x \notin \overline{\Pi A_s}$, then there exists an open set ΠO_s in X such that

$$x \in \Pi O_s \subseteq c\,\overline{\Pi A_s} \subseteq c\,\Pi A_s .$$

There is an index s such that $O_s \subseteq cA_s$ because if we had $O_s \cap A_s \neq \varnothing$ for every s, then ΠO_s and ΠA_s would intersect. Now if $O_s \subseteq cA_s$, then by $x_s \in O_s$ we get $x_s \notin \bar{A}_s$ and so $x \notin \Pi \bar{A}_s$. Therefore $\Pi \bar{A}_s \subseteq \overline{\Pi A_s}$ and the proposition is proved.

Let Y_s be a subspace of X_s for every $s \in S$. If the topology on Y_s is the topology induced by X_s, then $Y = \Pi Y_s$ can be considered as a topological space under the product of the topologies given on the subspace Y_s. A second topology can be defined on Y by considering it as a subspace of the product space X. We have:

Theorem 1. *If the topology on $Y_s \subseteq X_s$ $(s \in S)$ is the induced topology, then the topology of the product space $Y = \Pi Y_s$ is the topology induced on Y by the product topology of $X = \Pi X_s$.*

Proof. A subbase for the product topology of Y consists of the cylinders $Z(s, O_s \cap Y_s)$ lying in Y where O_s is open in X_s. Denote by $Z(s, O_s)$ the cylinder lying in X with base O_s. Then by $Z(s, O_s \cap Y_s) \subseteq Y$ we have $Z(s, O_s \cap Y_s) = Z(s, O_s) \cap Y$ and so $\{Z(s, O_s) \cap Y\}$ is a subbase for the product topology on Y. However, $\{Z(s, O_s)\}$ is a subbase for the product topology on X and so $\{Z(s, O_s) \cap Y\}$ is a subbase for the induced topology formed on Y. Therefore the two topologies on Y are identical.

Theorem 2. *Let $S = S_1 \cup S_2$ where S_1 and S_2 are disjoint sets and let X_s be a topological space for every $s \in S$. Choose a point x_s in X_s for every $s \in S_2$ and consider the set $A = \prod \{A_s : s \in S\}$ where $A_s = X_s$ for every $s \in S_1$ and $A_s = \{x_s\}$ for every $s \in S_2$. Then A as a subspace of X is homeomorphic to the product space $\prod \{X_s : s \in S_1\}$ and a homeomorphism is given by the natural map $\xi \to (\xi_s)$ where ξ_s is the* sth *coordinate of ξ for every $s \in S_1$.*

Proof. The open sets of A are the sets Q which can be represented in the form $Q = A \cap O$ where O is an open set in X. Therefore Q is open in A if and only if the following proposition holds for Q: For every $\xi \in Q$ and for every $s \in S_1$ there exist open sets O_s in X_s such that only finitely many of these are different from the corresponding X_s and

$$\xi \in \prod\{O_s : s \in S_1\} \times \prod\{\{x_s\} : s \in S_2\} \subseteq Q.$$

This proposition holds if and only if for every $\xi \in Q$, or equivalently for every $(\xi_s) \in \tilde{Q}$ where $\tilde{Q}$ denotes the image of Q under the map $\xi \to (\xi_s)$ $(s \in S_1)$, there exist open sets O_s in X_s such that only finitely many of these are different from the corresponding X_s and

$$(\xi_s) \in \prod\{O_s : s \in S_1\} \subseteq \tilde{Q}.$$

This last proposition is equivalent of saying that $\tilde{Q}$ is open in $\prod \{X_s : s \in S_1\}$.

Product spaces and product topology can be interpreted in the following way: The elements of $X = \prod \{X_s : s \in S\}$ are functions x which map S into $\bigcup \{X_s : s \in S\}$ and have the property that $x(s) \in X_s$ for every $s \in S$. The product topology is defined on the set of all these functions $x \in X$. The situation is particularly simple when the factor spaces are all equal to a fixed space Y. Then X is the set of all functions x mapping S into Y and the product topology is called the *weak topology* of the function space X or the *topology of pointwise convergence.*

The weak topology of a function space can be visualized by considering the example of real-valued functions and describing a base for the neighborhood filters of its points: Suppose for instance that S is the interval $[0, 1]$ and Y is the space of the reals. We associate a neighborhood of $x \in X$ with every $\epsilon > 0$ and every finite collection of distinct points $s_1, \ldots, s_n \in S$ as follows: A function $y \in X$ belongs to the $(\epsilon; s_1, \ldots, s_n)$-neighborhood of x if $|y(s_i) - x(s_i)| < \epsilon$ for every $i = 1, \ldots, n$, i.e., if y passes through each of the open intervals of length 2ϵ placed symmetrically around the functions values $x(s_1), \ldots, x(s_n)$. (See Figure 1, p. 62.)

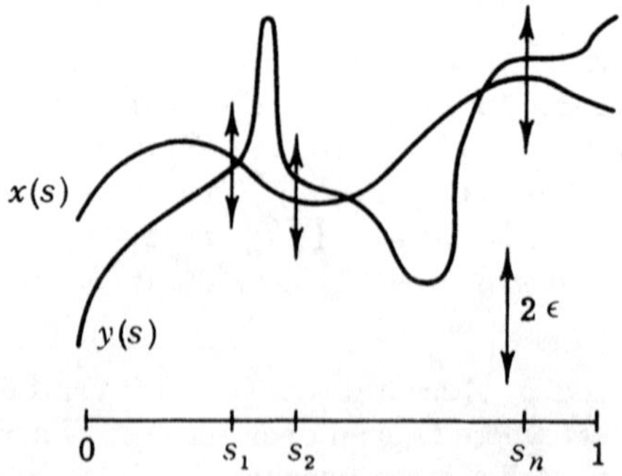

FIG. 1.

EXERCISES

1. Let X_1, X_2 be metric spaces with metrics d_1, d_2. Show that $d = d_1 + d_2$ is a metric on $X_1 \times X_2$ and the metric topology so obtained is the product topology.

(If $x \in O$ where O is open in $X_1 \times X_2$ relative to the product topology, then $x \in O_1 \times O_2 \subseteq O$ for some open sets O_1 and O_2 and so there is an $\epsilon > 0$ such that $x \in S_\epsilon^1[x_1] \times S_\epsilon^2[x_2] \subseteq O_1 \times O_2 \subseteq O$. Hence for every $x \in O$ there is an $\epsilon > 0$ such that $x \in S_\epsilon[x] \subseteq S_\epsilon^1[x_1] \times S_\epsilon^2[x_2] \subseteq O$ and so O is open in the metric sense. If O is open relative to the metric d, then for every $x \in O$ there is a ball satisfying $x \in S_\epsilon[x] \subseteq O$. Then $x \in S_{\epsilon/2}[x_1] \times S_{\epsilon/2}[x_2] \subseteq O$ and so O is open in the product topology.)

2. Let $\mathscr{B}_1$ and $\mathscr{B}_2$ be bases for the topologies of the spaces X_1 and X_2. Show that the family $\mathscr{B} = \{B_1 \times B_2\}$ $(B_1 \in \mathscr{B}_1$ and $B_2 \in \mathscr{B}_2)$ is a base for the product topology on $X_1 \times X_2$. Extend the result to infinite products.

(If the index set S is infinite then not all sets of the form $B = \Pi\{B_s : s \in S\}$ are in the base. The set B belongs to $\mathscr{B}$ if and only if $B_s = X_s$ for all but finitely many choices of $s \in S$. If S is infinite and no restriction is given on the B's we obtain a topology which is strictly finer than the product topology.)

3. Let X be the product space formed by the factor spaces X_1 and X_2 and let A_1 and A_2 be sets in X_1 and X_2, respectively. Show that

$$(A_1 \times A_2)^i = A_1{}^i \times A_2{}^i \qquad \text{and} \qquad (A_1 \times A_2)' = (A_1{}' \times \bar{A}_2) \cup (\bar{A}_1 \times A_2{}').$$

4. Determine the interior of all sets A of the form $A = \Pi\{A_s : s \in S\}$ when the index set S is infinite.

(We have $A^i = \Pi\{A_s^i : s \in S\}$ if and only if $A_s = X_s$ with at most finitely many exceptions. Otherwise $A^i = \varnothing$.)

5. Determine the relation between the isolated points of the factor spaces X_s and the isolated points of the product space X when the number of factor spaces is finite.

[Using the formula given in Exercise 3 we obtain $(X_1 \times X_2)' = (X_1' \times X_2) \cup (X_1 \times X_2')$. Hence $x \notin (X_1 \times X_2)'$ if and only if $x \in cX_1' \times cX_2'$ and so x is an isolated point if and only if its coordinates are isolated points of the factor spaces.]

6. Give an example of an infinite product space X containing a point x with the property that x is not isolated in X although each of its coordinates is an isolated point of the corresponding factor space.

[Take for X_s the subspace of the reals formed by the interval (1, 2) and the number 0.]

7. The *Boolean product topology* on $X = \Pi X_s$ is defined by the base $\mathscr{B} = \{\Pi O_s\}$ ($O_s \in \mathscr{O}_s$ and $s \in S$) where O_s for every s denotes an arbitrary open set in X_s. Show that a point is an isolated point of X relative to this topology if and only if each of its coordinates is an isolated point of the corresponding factor space.

8. Show the following proposition: If A_s is a subset of X_s for every index $s \in S$, then $\overline{\Pi A_s} = \Pi \bar{A}_s$ where on the left the closure is taken relative to the Boolean product topology.

(The proof given for Lemma 4 holds also in this case.)

9. A subbase for a closure operator on ΠX_s is given by the rule $\gamma(\Pi A_s) = \Pi \bar{A}_s$. Show that the topology defined by this operator is the product topology. Explain why the Boolean product topology is not obtained in this way.

(The sets of the form ΠC_s where C_s is closed in X_s form a subbase for the closed sets of the product topology. This subbase does not contain sufficiently many closed sets to define the finer Boolean product topology.)

10. Show that the cylinder Z formed by those points (x_1, x_2, x_3) which satisfy $x_1^2 + x_2^2 = r^2$ is a subspace of E_3 which is homeomorphic to the product of the real line and the circle.

(This is a direct application of Theorem 1: Let $E_3 = E_1 \times E_2$, let $Y_1 = E_1$, and let Y_2 be the circle $x_1^2 + x_2^2 = r^2$ in E_2. Then $Y_1 \times Y_2$ is the cylinder Z.)

11. Define what is meant by a base and a subbase for an interior operator and give the definition of the topology generated by a base and by a subbase for an interior operator.

(Compare Theorem 3.2 and Exercise 3.3.)

12. Show that both the product topology and the Boolean product topology can be introduced by a subbase for a suitable interior operator.

12. Products of Uniformizable Spaces

In the first half of this section we discuss a result on the product of metric spaces when there are at most denumerably many factors, then give a generalization of this result to arbitrary uniformizable spaces by showing that the product of such spaces is again a uniformizable space. In the second half of the section we consider product spaces and product structures in the special case when all the factors are equal and introduce a new topology and uniform structure.

Theorem 1. *The product topology of countably many pseudometric spaces is pseudometrizable and the product topology of countably many metric spaces is metrizable.*

Proof. Let $X_1, ..., X_n, ...$ be pseudometric spaces with pseudometrics $d_1, ..., d_n, ...$. We may assume that $d_n(x_n, y_n) \leqslant 2^{-n}$ for any two points x_n, y_n of the space X_n: In fact, if δ_n is an arbitrary pseudometric on X_n, then the pseudometric $d_n = 2^{-n}\,\delta_n/(1 + \delta_n)$ induces the same topology and has the required property. For any pair of points $x = (x_n)$ and $y = (y_n)$ in $X = \prod X_n$ the series $\sum d_n(x_n, y_n)$ is convergent and so $d = \sum d_n$ defines a pseudometric on the set X. The axioms of a pseudometric are clearly satisfied.

Now let $O \subseteq X$ be open in X with respect to the topology induced by the pseudometric d and let $x \in O$. Then there is an $\epsilon > 0$ such that $S_\epsilon[x] \subseteq O$. We choose a positive integer ν such that $\sum_{n>\nu} 2^{-n} < \epsilon/2$, and define $\epsilon_1, ..., \epsilon_\nu = \epsilon/2\nu$ and $\epsilon_n = 2^{-n}$ for every $n > \nu$. Then $\sum_{n\geqslant 1} \epsilon_n \leqslant \epsilon$ and so $\prod S_{\epsilon_n}[x_n] = \bigcap Z(n, S_{\epsilon_n}[x_n]) \subseteq S_\epsilon[x] \subseteq O$. Since $\epsilon_n = 2^{-n}$ for $n > \nu$ we have $S_{\epsilon_n}[x_n] = X_n$ for every $n > \nu$. Hence the set $\prod S_{\epsilon_n}[x_n] = \bigcap Z(n, S_{\epsilon_n}[x_n])$ is open relative to the product topology. We proved that if O is open relative to the pseudometric d, then O is an open set of the product space. Now let O be open relative to the product topology and let $x \in O$. Then $x \in \bigcap \{Z(n, O_n) : n \leqslant \nu\} \subseteq O$ for a sufficiently large ν and for suitable sets O_n ($n = 1, ..., \nu$) which are open relative to the pseudometric d_n. Hence for every $n \leqslant \nu$ there is a positive ϵ_n such that $S_{\epsilon_n}[x_n] \subseteq O_n$ and so

$$x \in \bigcap \{Z(n, S_{\epsilon_n}[x_n]) : n \leqslant \nu\} \subseteq \bigcap \{Z(n, O_n) : n \leqslant \nu\} \subseteq O.$$

We put $\epsilon = \min\{\epsilon_1, ..., \epsilon_\nu\}$ and obtain $x \in S_\epsilon[x] \subseteq O$. This shows that O is open relative to the pseudometric d. Therefore the pseudometric topology induced by d is identical with the product topology. Clearly, $d(x, y) = 0$ if and only if $d(x_n, y_n) = 0$ for every n and so if every X_n is a metric space then so is X.

The preceding theorem has an analog for uniform spaces. First we introduce the notion of the product structure:

Definition 1. *Let $X = \Pi X_s$ and for every index $s \in S$ let $\mathscr{U}_{Ss}$ be a subbase for a uniform structure for X_s. Denote by $\mathscr{U}_S$ the family of all sets*

$$Z(s, U_s) = \{(x, y) : x, y \in X \quad \text{and} \quad (x_s, y_s) \in U_s\}$$

where $U_s \in \mathscr{U}_{Ss}$. Then $\mathscr{U}_S$ is a subbase of a uniform structure $\mathscr{U}$ for X which is called the product structure.

The product of the uniform strctures $\mathscr{U}_s$ $(s \in S)$ is usually denoted by $\mathscr{U} = \Pi \mathscr{U}_s$. The family $\{Z(s, U_s)\}$ is only a subbase even if U_s varies over a base or over the whole uniform structure $\mathscr{U}_s$ for every s. Similarly, the family of all finite intersections $Z(s_1, U_{s_1}) \cap ... \cap Z(s_n, U_{s_n})$ is only a base of the product structure even if U_s varies over every element of $\mathscr{U}_s$ for every $s \in S$.

Theorem 2. *The product $X = \Pi X_s$ of an arbitrary family of uniform spaces is a uniform space, and if $\mathscr{U}_s$ $(s \in S)$ is a uniform structure for the topology of X_s, then the product structure $\mathscr{U} = \Pi \mathscr{U}_s$ is a uniform structure for the product topology.*

Note. The product need not be metrizable even if every X_s is metrizable.

Corollary. *If the uniform space X_s has a structure base of cardinality u_s, then the product space $X = \Pi X_s$ has a structure base of cardinality at most Σu_s.*

Proof. The family $\{Z(s, O_s)\}$ $(s \in S$ and $O_s \in \mathscr{O}_s)$ is a subbase for the product topology. Hence it is sufficient to verify the following statements:

(i) Given $x \in X$ and $Z(s, U_s) \in \mathscr{U}_S$ there is an open set O_s in X_s such that $x \in Z(s, O_s) \subseteq Z(s, U_s)[x]$.

(ii) Given $x \in X$ and an open set O_s in X_s which contains the projection x_s of x there is a $U_s \in \mathscr{U}_s$ such that $x \in Z(s, U_s)[x] \subseteq Z(s, O_s)$.

In order to satisfy (i) it is sufficient to choose as O_s the interior of $U_s[x_s]$. To see the other statement we choose U_s such that $U_s[x_s] \subseteq O_s$ where x_s is the sth coordinate of x. Then $Z(s, U_s)[x] \subseteq Z(s, O_s)$ and hence the proof is complete.

The present proof appears to be simpler than the proof of Theorem 1 but this is not surprising since for metric spaces the result is weaker.

Let us suppose now that all factor sets X_s $(s \in S)$ of the product $X = \prod\{X_s : s \in S\}$ are identical, say $X_s = Y$ for every $s \in S$. Let $\mathscr{U}_s$ $(s \in S)$ denote copies of the same uniform structure $\mathscr{V}$ for every $X_s = Y$. In this special case it is possible to make a slight modification in the definition of the product structure and of the product topology: By Definition 1, a base for $\prod \mathscr{U}_s$ consists of the sets $Z(s_1, U_{s_1}) \cap \dots \cap Z(s_n, U_{s_n})$ where $U_{s_1}, \dots, U_{s_n} \in \mathscr{V}$. Since $U = U_{s_1} \cap \dots \cap U_{s_n}$ is an element of $\mathscr{V}$ it follows that the subfamily consisting of the sets $Z(s_1, U) \cap \dots \cap Z(s_n, U)$ is also a base for the product structure $\prod \mathscr{U}_s$. This interpretation of the product structure makes it clear that the uniform structure which we are going to define now is finer than the ordinary product structure.

Definition 2. *Suppose that the same uniform structure $\mathscr{V}$ is associated with every copy X_s $(s \in S)$ of the fixed set X. Then the sets $\bigcap_s Z(s, V)$ $(V \in \mathscr{V})$ form a base of a uniform structure for $X = \prod X_s$ which is called the uniform product structure for X and is denoted by $\mathscr{V}^S$. The topology associated with $\mathscr{V}^S$ is the topology of uniform convergence relative to $\mathscr{V}$.*

Later we shall define what is meant by the uniform convergence of a sequence of functions $x_n \in \prod X_s$ relative to the uniform structure $\mathscr{V}^S$ or, what is the same thing, relative to the uniform structure $\mathscr{V}$. This will be a natural generalization of the notion of uniform convergence for real-valued functions of a real variable. We shall see then that a sequence of functions $x_n \in \prod X_s$ is uniformly convergent relative to $\mathscr{V}$ if and only if it is convergent relative to the topology of uniform convergence associated with $\mathscr{V}$.

EXERCISES

1. Let S be a noncountable index set (e.g., let $S = [0, 1]$) and let $X_s = Y = \{0, 1\}$ for every $s \in S$. The diagonal $I = \{(0, 0), (1, 1)\}$ of $Y \times Y$ is a base of the discrete structure $\mathscr{U}_s$ of $X_s = Y$. Determine a subbase for the product structure $\prod \mathscr{U}_s$ and show that there is no subbase for $\prod \mathscr{U}_s$ which consists of countably many elements.

[Let $U(s) = \{(x, y) : (x_s, y_s) \in I\} = \{(x, y) : x_s = y_s\}$. The family $\{U(s)\}$ $(s \in S)$ is a subbase for $\prod \mathscr{U}_s$ and no countable subfamily is a subbase: For let $\mathscr{U}$ be the uniform structure generated by the countable family $\{U(s_n)\}$ $(n = 1, 2, \dots)$ and let $s \neq s_1, s_2, \dots$. Consider a pair of distinct points $x, y \in X$ such that $x_{s_n} = y_{s_n}$ for every $n = 1, 2, \dots$

but $x_s \neq y_s$. On the one hand we have $(x, y) \in U$ for every $U \in \mathscr{U}$ and on the other hand there is a U in $\prod \mathscr{U}_s$, namely, $U = U(s)$, such that $(x, y) \notin U$. Hence the structures $\mathscr{U}$ and $\prod \mathscr{U}_s$ are distinct. It follows that $\prod \mathscr{U}_s$ cannot have any other countable subbase either.]

2. Show that every pseudometrizable structure has a countable structure base. Consequently the product structure $\prod \mathscr{U}_s$ constructed in the preceding exercise is not pseudometrizable though $X_s = Y$ is a metric space for every index $s \in S$.

3. Let X_s for every $s = 1, 2, \ldots$ be the set of positive integers topologized by the discrete topology. Show by using continued fractions that the product space $X = \prod \{X_s : s = 1, 2, \ldots\}$ is homeomorphic to the space formed by the set of irrational numbers exceeding one under the topology induced by the usual topology of the reals.

(A one-to-one correspondence between the elements of X and the irrationals $\xi > 1$ is given by the continued fraction development of ξ:

$$x = (x_s) \to \xi = x_1 + \cfrac{1}{x_2 + \cfrac{1}{x_3 + \cdots}}$$

A base in X is formed by the sets

$$B(x_1, \ldots, x_n) = \{y : y \in X \quad \textit{and} \quad y_1 = x_1, \ldots, y_n = x_n\}$$

where $x_1, \ldots, x_n$ are arbitrary positive integers. The image of $B(x_1, \ldots, x_n)$ is the set of irrational points in the interval whose end points are the rationals

$$x_1 + \cfrac{1}{x_2 + \cfrac{1}{x_3 + \cdots + \cfrac{1}{x_n + 1}}} \quad \text{and} \quad x_1 + \cfrac{1}{x_2 + \cfrac{1}{x_3 + \cdots + \cfrac{1}{x_n}}}$$

Hence the images of open sets of X are open relative to the topology induced on the irrationals. Now suppose that $\xi \in (\alpha, \beta)$ where $1 < \alpha < \beta$ are real numbers. If n is sufficiently large, then every irrational η satisfying $y_1 = x_1, \ldots, y_n = x_n$ also belongs to (α, β) and so the inverse image of (α, β) contains $B(x_1, \ldots, x_n)$. This shows that the inverse image of an open set is open in X.)

4. Let S be a noncountable index set. Suppose that every factor space X_s $(s \in S)$ contains a pair of elements which form a subspace with discrete induced topology. Prove that $\prod \{X_s : s \in S\}$ is not metrizable. (Use Theorem 11.1 and the result of Exercise 2.)

5. Determine the closure of $\prod A_s$ when X_s is the same set Y for every $s \in S$ and the topology on $X = \prod X_s$ is the topology of uniform convergence relative to a uniform structure $\mathscr{V}$ of Y.

(The set $\prod \bar{A}_s$ is closed since it is closed in the product topology. Hence the closure of $\prod A_s$ is contained in $\prod \bar{A}_s$. Since the closure is the intersection of all closed sets containing $\prod A_s$ and the closure of $\prod A_s$ in the Boolean product topology is $\prod \bar{A}_s$, its closure in the topology of uniform convergence contains $\prod \bar{A}_s$. Hence it is $\prod \bar{A}_s$.)

6. Suppose that Y is a pseudometric space with pseudometric d where we assume $d \leqslant 1$. Show that the topology of uniform convergence of $\prod X_s$ where $X_s = Y$ for every s is a metrizable topology and if $\mathscr{V}$ is the uniform structure generated by d, then $\mathscr{V}^S$ is generated by the pseudometric δ defined as follows: $\delta(x, y) = \text{lub}\{d(x_s, y_s) : s \in S\}$.

13. Inverse and Direct Images of Topologies

Here we describe two more methods which are commonly used to construct new topological spaces by using some known topologies. The first of these occurs when we have a set X, a topological space Y, and a function f which maps X into Y. The object is to introduce a topology on X by using the topology of Y and the function f. The second method concerns the case when X is a topological space, Y is a set, and f maps X onto Y. Using the topology of X we introduce a topology on Y.

Theorem 1. *Let f map the set X into the topological space Y with topology $\mathscr{T}$. Then there is a weakest topology on X having the property that the inverse image of every open set in Y is an open set in X. This topology is called the inverse image of $\mathscr{T}$ under f and is denoted by $f^{-1}(\mathscr{T})$. The family of open sets of $f^{-1}(\mathscr{T})$ consists of the sets $f^{-1}(O)$ where O is open in Y.*

Note. We define $f^{-1}(A) = \{x : f(x) \in A\}$ for any $A \subseteq Y$ without regard whether A is in the range of f or not.

Proof. There exist topologies such that the inverse image of open sets is open in X. For instance, if X is topologized by the discrete topology, then $f^{-1}(O)$ is open for every open O of Y. If a topology on X has this property, then the family $\{f^{-1}(O)\}$ $(O \in \mathscr{O})$ is a subcollection of the family of its open sets. Since

$$\bigcap f^{-1}(O_i) = f^{-1}(\bigcap O_i) \quad \text{and} \quad \bigcup f^{-1}(O_i) = f^{-1}(\bigcup O_i)$$

the family $\{f^{-1}(O)\}$ $(O \in \mathcal{O})$ satisfies the axioms for the open sets of a topology on X. Hence $f^{-1}(\mathcal{T})$ exists and its open sets are the sets $f^{-1}(O)$ where $O \in \mathcal{O}$.

The theorem can be extended to the more general situation when a whole family of functions is given on X:

Theorem 2. *Let $\{f_i\}$ $(i \in I)$ be a family of functions mapping the set X into the topological space Y. Then there is a weakest topology on X such that the inverse images $f_i^{-1}(O)$ are open in X for every f_i and every open set O in Y.*

Proof. We consider the inverse images $f_i^{-1}(\mathcal{T})$ $(i \in I)$ and take $\text{lub}\{f_i^{-1}(\mathcal{T}) : i \in I\}$ where as usual the least upper bound denotes the weakest topology which is at least as strong as any $f_i^{-1}(\mathcal{T})$. This topology has the required properties.

Later when discussing functions and continuity we shall see that a function f is defined as *continuous*, or *everywhere continuous*, relative to a topology given on its domain and a topology given on its range if the inverse image of open sets is open. The topology $f^{-1}(\mathcal{T})$ has the property that the inverse image $f^{-1}(O)$ of open sets is open. Hence $f^{-1}(\mathcal{T})$ is the weakest topology on X such that f is continuous relative to $f^{-1}(\mathcal{T})$ and $\mathcal{T}$. The topology $\text{lub}\{f_i^{-1}(\mathcal{T}) : i \in I\}$ whose existence is stated in Theorem 2 is the weakest topology on X such that every f_i $(i \in I)$ is continuous.

Another interesting property of $f^{-1}(\mathcal{T})$ is that the image $f(O)$ of every open set O in X is an open set of $f(X)$. For if O is open in $f^{-1}(\mathcal{T})$, then $O = f^{-1}(Q)$ for some open set Q in Y and so $f(O) = Q \cap f(X)$. If X and Y are topological spaces and f maps X into Y such that the image of open sets is open, we speak about an *open map*. Hence $f^{-1}(\mathcal{T})$ is a topology on X such that f is an open map relative to $f^{-1}(\mathcal{T})$ and the relativization of $\mathcal{T}$ to $f(X)$. If C is closed in $f^{-1}(\mathcal{T})$, then $C = cf^{-1}(Q)$ for some open set Q in Y, so $f(C) = cQ \cap f(X)$ and $f(C)$ is closed in the range of f under the induced topology. If a function f maps a topological space X into another Y such that closed sets are mapped into closed sets we say that f is a *closed map*. Hence $f^{-1}(\mathcal{T})$ is a topology on X such that f is a closed map relative to $f^{-1}(\mathcal{T})$ and the relativization of $\mathcal{T}$ to $f(X)$. The formation of the inverse images is a *transitive process*; that is, $f^{-1}(g^{-1}(\mathcal{T})) = (f \circ g)^{-1}(\mathcal{T})$. This follows from the fact that $(f \circ g)^{-1} = g^{-1} \circ f^{-1}$ for any pair of functions $f : X \to Y$ and $g : Y \to Z$.

Let us now consider a topological space X with topology $\mathcal{T}$ and a function f mapping X into a set Y. There is at least one topology on Y, namely, the nondiscrete topology, such that f is continuous. If f is continuous with respect to some topology on Y, then it is continuous also with respect to every weaker topology. It is reasonable to ask whether a strongest topology exists on Y such that f is continuous. Suppose f is continuous relative to some topology on Y having open sets $Q \in \mathcal{Q}$. Then $f^{-1}(Q)$ is open for every $Q \in \mathcal{Q}$ and so $\mathcal{Q}$ is a subfamily of

$$\{Q : Q \subseteq Y \quad \text{and} \quad f^{-1}(Q) \in \mathcal{O}\}$$

where $\mathcal{O}$ is the family of open sets of X. This family is closed under the formation of finite intersections and arbitrary unions. Hence it defines a topology on Y which has the required property: It is the strongest among those topologies of Y which make f continuous.

We proved the following:

Theorem 3. *Let X be a topological space with topology $\mathcal{T}$ and let f map X into the set Y. Then there is a strongest topology on Y which makes f a continuous map. This topology is called the direct image of $\mathcal{T}$ under f and is denoted by $f(\mathcal{T})$. A set Q is open relative to $f(\mathcal{T})$ if and only if $f^{-1}(Q)$ is an open set of X.*

Since every family of topologies admits a greatest lower bound we have the following extension of the foregoing result:

Theorem 4. *Let $\{f_i\}$ $(i \in I)$ be a family of functions mapping the topological space X into the set Y. Then there is a strongest topology on Y which makes every f_i continuous.*

An important application of the direct image topology will be studied in the next section.

EXERCISES

1. Determine $f^{-1}(\mathcal{T})$ when $X = Y$ is the set of reals, Y is topologized in the usual way, and $f : x \to x^2$.

(A set O is open in X relative to $f^{-1}(\mathcal{T})$ if it is open in the usual sense and is symmetric with respect to zero.)

2. Determine the greatest lower bound of the topologies $f^{-1}(\mathcal{T})$ and $g^{-1}(\mathcal{T})$ when $f : x \to (x - \frac{1}{2})^2$ and $g : x \to (x + \frac{1}{2})^2$, $X = Y$ is the set of reals, and $\mathcal{T}$ is the usual topology of the reals.

[By the solution of Exercise 1 a set O is open relative to $f^{-1}(\mathcal{T})$ if it is open in the usual sense and is symmetric with respect to $\frac{1}{2}$. Hence O is open in the greatest lower bound if it is of the form

$$O = \bigcup \{(2n + 1 - Q) \cup (2n + Q) : n = 0, \pm 1, \pm 2, \ldots\}$$

where Q is open in the usual sense.]

3. Let f be a one-to-one map of the set X onto the topological space Y with topology $\mathscr{T}$. Show that if X is topologized by $f^{-1}(\mathscr{T})$, then X and Y are homeomorphic spaces.

4. Prove that the product topology is the weakest topology on $X = \prod X_s$ such that the projection maps $\pi_s : X \rightarrow X_s$ are continuous.

5. If $X = \prod X_s$ has the product topology, is $\pi_s : X \rightarrow X_s$ an open map? Is it closed?

14. Quotient Spaces

Let X be a topological space with topology $\mathscr{T}$ and let R be an equivalence relation for X. We denote by X/R the set of equivalence classes; $\{x\}$ is the equivalence class containing the element $x \in X$. Let f be the function which associates with each $x \in X$ the class $\{x\} \in X/R$. This function f is called the *natural map* of X onto X/R. Thus $f : X \rightarrow X/R$ where X is a topological space and X/R is a set, a situation studied in the preceding section. The direct image $f(\mathscr{T})$ of $\mathscr{T}$ is called the *quotient topology* and is denoted by $\mathscr{T}/R$. The set X/R endowed with the topology $\mathscr{T}/R$ is called a *quotient space*. By Theorem 13.3 $\mathscr{T}/R$ is the strongest topology on X/R which makes the natural map $f : X \rightarrow X/R$ a continuous function.

The process of forming a quotient space is called a *topological identification*. This is so because it corresponds to the intuitive idea of identifying certain points of the space. The points to be identified with each other form an equivalence class and the equivalence relation R is determined by these equivalence classes.

We discuss a few instances of topological identification without actually proving that the spaces so obtained are homeomorphic to some other spaces already defined. First let X be the closed interval $[0, 1]$ under the induced topology of the reals. We identify the end points 0 and 1 and obtain a quotient space which is the homeomorphic image of a circle S^1. The same result can be obtained by choosing for X the whole real line and identifying any two points which are congruent modulo 1; that is, $x \sim y$ if and only if $x - y$ is an integer.

The same method can be applied in the plane to obtain the cylinder and the torus: Let X be the plane with its usual topology and let $x = (x_1, x_2) \sim y = (y_1, y_2)$ if $x_1 = y_1$ modulo 1 and $x_2 = y_2$. The resulting space is a cylinder and intuitively the process corresponds to rolling up an infinitely large sheet of paper to form a cylinder. The same quotient space can be formed by starting from the infinite strip

$X = \{x : x = (x_1, x_2)$ *and* $0 \leqslant x_1 \leqslant 1\}$ and identifying the points $(0, x_2)$ and $(1, x_2)$ for every x_2. The intuitive process is to glue together two parallel edges of a paper strip to form a cylinder.

Again let X be the plane and let $x \sim y$ if $x_1 = y_1$ modulo 1 and $x_2 = y_2$ modulo 1. The quotient space can be easily visualized if we assume that this identification process can be performed in two successive steps: First let $x \sim y$ if $x_1 = y_1$ modulo 1 and $x_2 = y_2$. Then the quotient space X/R is the homeomorphic image of a cylinder. Next we introduce an equivalence relation in X/R: $\{x\} \sim \{y\}$ if $x_2 = y_2$ modulo 1 where $x = (x_1, x_2)$ and $y = (y_1, y_2)$. This second identification corresponds to inserting one end of the infinitely long cylinder into its other end and rolling it up by sliding it within itself. The end result is a torus.

In what follows we are going to prove a theorem which states that a single complicated identification can be replaced by a sequence of successive simpler identifications.

Let R_1 and R_2 be equivalence relations on a set X such that R_1 implies R_2. Using these relations we can introduce an equivalence relation R_2/R_1 on the set X/R_1 of equivalence classes modulo R_1. We say that $\{x\}_1 \sim \{y\}_1$ modulo R_2/R_1 if $x \sim y$ modulo R_2. This is a correct definition because it is independent of the choice of the representatives x and y: If $x_1, x_2 \in \{x\}_1$ and $y_1, y_2 \in \{y\}_1$, then $x_1 \sim x_2$ and $y_1 \sim y_2$ modulo R_1 and so $x_1 \sim x_2$ and $y_1 \sim y_2$ modulo R_2. Hence if $x_1 \sim y_1$ modulo R_2, then also $x_2 \sim y_2$ modulo R_2.

There is a natural map of the set X/R_2 onto $(X/R_1)/(R_2/R_1)$, namely, one associates with $\{x\}_2 \in X/R_2$ the class $\{\{x\}_1\}$ of those classes in X/R_1 which are equivalent to $\{x\}_1$ modulo R_2/R_1. We prove that this natural map is a one-to-one correspondence between the elements of X/R_2 and of $(X/R_1)/(R_2/R_1)$: For if $\{\{x\}\}_1 = \{\{y\}\}_1$, then $\{x\}_1 \sim \{y\}_1$ modulo R_2/R_1, then $x \sim y$ modulo R_2, and so $\{x\}_2 = \{y\}_2$.

Theorem 1. *Let R_1 and R_2 be equivalence relations on the topological space X such that R_1 implies R_2. Then X/R_2 and $(X/R_1)/(R_2/R_1)$ are homeomorphic spaces and a homeomorphism is given by the natural map of X/R_2 onto $(X/R_1)/(R_2/R_1)$.*

Proof. We have four natural maps:

$$f : x \in X \to \{x\}_1 \in X/R_1,$$
$$g : \{x\}_1 \in X/R_1 \to \{\{x\}_1\} \in \{X/R_1)/(R_2/R_1),$$
$$h : x \in X \to \{x\}_2 \in X/R_2$$
$$k : \{x\}_2 \in X/R_2 \to \{\{x\}_1\} \in (X/R_1)/(R_2/R_1).$$

The following *diagram* helps to visualize the situation:

$$\begin{array}{ccc} X & \xrightarrow{\;f\;} & X/R_1 \\ {\scriptstyle h}\downarrow & & \downarrow{\scriptstyle g} \\ X/R_2 & \xrightarrow{\;k\;} & X/R_1/(R_2/R_1) \end{array}$$

These maps have the property that $g(f(x)) = k(h(x))$, for every x in X, written in our notation as $f \circ g = h \circ k$. Mapping diagrams with this property are called *commutative*. Now Q is open in $(X/R_1)/(R_2/R_1)$ if and only if $(g^{-1} \circ f^{-1})(Q)$ is open in X. However,

$$(g^{-1} \circ f^{-1})(Q) = (f \circ g)^{-1}(Q) = (h \circ k)^{-1}(Q) = (k^{-1} \circ h^{-1})(Q) = h^{-1}(k^{-1}(Q))$$

and $k^{-1}(Q)$ is open in X/R_2 if and only if $h^{-1}(k^{-1}(Q))$ is open in X. Therefore Q is open in $(X/R_1)/(R_2/R_1)$ if and only if $k^{-1}(Q)$ is open in X/R_2.

Those open sets Q of X which are unions of equivalence classes modulo R are called *saturated open sets*. We can easily check that the family $\mathcal{Q}$ of saturated open sets satisfies the axioms for the open sets of a topological space and so it determines a topology $\mathcal{T}_R$ on X. Since $\mathcal{Q} \subseteq \mathcal{O}$ this topology is coarser than $\mathcal{T}$ and clearly $(\mathcal{T}_R)_R = \mathcal{T}_R$. The topologies $\mathcal{T}_R$ on X and $\mathcal{T}/R$ on X/R are closely related, namely, $\mathcal{T}_R$ is the inverse image of $\mathcal{T}/R$ relative to the natural map f of X onto X/R. Often it is more convenient to deal with the original set X under the topology $\mathcal{T}_R$ instead of the set of equivalence classes X/R under the quotient topology $\mathcal{T}/R$. This is the case for instance when an algebraic structure is given on X which is not compatible with the equivalence relation R.

We defined the torus by identifying some points of the plane. It can be obtained also in another way by forming the product of two circles which as we saw earlier, can be obtained from the real line by identification. It is no accident that these two definitions lead to the same space but rather an instance of a general principle on identification in product spaces:

If R_1 and R_2 are equivalence relations for the sets X_1 and X_2, then an equivalence relation $R_1 \times R_2$ can be defined on $X_1 \times X_2$ as follows: $x = (x_1, x_2) \sim y = (y_1, y_2)$ modulo $R_1 \times R_2$ if $x_1 \sim y_1$ and $x_2 \sim y_2$. There is a natural one-to-one correspondence between the elements of $(X_1/R_1) \times (X_2/R_2)$ and of $(X_1 \times X_2)/(R_1 \times R_2)$. The map $(\{x_1\}, \{x_2\}) \to \{(x_1, x_2)\}$ gives such a one-to-one correspondence because if $x_1' \sim x_1''$ and $x_2' \sim x_2''$, then $(x_1', x_2') \sim (x_1'', x_2'')$ and conversely.

If X and Y are sets and $f : X \to Y$ is an arbitrary map we call $f^{-1}(f(A)) = (f \circ f^{-1})(A)$ the *saturation of A* with respect to $f : X \to Y$.

In particular, if X is a set, R is an equivalence relation for X and $f: X \to X/R$ the natural map then $f^{-1}(f(A))$ is called the *saturation of A with respect to R*. If X is a topological space and R is such that the saturation of every open set of X with respect to R is again open, then the equivalence relation R is said to be compatible with the topology of X.

Theorem 2. *Let X_1 and X_2 be topological spaces and let R_1 and R_2 be equivalence relations on X_1 and X_2, respectively. Then the natural map*

$$g: (\{x_1\}, \{x_2\}) \in (X_1/R_1) \times (X_2/R_2) \to \{(x_1, x_2)\} \in (X_1 \times X_2)/(R_1 \times R_2)$$

is an open map. Moreover, if R_i is compatible with the topology of X_i $(i = 1, 2)$ then g is a homeomorphism.

Proof. Let $f_i: X_i \to X_i/R_i$ be the natural map. A set $\tilde{Q}_i$ is open in X_i/R_i if and only if $Q_i = \overset{-1}{f_i}(\tilde{Q}_i)$ is a saturated open set in X_i. The image of

$$f_1(Q_1) \times f_2(Q_2) = \{\{x_1\} : x_1 \in Q_1\} \times \{\{x_2\} : x_2 \in Q_2\}$$

under the natural map g is

$$h(Q_1 \times Q_2) = \{\{(x_1, x_2)\} : (x_1, x_2) \in Q_1 \times Q_2\}$$

where h denotes the natural map of $X_1 \times X_2$ onto $(X_1 \times X_2)/(R_1 \times R_2)$. Since Q_1 and Q_2 are saturated the inverse image of $h(Q_1 \times Q_2)$ under h is $Q_1 \times Q_2$ which is an open set of $X_1 \times X_2$. Therefore $h(Q_1 \times Q_2)$ is an open set of $(X_1 \times X_2)/(R_1 \times R_2)$. The sets $f_1(Q_1) \times f_2(Q_2)$ form a base for the open sets of $(X_1/R_1) \times (X_2/R_2)$ and as we just proved they are mapped into open sets of $(X_1 \times X_2)/(R_1 \times R_2)$ by the natural map g. Therefore g is an open map. Moreover, if R_1 and R_2 are compatible with the corresponding topologies, then the image sets $h(Q_1 \times Q_2)$ also form a base in $(X_1 \times X_2)/(R_1 \times R_2)$ and so g is a homeomorphism.

EXERCISES

1. Consider the rectangle in the plane with vertices $(\pm 1, \pm 1)$. Try to visualize the spaces obtained by the following identifications:

(a) $(-x, -1) \sim (1, x)$ and $(-1, -y) \sim (y, 1)$.
(b) $(x, -1) \sim (x, 1)$ and $(-1, y) \sim (1, y)$.
(c) $(x, -1) \sim (x, 1)$
(d) $(x, -1) \sim (-x, 1)$.
(e) $(x, -1) \sim (-x, 1)$ and $(-1, y) \sim (1, -y)$.
(*f*) $(x, -1) \sim (x, 1)$ and $(-1, y) \sim (1, -y)$.

(Sphere, torus, cylinder, Möbius band, projective plane, Klein bottle.)

2. Let X be the three-dimensional Euclidean space, the origin being omitted, and let $x \sim y$ if there is a real number λ such that $x_i = \lambda y_i$ $(i = 1, 2, 3)$. Show that the quotient space is homeomorphic to the space defined in Exercise 1(e).

3. Let X be the set of ordered pairs (z_1, z_2) of complex numbers, the pair (0, 0) being omitted. Define $u \sim v$ if there is a complex number λ such that $u_i = \lambda v_i$ $(i = 1, 2)$. Show that the quotient space is homeomorphic to the sphere.

NOTES

For further literature on regular open and regular closed sets see Kuratowski [1].

One says that A is regularly contained in B if $\bar{A} \subseteq B^i$.

Axioms (K.1)–(K.4) are equivalent to the following pair of axioms: (a) $\bar{\varnothing} = \varnothing$; (b) $\bar{\bar{X}} \cup \bar{Y} \cup Y = \overline{X \cup Y}$. These axioms were introduced by Monteiro [2]. As an exercise the reader should show the equivalence of axioms (a) and (b) and of axioms (K.1)–(K.4). Another set of axioms for a topology can be given in terms of the binary operation $A * B = (A \cap \bar{B}) \cup (B \cap \bar{A})$ (see Murti [3]). An extensive study of the definition of abstract spaces by derivative, closure, boundary, interior, exterior, and other operators was published by Monteiro and Ribeiro [4].

Axioms (B.1) and (B.2) of a base for a topology can be reformulated as follows:

(B.1′) $$\bigcup \{B : B \in \mathscr{B}\} = X$$

(B.2′) *For every* $B_1, B_2 \in \mathscr{B}$ *we have* $B_1 \cap B_2 = \bigcup \{B : B \subseteq B_1 \cap B_2\}$.

Topologies and uniform structures for linearly ordered sets were studied in two notes by Nachbin [5].

The axioms of a metric are equivalent to the following two axioms:

(M′) $d(x, y) = 0$ *if and only if* $x = y$.

(M″) $d(x, y) + d(x, z) \geqslant d(y, z)$ *for any* $x, y, z \in X$.

These axioms were introduced by Lindenbaum [6].

The condition given in Lemma 6.2 expresses a continuity property of d_2 relative to the topology $\mathscr{T}_1$ generated on X by the pseudometric d_1: For each fixed value of x the pseudometric d_2 determines the function

$d_2^x : y \rightarrow d_2(x, y)$. The lemma states that $\mathcal{T}_1 \geqslant \mathcal{T}_2$ if and only if every d_2^x ($x \in X$) is continuous at x relative to $\mathcal{T}_1$ and the usual topology of the reals.

The result discussed in Exercise 6.4 is connected with a result of Sreenivason [7]: A necessary and sufficient condition that $d \circ f = f(d)$ be a metric whenever d is a metric on X is (i) $f(0) = 0$; (ii) $f(d) \geqslant 0$ *for every* $d \geqslant 0$; (iii) $f(d) \leqslant f(u) + f(v)$ *whenever* $0 \leqslant v - u \leqslant d \leqslant u + v$. If f is strictly increasing and concave, the conditions are satisfied. The topologies generated by d and $d \circ f$ need not be the same.

An equivalent definition of product topologies which is not discussed in the text is due to Efremovič [8]: Define the closure of every set $A : x \in \tilde{A}$ if given a decomposition $A = A^1 \cup \ldots \cup A^n$ of A there is an A^k $(1 \leqslant k \leqslant n)$ such that $x_s \in \overline{A_s^k}$ for every $s \in S$. The product topology is defined by the closure operator $\tilde{A}$.

An extensive study of Boolean product spaces and of Boolean product of uniform structures was made by Nöbeling [9].

REFERENCES

1. C. Kuratowski, "Topologie I," 4th ed., in French. Hafner, New York, 1958.
2. A. Monteiro, Caractérisation de l'opération de fermeture par un seul axiome. *Portugal. Mat.* **4**, 158–160 (1945).
3. K. Murti, A set of axioms for topological algebra, *J. Indian Math. Soc.* [N.S.] **4**, 116–119 (1940).
4. A. Monteiro and H. Ribeiro, Sur l'axiomatique des espaces, V. *Portugal. Math.* **1**. 275–288 (1940).
5. L. Nachbin, Sur les espaces topologiques ordonnés. *C. R. Acad. Sci. Paris* **226**, 381–382 (1948); Sur les espaces uniformes ordonnés. *Ibid.* **226**, 774-775 (1948).
6. A. Lindenbaum, Contributions à l'étude de l'espace métrique, I. *Fund. Math.* **8**, 209–222 (1926).
7. T. K. Sreenivason, Some properties of distance functions. *J. Indian Math. Soc.* [N.S.] **11**, 38–43 (1947).
8. V. A. Efremovič, Invariant definition of topological product. *Uspehi Mat. Nauk* [N.S.] **7** (47), 159–161 (1952).
9. G Nöbeling, "Grundlagen der analytischen Topologie." Springer-Verlag, Berlin, 1954.

CHAPTER II

Separation Properties

1. (T_0) and (T_1) Axioms, Hausdorff Spaces

The most primitive axiom of separation states that given any pair of distinct points a and b in a topological space X there exists an open set O containing one of these points and not containing the other. We do not care which of the possibilities $a \in O$ and $b \notin O$ or $a \notin O$ and $b \in O$ can be realized; the axiom states only that at least one of these alternatives can be secured by choosing a suitable $O \in \mathcal{O}$. The axiom is not a consequence of the axioms of a topological space, as is shown for instance by the nondiscrete topology of any set X consisting of more than a single element. This axiom is called (T_0) *axiom* or *Kolmogorov's axiom of separation*. If it holds we speak about a (T_0) space. One can immediately see that a topological space X is a (T_0) space if and only if for any pair of distinct points a and b at least one of the relations $a \notin \overline{\{b\}}$ and $b \notin \overline{\{a\}}$ takes place.

All other axioms of separation are of a similar nature: Two distinct points or two disjoint sets A and B are given subject to various conditions, and the axiom states the existence of an open set O or of a pair of open sets O_A and O_B which in some way or another separate these points or point sets. For example, a topological space X is called a (T_1) *space* if given any pair of distinct points a, b there exist open sets O_a and O_b such that $a \in O_a$, $b \notin O_a$ and $b \in O_b$, $a \notin O_b$. If X is a (T_1) space we say that the (T_1) *axiom of separation* holds in X. One speaks also about *Fréchet's axiom* or *Riesz's axiom of separation*.

It is clear that every (T_1) space is necessarily a (T_0) space. On the other hand, there exist (T_0) spaces which are not (T_1) spaces. For example, we can choose X as a set of two elements $X = \{a, b\}$ and define the family of open sets as $\mathcal{O} = \{\emptyset, \{a\}, \{a, b\}\}$.

Theorem 1. *The space X is a (T_1) space if and only if any—and thus each—of the following statements holds:*

(1) *Every set $\{a\}$ consisting of a single point a is closed.*

(2) *For every A the intersection of all open sets O_A containing A is the set A.*

(3) *The intersection of all open sets O_a containing the point a is the set $\{a\}$.*

(4) *Every subset of X is a union of closed sets.*

(5) *Every nonvoid set contains some nonvoid closed subset.*

Proof. We prove the implications $(T_1) \to (1) \to (2) \to (3) \to (T_1)$ and thus obtain the equivalence of (T_1), (1), (2), and (3): If (T_1) holds, $\{a\}$ is closed because the complement of $\{a\}$ is the union of open sets O_b where $b \in O_b$. Next, if (1) holds, then given a set A and a point $b \notin A$ there are open sets containing A and not containing b, e.g., $X - \{b\}$ is such a set. Hence the intersection of all open sets containing A is A itself. It is plain that (2) implies (3). Finally, (3) implies the existence of an open set O_a for each pair of distinct points $a, b \in X$ such that $a \in O_a$ and $b \notin O_a$. Hence (3) implies (T_1). The equivalence of (1), (4), and (5) is obvious.

Theorem 2. *If X is a (T_1) space, then every derived set is closed.*

Note. There are (T_0) spaces such that not every derived set is closed. On the other hand, there are spaces for which (T_0) fails but nonetheless every derived set is closed. The nondiscrete topology of a set yields such a space.

Proof. If $a'' \in A''$, then in every open set O containing a'' there is an $a' \in O$ such that $a' \in A'$ and $a' \neq a''$. Since X is a (T_1) space we can find an open set Q containing a' and not containing a''. The set $S = O \cap Q$ is open, $a' \in S$ and $a'' \notin S$. By $a' \in A'$ there exists an a in A which belongs to S. By $a'' \notin S$ we have $a'' \neq a$. Thus for every $a'' \in A''$ and every open O containing a'' there is an a in A, distinct from a'' which belongs to O. Hence a'' is an accumulation point of A.

The best-known separation axiom was introduced by Hausdorff: *Given any pair of distinct points $a, b \in X$ there exist disjoint open sets O_a and O_b in X such that $a \in O_a$ and $b \in O_b$.* A topological space satisfying this axiom is called a *Hausdorff space*, a (T_2) *space*, or a *separated space*. The words separated and separable should not be confused, for the expression *separable space* occurs in the literature with a completely different meaning. If X is a Hausdorff space we say that the (T_2) axiom or Hausdorff's axiom of separation holds in X. Since O_a and O_b are disjoint, $b \notin O_a$ and $a \notin O_b$. Hence (T_2) implies (T_1) and so all the properties discussed above also hold for Hausdorff spaces. As an additional result we have:

Theorem 3. *A topological space is a Hausdorff space if and only if for each point a the intersection of all closed neighborhoods of a is the set* $\{a\}$.

Proof. Let X be a Hausdorff space and let $b \neq a$. Choose disjoint open sets O_a and O_b such that $a \in O_a$ and $b \in O_b$. Then $a \in O_a \subseteq cO_b$ but $b \notin cO_b$ and so cO_b is a closed neighborhood of a not containing b. Hence the intersection of all closed neighborhoods of a does not contain any $b \neq a$. Conversely, if this intersection is $\{a\}$, then given any $b \neq a$ there exist an open O_a and a closed C_a such that $a \in O_a \subseteq C_a$ and $b \notin C_a$. Then $O_b = cC_a$ is open, $b \in O_b$, and $O_a \cap O_b = \emptyset$. Hence Hausdorff's axiom holds in X.

There exist (T_1) spaces which are not (T_2) spaces. For instance, let X be an infinite set and let the topology on X be the topology of finite complements. For any pair of distinct points a, b the open sets $O_b = X - \{a\}$ and $O_a = X - \{b\}$ fulfill the requirements of axiom (T_1) but not of (T_2). Since any two nonvoid open sets have a nonvoid intersection, there exists no pair O_a, O_b which would satisfy the requirements of axiom (T_2).

It is clear that every metric space is a Hausdorff space and it is easy to see that every pseudometric space satisfying axiom (T_0) is a Hausdorff space. This result can be extended to uniform spaces:

Theorem 4. *If axiom* (T_0) *holds in a uniform space, then it is a Hausdorff space.*

Note. If the uniform topology associated with the uniform structure $\mathcal{U}$ is a Hausdorff topology, then $\mathcal{U}$ is called a *separated structure*.

Proof. Let $\mathcal{U}$ be a uniform structure for the uniform topology on X and let a and b be distinct points in X. If X is a (T_0) space, there is a $U \in \mathcal{U}$ such that $a \notin U[b]$ or $b \notin U[a]$. According to Lemma I.8.1 we can choose a symmetric uniformity $V \in \mathcal{U}$ such that $V \circ V \subseteq U$. Then $V[a]$ and $V[b]$ are disjoint: For if $x \in V[a]$ and $x \in V[b]$, then $(a, x) \in V$ and $(b, x) \in V$, so by the symmetry of V we have $(a, b) \in V \circ V \subseteq U$ and $(b, a) \in V \circ V \subseteq U$ which contradicts "$a \notin U[b]$ or $b \notin U[a]$." Therefore the interiors $O_a = V[a]^i$ and $O_b = V[b]^i$ are disjoint open sets containing a and b, respectively.

EXERCISES

1. Let f be a real-valued function on some set X. Show that the pseudometric $d : d(x, y) = |f(x) - f(y)|$ generates a (T_0)-space on X if and only if f is invertible.

2. Show that a pseudometric (T_0) space is necessarily a Hausdorff space.

3. Define the equivalence relation R on the topological space X as follows: $a \sim b$ if and only if $a \in O_b$ and $b \in O_a$ for every $O_a \in \mathcal{O}(a)$ and $O_b \in \mathcal{O}(b)$. Show that X/R is a (T_0) space and X and X/R are topologically isomorphic, i.e., there is a one-to-one correspondence between their open sets which is an isomorphism with respect to the operations $\cup$ and $\cap$.

4. Show that the properties (T_0), (T_1), and (T_2) are topological invariants, i.e., if X and Y are homeomorphic spaces, then either both or neither of them have property (T_i) ($i = 0, 1$, or 2).

5. The spaces X and X/R are homeomorphic if and only if the equivalence relation constructed in Exercise 3 is the identity relation.

6. If X is a (T_1) space and if $\mathcal{O} = \mathcal{C}$, then the topology on X is discrete.

7. Show that if X is a finite set and its topology is (T_1), then it is discrete.

8. The following topological space is an example of a (T_1) space which is not a Hausdorff space: Let X be the set of real numbers and let $C \subseteq X$ be closed if it is bounded and closed in the usual sense.

(Any two nonvoid open sets have a nonvoid intersection and so axiom (T_2) cannot hold. Similar topologies can be constructed by starting from an arbitrary topology on a set X. This is the *topology of compact complements.*)

9. Show the following result: *If a* (T_0) *space* X *has a countable base, then the cardinality of the set* X *is at most the cardinality of the continuum.*

(Let $\mathcal{B}$ with card $\mathcal{B} \leqslant \omega$ be a base. For every $x \in X$ let

$$\mathcal{B}_x = \{B : B \in \mathcal{B} \quad \textit{and} \quad x \in B\}.$$

Then $\mathcal{B}_x \in \mathfrak{P}(\mathcal{B})$ where $\mathfrak{P}(\mathcal{B})$ denotes the power set of $\mathcal{B}$ and by the (T_0) axiom $\mathcal{B}_{x_1} \neq \mathcal{B}_{x_2}$. Hence card $X =$ card$\{\mathcal{B}_x\} \leqslant$ card $\mathfrak{P}(\mathcal{B}) \leqslant 2^\omega$.)

2. (T_3) Spaces, Regular and Semiregular Spaces

We call a topological space X a (T_3) *space* if given any closed set A and any point $b \notin A$ there exist disjoint open sets O_A and O_b such that $A \subseteq O_A$ and $b \in O_b$. Two simple examples show that axiom (T_3) is independent of axioms (T_0), (T_1), and (T_2): First let $X = \{a, b, c\}$ and let the open sets of X be $\varnothing$, $\{a\}$, $\{b, c\}$, $\{a, b, c\}$. These sets are also closed and they are the only closed sets of X. Axiom (T_3) is clearly satisfied in X but axiom (T_0) fails because b and c cannot be separated by open sets. Consequently, axiom (T_3) does not imply axiom (T_0).

Next consider the following example of a Hausdorff space in which axiom (T_3) does not hold.

Example 1. Let X be the set of reals and let $A = \{1/n : n = 1, 2, ...\}$. We call a set O open if $O = Q - B$ where $B \subseteq A$ and Q is open in the usual sense. This defines a topology on X which makes X a (T_2) space because it contains more open sets than the usual Hausdorff topology. However, X is not a (T_3) space as can be seen by choosing $b = 0$ and the closed set $A = \{1/n : n = 1, 2, ...\}$: Indeed, if $A \subseteq O_A$, then O_A is open in the usual sense and so it contains intervals arbitrarily close to O. On the other hand, if $b = 0$ is contained in O_b, then $Q_b = O_b \cup B_b$ contains an interval $(-\delta, \delta)$ for some $\delta > 0$. Hence O_A and O_b intersect and the (T_3) property fails.

Theorem 1. *If a topological space satisfies axioms* (T_0) *and* (T_3), *then it is a Hausdorff space.*

Proof. Given any pair of distinct points $a, b \in X$, by axiom (T_0) there exists an open set O which separates these points. We may assume that $a \notin O$ and $b \in O$. We apply axiom (T_3) with $A = cO$ and find open sets O_A and O_b such that $A \subseteq O_A$, $b \in O_b$, and $O_A \cap O_b = \emptyset$. Therefore $a \in A \subseteq O_A$ and $b \in O_b$. This shows that X is a Hausdorff space.

A topological space satisfying axioms (T_0) and (T_3) is called a *regular topological space*. These spaces were first identified by Vietoris. In view of the foregoing remarks every regular topological space is a Hausdorff space but not every Hausdorff space is regular. The reader should be careful because there are several alternative terminologies in the literature all using the same symbol (T_3) and the expression "regular." The best known among these alternatives is a simple interchange in terminology: Our (T_3) spaces are called regular and our regular spaces are termed (T_3). The strength of the axioms (T_0), (T_1), (T_2), $(T_0) + (T_3)$ is increasing in their order.

Theorem 2. *A topological space* X *is a* (T_3) *space if and only if any, and thus each, of the following propositions holds in* X:

(1) *For every nonvoid open set* O *and for every* $x \in O$ *there is an open set* Q_x *such that* $x \in Q_x \subseteq \bar{Q}_x \subseteq O$.

(2) *For every closed* A *the intersection of all closed neighborhoods of* A *is the set* A.

(3) *For every* A *and open* B *satisfying* $A \cap B \neq \emptyset$ *there exists an open* O *such that* $A \cap O \neq \emptyset$ *and* $\bar{O} \subseteq B$.

(4) *For every nonvoid A and closed B satisfying $A \cap B = \emptyset$ there exist disjoint open sets O_A and O_B such that* $A \cap O_A \neq \emptyset$ and $B \subseteq O_B$.

Note. If the set A is contained in the interior of the set N_A then N_A is called a *neighborhood* of A.

Proof. We prove the following sequence of implications: $(T_3) \rightarrow (1) \rightarrow (2) \rightarrow (3) \rightarrow (4) \rightarrow (T_3)$. This will show the equivalence of any two of the statements (1), (2), (3), (4), and (T_3).

$(T_3) \rightarrow (1)$: Let O be open and $x \in O$. Then cO is closed and $x \notin cO$. Hence by (T_3) there exist open sets Q_x and Q such that $x \in Q_x$, $cO \subseteq Q$, and $Q \cap Q_x = \emptyset$. Consequently, $Q_x \subseteq cQ \subseteq O$ and since cQ is closed $\bar{Q}_x \subseteq cQ \subseteq O$. Therefore $x \in Q_x \subseteq \bar{Q}_x \subseteq O$ and (1) holds for x and O.

$(1) \rightarrow (2)$: If A is closed and $x \notin A$, then $O = cA$ is open and $x \in O$. Hence by (1) we have $x \in Q_x \subseteq \bar{Q}_x \subseteq O$ for some open set Q_x. The set cQ_x is closed and $cQ_x \supseteq c\bar{Q}_x \supseteq cO = A$ showing that cQ_x is a closed neighborhood of A which does not contain x. Therefore the intersection of all closed neighborhoods of A contains no x lying outside of A.

$(2) \rightarrow (3)$: Let $A \cap B \neq \emptyset$ and let B be open. Then there is an x such that $x \in A$ and $x \in B$. Since x does not belong to the closed set cB, by (2) there is a closed neighborhood of cB not containing x: Let us say $cB \subseteq Q \subseteq C$ where Q is open, C is closed, and $x \notin C$. Then $O = cC$ is an open set which contains x and therefore $A \cap O \neq \emptyset$. Moreover, cQ being closed $\bar{O} = \overline{cC} \subseteq cQ \subseteq B$. This proves (3) for the sets A and B.

$(3) \rightarrow (4)$: If $A \cap B = \emptyset$ where A is not void and B is closed, then $A \cap cB \neq \emptyset$ and cB is open. Using (3) with cB in place of B we obtain an open set O such that $A \cap O \neq \emptyset$ and $O \subseteq \bar{O} \subseteq cB$. Hence if we put $O_A = O$ and $O_B = c\bar{O}$, then O_A and O_B are open sets satisfying $A \cap O_A \neq \emptyset$ and $B \subseteq O_B$. Furthermore, by $O_B = c\bar{O} \subseteq cO = cO_A$ the sets O_A and O_B are disjoint.

$(4) \rightarrow (T_3)$: Applying (4) in the special case when A consists of a single point a we obtain a pair of disjoint open sets O_a and O_B such that $a \in O_a$ and $B \subseteq O_B$. Hence if (4) holds, then axiom (T_3) is also satisfied in X. The proof of the theorem is complete.

Perhaps the most useful form of axiom (T_3) is form (1) above. Two simple variants of (1) are also common. The first of these states the following: "*For every x and every neighborhood N_x there is a neighborhood Q_x such that $x \in Q_x \subseteq \bar{Q}_x \subseteq N_x$.*" A base $\mathscr{B}$ for the open sets of X is called a *regular base* if for each $B \in \mathscr{B}$ and $x \in B$ there is an open set Q_x such that $x \in Q_x \subseteq \bar{Q}_x \subseteq B$. The second variant of (1) states: "*Every base for the open sets of X is a regular base.*" It is easy to see that the existence of one regular base implies that every base is regular.

A topological space X is called a ($T_3{}^s$) space if its regular open sets form a base for the open sets of X. If in addition X is a Hausdorff space, then it is called a *semiregular* topological space. It follows that X can be a ($T_3{}^s$) space without satisfying axiom (T_0). We can easily show that every (T_3) space is a ($T_3{}^s$) space and consequently every regular space is semiregular: First we prove that the regular open sets of a (T_3) space form a base. Since X itself is regular open, only axiom (B. 2) has to be verified. We assume that O_1 and O_2 are regular open sets and $x \in O_1 \cap O_2$. By axiom (T_3) there is an open set Q such that $x \in Q \subseteq \bar{Q} \subseteq O_1 \cap O_2$. If we let $\tilde{O} = \bar{Q}^i$, then by Theorem I.2.9 the set $\tilde{O}$ is regular open and we have $x \in Q \subseteq \tilde{O} \subseteq \bar{Q} \subseteq O_1 \cap O_2$. Hence $x \in \tilde{O} \subseteq O_1 \cap O_2$ and so (B. 2) holds for O_1 and O_2. The same argument shows that if O is open and $x \in O$, then there is a regular open set $\tilde{O}$ such that $x \in \tilde{O} \subseteq O$. Hence the regular open sets generate the original topology of the space.

Another separation property which is actually stronger than (T_2) but weaker than regularity was first noticed by Urysohn: A space is usually called a *Urysohn space* if any pair of distinct points a, b can be separated by disjoint closed neighborhoods. On the one hand, this requirement is clearly at least as strong as Hausdorff's separation axiom. On the other hand, if a and b are distinct points in a regular space, then a and b have disjoint open neighborhoods O_a and O_b and so by form (1) of axiom (T_3) there are open sets Q_a and Q_b such that $a \in Q_a \subseteq \bar{Q}_a \subseteq O_a$ and $b \in Q_b \subseteq \bar{Q}_b \subseteq O_b$. Hence every regular space is a Urysohn space. Semiregularity and Urysohn's axiom are not comparable: There exist Urysohn spaces which are not semiregular and also semiregular spaces which do not satisfy Urysohn's axiom.

A pair of sets A and B satisfying $A \subseteq c\bar{B}$ and $B \subseteq c\bar{A}$ is called *separated.* Axioms (T_4) and (T_5) which will be discussed in the following sections concern the separation of separated sets by disjoint open sets. The next theorem shows that in order to find Hausdorff spaces having property (T_3) but not (T_4) or else having property (T_4) but not (T_5), we must look for rather sophisticated examples.

Theorem 3. *If X is regular and if A and B are countable separated sets, then there exist disjoint open sets O_A and O_B containing A and B, respectively.*

Proof. The only interesting case is when both A and B are denumerable sets. We suppose that $A = \{a_n\}$ $(n = 1, 2, \ldots)$ and $B = \{b_n\}$ $(n = 1, 2, \ldots)$. Since X is a (T_2) space we can find disjoint open sets O_{a_1} and O_{b_1} such that $a_1 \in O_{a_1}$ and $b_1 \in O_{b_1}$. By hypothesis $O_{a_1} \cap c\bar{B}$ is an open set containing a_1 and so by axiom (T_3) there is an open neighborhood Q_{a_1} of a_1 such that $Q_{a_1} \subseteq \bar{Q}_{a_1} \subseteq O_{a_1} \cap c\bar{B}$. Similarly, there is an open neighborhood Q_{b_1} of b_1 such that $Q_{b_1} \subseteq \bar{Q}_{b_1} \subseteq O_{b_1} \cap c\bar{A}$. It is obvious from the construction that $\bar{Q}_{a_1} \cap \bar{B} = \emptyset$, $\bar{Q}_{b_1} \cap \bar{A} = \emptyset$, and

$$\bar{Q}_{a_1} \cap \bar{Q}_{b_1} = \emptyset.$$

Now we use induction to construct sequences (Q_{a_n}) $(n = 1, 2, \ldots)$ and (Q_{b_n}) $(n = 1, 2, \ldots)$ of open neighborhoods such that $\bar{Q}_{a_n} \cap \bar{B} = \emptyset$, $\bar{Q}_{b_n} \cap \bar{A} = \emptyset$, and $\bar{Q}_{a_m} \cap \bar{Q}_{b_n} = \emptyset$ for every $m, n = 1, 2, \ldots$; for instance, if $Q_{a_1}, \ldots, Q_{a_n}$ and $Q_{b_1}, \ldots, Q_{b_n}$ are already determined in a proper way, then $a_{n+1} \notin \bar{Q}_{b_1} \cup \ldots \cup \bar{Q}_{b_n}$ and so $a_{n+1} \in c(\bar{B} \cup \bar{Q}_{b_1} \cup \ldots \cup \bar{Q}_{b_n})$. Similarly, $b_{n+1} \in c(\bar{A} \cup \bar{Q}_{a_1} \cup \ldots \cup \bar{Q}_{a_n})$. By the Hausdorff property we can determine disjoint open neighborhoods $O_{a_{n+1}}$ and $O_{b_{n+1}}$ of a_{n+1} and b_{n+1}, respectively. Next, by axiom (T_3) there is an open neighborhood $Q_{a_{n+1}}$ of a_{n+1} such that

$$a_{n+1} \in Q_{a_{n+1}} \subseteq \bar{Q}_{a_{n+1}} \subseteq c(\bar{B} \cup \bar{Q}_{b_1} \cup \ldots \cup \bar{Q}_{b_n}) \cap O_{a_{n+1}}.$$

The neighborhood $Q_{b_{n+1}}$ can be determined similarly. The sets $O_A = \bigcup\{Q_{a_n} : n = 1, 2, \ldots\}$ and $O_B = \bigcup\{Q_{b_n} : n = 1, 2, \ldots\}$ are disjoint open sets containing A and B, respectively.

EXERCISES

1. Give a simple proof for the implication $(T_1) + (T_3) \rightarrow (T_2)$.

2. Prove directly without using any other statements of Theorem 2 that $(1) \rightarrow (T_3)$ and also that $(1) \rightarrow (3) \rightarrow (1)$.

3. Let X be any linearly ordered set. Show that the interval topology of X and also its half-open interval topologies are regular.

(Let x be a point in the open interval (a, b). If $\{x\} = (a, b)$, then $\{x\}$ is both open and closed because the interval topology satisfies axiom (T_1). Hence $Q_x = \{x\}$ is an open set containing x and $Q_x \subseteq \bar{Q}_x \subseteq (a, b)$. If $(a, x) \neq \emptyset$ but $(x, b) = \emptyset$, then we can choose a point c in (a, x) and we see that $[c, b) = [c, x]$ is a closed set. Hence $Q_x = (c, b)$ has the property that $x \in Q_x \subseteq \bar{Q}_x \subseteq (a, b)$. If $(a, x) \neq \emptyset$ and $(x, b) \neq \emptyset$, then we choose a point c in (a, x) and a point d in (x, b). The open set (c, d) has the required properties. If X is topologized by the half-open interval topology, the situation is similar but much simpler.)

4. Show that every uniformizable space is a (T_3) space.

(Let $\mathscr{U}$ be a uniform structure and let x be a point in the open set O. By Lemma I.9.2 there is a $U \in \mathscr{U}$ satisfying $x \in U[x] \subseteq O$. If $V \in \mathscr{U}$ is chosen such that $V \circ V \subseteq U$, then by Lemma I.9.7 we have

$$x \in V[x]^i \subseteq \overline{V[x]} \subseteq U[x] \subseteq O.)$$

5. Show that the topology of finite complements of an infinite set is not a $(T_3{}^s)$ topology.

(The only regular open sets are ø and X.)

6. On the set of reals let $\mathscr{T}$ be the least upper bound of the usual topology and of the topology of countable complements. Show that $\mathscr{T}$ is not semiregular.

[By Exercise I.5.4 a set O is open relative to $\mathscr{T}$ if and only if $O = Q \cap cA$ where Q is open in the usual sense and A is countable. We determine the closure of O relative to $\mathscr{T}$: The closed sets containing O are of the form $C \cup B$ where C is closed in the usual sense and B is countable. If $a \notin C$ then there is an open interval (α, β) such that $a \in (\alpha, \beta) \subseteq cC$. It follows that $a \notin O$; for $(\alpha, \beta) \cap Q \cap cA \subseteq B$ and $A \cup B$ being countable $(\alpha, \beta) \cap Q = ø$. Hence if $O = Q \cap cA \subseteq C \cup B$, then $O \subseteq C$ and so $\bar{O} = \bigcap\{C : O \subseteq C\}$, i.e., the closure of O relative to $\mathscr{T}$ is the same as its closure in the usual sense. Since $C^i = c\overline{C^e}$ we see that if C is closed relative to $\mathscr{T}$, then its interior is the same as its interior in the usual sense. Consequently a set O is regular open relative to $\mathscr{T}$ if and only if it is regular open in the usual sense. The complement of the rationals for instance is not the union of regular open sets though it is open relative to $\mathscr{T}$.]

7. Give an example of a set X and two topologies $\mathscr{T}_1 < \mathscr{T}_2$ such that $\mathscr{T}_1$ is regular but $\mathscr{T}_2$ is not.

(Let X be the set of reals, let $\mathscr{T}_1$ be the usual topology of the reals, and let $\mathscr{T}_2$ be the topology given in the preceding exercise.)

8. Show the following proposition: *If* $\{\mathscr{T}_i\}$ $(i \in I)$ *is a family of* (T_3) *topologies on a fixed set* X, *then* $\operatorname{lub}\{\mathscr{T}_i\}$ *is also a* (T_3) *topology.*

(Let O be open relative to $\operatorname{lub}\{\mathscr{T}_i\}$ and let $x \in O$. Then there are indices $i_1, \ldots, i_n \in I$ and sets O_{i_k} open relative to $\mathscr{T}_{i_k}$ such that

$$x \in O_{i_1} \cap \ldots \cap O_{i_n} \subseteq O.$$

Let Q_{i_k} be open relative to $\mathscr{T}_{i_k}$ and such that $x \in Q_{i_k} \subseteq \bar{Q}_{i_k} \subseteq O_{i_k}$. The set $Q = Q_{i_1} \cap \ldots \cap Q_{i_n}$ is open relative to $\operatorname{lub}\{\mathscr{T}_i\}$, its closure is contained in $\bar{Q}_{i_1} \cap \ldots \cap \bar{Q}_{i_n}$, and so $x \in Q \subseteq \bar{Q} \subseteq O$.)

3. (T_4) Spaces and Normal Spaces

The (T_4) axiom was introduced by Tietze in one of the classic papers on separation axioms. A topological space X is called a (T_4) *space* if for any pair of disjoint closed sets A and B there exist disjoint open sets O_A and O_B such that $A \subseteq O_A$ and $B \subseteq O_B$. It is clear that for (T_1)

spaces the (T_4) axiom is stronger than the (T_3) axiom. However, if the space does not satisfy the (T_1) axiom, then axiom (T_4) does not imply axiom (T_3): For example, let X be a set of five elements, say $X = \{a, b, c, d, e\}$ and let the open sets of X be $\{a, b, c, d, e\}$, $\{a, b, c, d\}$, $\{a, b, c\}$, $\{b, c, d, e\}$, $\{b, c, d\}$, $\{b, c\}$, $\{b\}$, $\{b, d, e\}$, $\{b, d\}$, $\{d, e\}$, $\{d\}$, and $\emptyset$. One can easily see that these sets define a topology on X which satisfies axiom (T_0) but not (T_1). The closed sets of X are $\emptyset$, $\{e\}$, $\{d, e\}$, $\{a\}$, $\{a, e\}$, $\{a, d, e\}$, $\{a, c, d, e\}$, $\{a, c\}$, $\{a, c, e\}$, $\{a, b, c\}$, $\{a, b, c, e\}$, and X. Let us now suppose that A and B are disjoint closed proper subsets of X. Since the only proper closed sets not containing a are $\{d, e\}$ and $\{e\}$, one of A and B must contain a while the other must contain e. Hence A is $\{a, b, c\}$, $\{a, c\}$, or $\{a\}$ and B is $\{d, e\}$ or $\{e\}$. In any case we can choose $O_A = \{a, b, c\}$ and $O_B = \{d, e\}$ so that the requirements of axiom (T_4) are satisfied. However, X is not a (T_3) space because the only open set containing the closed set $\{a, d, e\}$ is X itself. Therefore (T_0) and (T_4) do not imply (T_3).

Theorem 1. *If a topological space satisfies axioms* (T_1) *and* (T_4), *then it also satisfies axioms* (T_2) *and* (T_3).

Proof. This is a simple consequence of Theorems 1.1 and 2.1: In fact, if X is a (T_1) space, then the set $\{b\}$ is closed for every $b \in X$ and so for such a space axiom (T_3) is a special case of axiom (T_4). By the second theorem, axiom (T_2) is a consequence of (T_1) and (T_3).

Example 1. Now we construct a topological space in which axioms (T_2) and (T_3) hold but (T_4) fails. This will show that (T_4) is not a consequence of the preceding axioms (T_0), (T_1), (T_2), and (T_3). Let X be the "closed upper half plane," that is, the set of ordered pairs (x_1, x_2) of real numbers satisfying $x_2 \geqslant 0$. We put the set O in the family $\mathcal{O}$ if it contains with each $x = (x_1, x_2)$; $x_2 > 0$ a disk

$$\{\xi : (\xi_1 - x_1)^2 + (\xi_2 - x_2)^2 \leqslant \epsilon^2\}$$

and if it contains with each $(x_1, 0)$ a disk

$$D_\epsilon[x] = \{\xi : (\xi_1 - x_1)^2 + (\xi_2 - \epsilon)^2 \leqslant \epsilon^2\}.$$

The family $\mathcal{O}$ satisfies the axioms of a topological space and it is obvious that X is a Hausdorff space in this topology. Using the fact that every closed disk is a closed set in the present topology we see that form (1) of axiom (T_3) holds.

Now we show that axiom (T_4) fails in this space X. For let $A = \{(x_1, 0) : x_1 = \textit{rational}\}$ and $B = \{(x_1, 0) : x_1 = \textit{irrational}\}$. Every subset of the real line is closed relative to this topology because its complement is open. Hence A and B are closed sets. Suppose that O_A and O_B are open sets containing A and B, respectively. To each $x \in B$ there corresponds an $\epsilon = \epsilon_x > 0$ such that $D_\epsilon[x] \subseteq O_B$. Let B_n denote the set of those irrational numbers x_1 for which $\epsilon_x \geqslant 1/n$ where $x = (x_1, 0)$.

Let us suppose that for each n and for each interval I on the real line $A \cup B$ there is a subinterval J in I such that $B_n \cap J = \emptyset$. Let the rational numbers be ordered in a single sequence $(r_0, r_1, ..., r_n, ...)$. Then we can construct a sequence of closed intervals I_n such that $I_{n+1} \subseteq I_n$, $r_n \notin I_n$, and $B_n \cap I_n = \emptyset$. By the "principle of nested intervals" there is a real number ρ which belongs to all of these intervals. However, by $r_n \notin I_n$ this number is not rational and so it belongs to B_n for a sufficiently large value of n. Since $\rho \in I_n$ and $B_n \cap I_n = \emptyset$ this leads to a contradiction. Hence we proved that for some n there is an interval I on the real line with the property that every subinterval of I contains points of B_n. This shows that every point of I is an accumulation point of B_n. Consequently B_n has a rational accumulation point r. A disk $D_\delta[(r, 0)]$ can be given such that $D_\delta[(r, 0)] \subseteq O_A$. To each $x_1 \in B_n$ there corresponds a disk $D_\epsilon[(x_1, 0)] \subseteq O_B$ with radius $\epsilon = \epsilon_x \geqslant 1/n$. Hence, if x_1 is sufficiently close to the point r, the disks $D_\delta[(r, 0)]$ and $D_\epsilon[(x_1, 0)]$ intersect. Thus O_A and O_B are not disjoint and so the requirements of axiom (T_4) cannot be fulfilled.

A topological space satisfying axioms (T_1) *and* (T_4) *is called a normal space*. The foregoing results and examples show that every normal space is regular but there exist regular topological spaces which are not normal. Therefore axiom (T_1) + (T_4) is stronger than any of the axioms (T_0), (T_1), (T_2), (T_0) + (T_3). In the literature (T_4) spaces are often called normal but then a normal space need not be regular nor a Hausdorff space. If this alternative terminology is used, our normal spaces are identified by the symbol (T_4).

Theorem 2. *A topological space X is a* (T_4) *space if and only if for every open set O and every closed subset A of O there exists an open set O_A such that $A \subseteq O_A \subseteq \bar{O}_A \subseteq O$.*

Proof. First we prove that axiom (T_4) implies the present criterion: Indeed if we set $B = cO$, then A and B are disjoint closed sets. Hence by axiom (T_4) there exist disjoint open sets O_A and O_B such that $A \subseteq O_A$ and $B \subseteq O_B$. Therefore by $O_A \subseteq cO_B$ we have $A \subseteq O_A \subseteq \bar{O}_A \subseteq cO_B \subseteq cB = O$. Next we prove axiom ($T_4$) by using the statement of the criterion: Given any pair of disjoint closed sets A and B we can apply the statement with A and $O = cB$. We obtain an open set O_A such that

$A \subseteq O_A \subseteq \bar{O}_A \subseteq O$. The open sets O_A and $O_B = c\bar{O}_A$ are disjoint and $A \subseteq O_A$ and $B = cO \subseteq c\bar{O}_A = O_B$. Therefore axiom ($T_4$) holds in X.

EXERCISES

1. Let $X = (-1, +1)$ and let the closed sets of X be ø and the intervals $[\alpha, \beta] = \{x : \alpha \leqslant x \leqslant \beta\}$ where $-1 < \alpha \leqslant 0 \leqslant \beta < +1$. Show that the axioms for the open sets of a topology are satisfied and the space is a (T_0) + (T_4) space but it is not a (T_1) space nor a (T_3) space.

2. Let X be linearly ordered and topologized by the order topology. Show that X is neither (T_1) nor (T_3) but it is (T_0) and (T_4).

3. Show that if Y is a nonvoid closed set in a (T_4) space X, then Y is a (T_4) space in the induced topology.

(If $A \subseteq Y$ is closed relative to the induced topology, then A is a closed set in X. If $A \subseteq Y$ and $B \subseteq Y$ are disjoint closed sets in X, then there are disjoint open sets O_A and O_B containing A and B, respectively. The sets $O_A \cap Y$ and $O_B \cap Y$ are open neighborhoods of A and B in Y.)

4. Let $C_1, C_2, \ldots$ be closed sets in a (T_4) space such that for every $m = 1, 2, \ldots$ the closure of $\bigcup\{C_n : n \neq m\}$ does not intersect C_m. Show the existence of disjoint open sets $O_1, O_2, \ldots$ such that $C_m \subseteq O_m$ for every $m = 1, 2, \ldots$.

(Construct the sets O_m step by step using Theorem 2.)

5. Modify the space described in Example 1 as follows: O is open in X if it contains with each $x = (x_1, x_2)$ where $x_2 > 0$ a disk

$$\{\xi : (\xi_1 - x_1)^2 + (\xi_2 - x_2)^2 \leqslant \epsilon^2\}$$

and if it contains with each $(x_1, 0)$ a sector

$$\{\xi : (\xi_1 - x_1)^2 + \xi_2^2 \leqslant \epsilon^2 \quad \text{and} \quad |\xi_1 - x_1| \leqslant \xi_2\}.$$

Show that the topological space defined in this manner is regular but not normal.

6. Prove that if X is finite, then $(T_3) \rightarrow (T_4)$.

4. Point-Finite and Star-Finite Open Coverings

The criterion given in Theorem 3.2 can be formulated such that it leads to useful generalizations: Given any closed set A and a covering of A by a single open set O there is another covering of A consisting of one open set Q such that $Q \subseteq \bar{Q} \subseteq O$. The question arises whether a similar statement holds for coverings consisting of more than one open set. For finite covers the answer is affirmative.

Theorem 1. *If X is a (T_4) space, A a closed set in X, and $\{O_1, O_2, ..., O_n\}$ an open covering of A, then there is another open covering $\{Q_1, Q_2, ..., Q_n\}$ of A such that $\bar{Q}_i \subseteq O_i$ for each $i = 1, 2, ..., n$.*

Note. If the conclusion of the theorem holds in the special case when $A = X$ and $n = 2$, then X is necessarily a (T_4) space. This is an easy consequence of Theorem 3.2: Given a closed set C and an open set O containing C it is sufficient to consider the open cover of X which consists of cC and O.

Proof. First we restrict ourselves to the special case when $A = X$: Let $\{O_1, O_2, ..., O_n\}$ be an open covering of X and let $A_1 = c(O_2 \cup ... \cup O_n)$. Since we are dealing with a covering of X we see that $A_1 \subseteq O_1$. We use axiom (T_4) in the form which is given in Theorem 3.2 and find an open set Q_1 such that $A_1 \subseteq Q_1 \subseteq \bar{Q}_1 \subseteq O_1$. Then $\{Q_1, O_2, ..., O_n\}$ is an open covering of X because $\{A_1, O_2, ..., O_n\}$ is a covering of X and $A_1 \subseteq Q_1$. Therefore any open covering $\{O_1, O_2, ..., O_n\}$ of X can be replaced by another one $\{Q_1, O_2, ..., O_n\}$ such that $\bar{Q}_1 \subseteq O_1$. If we apply this principle n times we obtain an open covering $\{Q_1, Q_2, ..., Q_n\}$ such that $\bar{Q}_i \subseteq O_i$ for every $i = 1, 2, ..., n$.

Now we prove the theorem in the general case: Given a closed set A and an open covering $\{O_1, ..., O_n\}$ of A we put $O_0 = cA$. Then $\{O_0, O_1, ..., O_n\}$ is an open covering of X and in this special case we may apply the theorem. We obtain an open covering $\{Q_0, Q_1, ..., Q_n\}$ of X satisfying $\bar{Q}_0 \subseteq O_0$ and $\bar{Q}_i \subseteq O_i$ for each $i = 1, ..., n$. Since $Q_0 \subseteq \bar{Q}_0 \subseteq O_0 = cA$ we have $A \subseteq cQ_0$ so that the points of A are covered by the open sets $Q_1, ..., Q_n$. Therefore $\{Q_1, ..., Q_n\}$ is an open covering of A which satisfies $\bar{Q}_i \subseteq O_i$ for each $i = 1, ..., n$. This completes the proof.

A cover A_i $(i \in I)$ of a set X is called *point-finite* if each point of X is covered only by finitely many sets A_i. The last theorem can be extended to point-finite open covers by using Zorn's lemma:

Theorem 2. *Let X be a (T_4) space and let A be a closed set. Consider an open covering $\{O_i\}$ $(i \in I)$ of A such that every point $x \in X$ is covered only by finitely many O_i's. Then there is an open covering $\{Q_i\}$ $(i \in I)$ of A such that $\bar{Q}_i \subseteq O_i$ for each $i \in I$.*

Note. There is one special case which is worth mentioning separately: A cover $\{A_i\}$ $(i \in I)$ is called *star-finite* if every A_i intersects only finitely many sets of the family $\{A_i\}$. Clearly every star-finite cover is point-finite and so the above statement holds in particular for star-finite open covers of the closed set A.

First we restrict ourselves to the case when $A = X$. Let $\{O_i\}$ $(i \in I)$ be an open covering of X which satisfies the hypotheses of the theorem. We denote by $\mathcal{Q}$ the family of all open coverings $\{Q_i\}$ $(i \in I)$ of X which have the property that $\bar{Q}_i \subseteq O_i$ or $Q_i = O_i$ for every $i \in I$. Let J be the set of those indices $i \in I$ for which $\bar{Q}_i \subseteq O_i$. Of course, J depends on the particular covering $\{Q_i\}$ in question. We introduce a reflexive ordering relation in $\mathcal{Q}$ by saying that $\{Q_i^1\} \leqslant \{Q_i^2\}$ if $Q_i^1 = Q_i^2$ for every $i \in J^1$. The transitivity of the ordering follows from the remark that $J^1 \subseteq J^2$ whenever $\{Q_i^1\} \leqslant \{Q_i^2\}$.

Let S be an index set and let $\{\{Q_i^s\}\}$ $(s \in S)$ be a linearly ordered family of open coverings $\{Q_i^s\} \in \mathcal{Q}$. We show that the family has an upper bound in $\mathcal{Q}$ and so the hypothesis of Zorn's lemma is satisfied. In fact, let $J = \bigcup J^s$ and for every $i \in J$ let $Q_i = Q_i^s$ where s is arbitrarily chosen such that $i \in J^s$. Since $\{\{Q_i^s\}\}$ $(s \in S)$ is linearly ordered Q_i is independent of the choice of s. For every $i \notin J$ let $Q_i = O_i$. We show that $\{Q_i\}$ is an open covering of X: For let $x \in X$ be arbitrary. By the hypothesis of the theorem there exist only finitely many of indices, say $i_1, ..., i_n$ such that $x \in O_{i_\nu}$. Since the number of these indices is finite and $\{J^s\}$ is a linearly ordered family there is an index $s \in S$ such that if $i_\nu \in J$ then $i_\nu \in J^s$. The family $\{Q_i^s\}$ is a covering of X and so $x \in Q_i^s$ for some $i \in I$, where necessarily $i = i_\nu$ for some index ν. Thus, if $i \in J$, then $i \in J^s$ and so $Q_i = Q_i^s$ and $x \in Q_i$. If $i \notin J$, then by definition $Q_i = O_i$ and so $x \in Q_i^s \subseteq O_i = Q_i$. Hence in any case x is covered by $\{Q_i\}$. According to the definition of the ordering relation $\leqslant$ we have $\{Q_i^s\} \leqslant \{Q_i\}$ for every $s \in S$ because for each $i \in J^s$ we have $Q_i^s = O_i$. Therefore the hypotheses of Zorn's lemma are satisfied.

By Zorn's lemma there exists an open covering $\{Q_i\} \in \mathcal{Q}$ which is maximal with respect to the given ordering relation. Using axiom (T_4) we can now show that this particular covering satisfies the requirements of the theorem. Indeed if $\bar{Q}_i \subseteq O_i$ is not true for some index $i \in I$, then $Q_i = O_i$ and so C, the complement of $\bigcup\{Q_j : j \neq i\}$ is a closed subset of the open set O_i. Hence by Theorem 3.2 we can replace $Q_i = O_i$ by another open set Q_i^* such that $C \subseteq Q_i^* \subseteq \bar{Q}_i^* \subseteq O_i$. For every $j \neq i$ we define $Q_j^* = Q_j$. The new family $\{Q_i^*\}$ is an open covering of X because $C \subseteq Q_i^*$ and $X = C \cup \bigcup\{Q_j^*: j \neq i\}$. Morever, $\{Q_i\} < \{Q_i^*\}$ whence $\{Q_i\}$ is not maximal. This completes the proof for the case when $A = X$.

If A is an arbitrary closed set in X and $\{O_i\}$ $(i \in I)$ is an open covering of A, then we can complete this to a covering of the entire space X by joining $O_0 = cA$ to the family $\{O_i\}$. Using the theorem in the special case for which it was already proved we find an open covering $\{Q_i\}$ ($i = 0$ or $i \in I$) of X such that $\bar{Q}_i \subseteq O_i$ for $i = 0$ and for every $i \in I$. The family $\{Q_i\}$ ($i \in I$) gives an open covering of A which has the required property.

We give one more characterization of (T_4) spaces. This one depends on the following notion of similarity: *An ordered n-tuple* $(A_1, ..., A_n)$ *of subsets* $A_1, ..., A_n$ *of a set* X *is called similar to an ordered n-tuple* $(B_1, ..., B_n)$ *of subsets* $B_1, ..., B_n$ *of a set* Y *if for any selection of the indices* $A_{i_1} \cap ... \cap A_{i_k} = \emptyset$ *if and only if* $B_{i_1} \cap ... \cap B_{i_k} = \emptyset$. We have the following:

Theorem 4. *A topological space X is a* (T_4) *space if and only if for every ordered n-tuple* $(A_1, ..., A_n)$ *of closed sets in X there exists an ordered n-tuple* $(O_1, ..., O_n)$ *of open sets in X such that* $(A_1, ..., A_n)$ *is similar to* $(\bar{O}_1, ..., \bar{O}_n)$ *and* $A_i \subseteq O_i$ *for every* $i = 1, ..., n$.

Proof. First we show that the condition is necessary. Let $(A_1, ..., A_n)$ be given and let A be the union of all those intersections $A_{i_1} \cap ... \cap A_{i_k}$ which do not meet A_1. Then A is closed and $A \cap A_1 = \emptyset$, so cA is open and $A_1 \subseteq cA$. Hence by Theorem 3.2 there is an open set O_1 such that $A_1 \subseteq O_1 \subseteq \bar{O}_1 \subseteq cA$. The ordered family $(\bar{O}_1, A_2, ..., A_n)$ is similar to $(A_1, ..., A_n)$: First, if we consider intersections which do not involve A_1, then the corresponding sets are identical. Next, if $A_1 \cap A_{i_1} \cap ... \cap A_{i_k} \neq \emptyset$, then by $A_1 \subseteq \bar{O}_1$ also $\bar{O}_1 \cap A_{i_1} \cap ... \cap A_{i_k} \neq \emptyset$. If $A_1 \cap A_{i_1} \cap ... \cap A_{i_k} = \emptyset$, then $A_{i_1} \cap ... \cap A_{i_k} \subseteq A$ and so

$$c(A_{i_1} \cap ... \cap A_{i_k}) \supseteq cA \supseteq \bar{O}_1$$

whence $\bar{O}_1 \cap A_{i_1} \cap ... \cap A_{i_k} = \emptyset$.

We proved that if X is a (T_4) space and if $A_1, ..., A_n$ are closed sets in X, then there is an open set O_1 such that $(A_1, ..., A_n)$ is similar to $(\bar{O}_1, A_2, ..., A_n)$ and $A_1 \subseteq O_1$. If we start from $(A_1, ..., A_n)$ and apply this principle n times we obtain an n-tuple $(\bar{O}_1, ..., \bar{O}_n)$ which is similar to $(A_1, ..., A_n)$ and is such that $A_i \subseteq O_i$ for every $i = 1, ..., n$. This shows the necessity of the condition. The sufficiency is obvious because if we choose $n = 2$ and A_1, A_2 to be disjoint then the condition implies axiom (T_4).

5. (T_5) Spaces and Completely Normal Spaces

It is simple to show that every subspace Y of a (T_i) space X where $i = 0, 1, 2$, or 3 is itself a (T_i) space. In particular, every subspace of a Hausdorff space is a Hausdorff space and every subspace of a regular topological space is regular. For (T_4) spaces and normal spaces the situation is different: There exist normal topological spaces such that not every subspace satisfies the (T_4) axiom. For instance, let Y be the set of all ordered pairs (x_1, x_2) of real numbers x_1, x_2 satisfying $x_2 \geqslant 0$ and let $X = Y \cup \{\infty\}$ where ∞ is not an element of Y. We introduce a topology on X: If $\infty \in O$ and if $O - \{\infty\}$ is open in the space of Example 3.1 and if O contains all but a finite number of pairs $(x_1, 0)$, then we say that O is open. If $\infty \notin O$, i.e., if $O \subseteq Y$ and if O is open in the same sense as in Example 3.1, then again we say that O is open. Under this topology X is a Hausdorff space and it also satisfies axiom (T_4). However, the topology of the subspace Y is the same topology which was described in Example 3.1 and therefore Y is not a normal space.

Topological spaces which have the property that every one of their subspaces is a (T_4) space are of special interest. These were first systematically studied by Urysohn who found that they can be characterized by a separation axiom which we call the (T_5) *axiom* of separation: *A space X is called a* (T_5) *space if given any pair of separated sets A and B there exist disjoint open sets O_A and O_B such that $A \subseteq O_A$ and $B \subseteq O_B$.*

It is clear that axiom (T_5) implies axiom (T_4) so that a $(T_1) + (T_5)$ space is necessarily a normal space and as such is a Hausdorff space. *A topological space satisfying axioms* (T_1) *and* (T_5) *is called completely normal.*

The main result concerning (T_5) spaces and completely normal topological spaces is:

Theorem 1. *A topological space X is a* (T_5) *space if and only if every subspace of X is a* (T_4) *space. X is completely normal if and and only if every subspace of X is normal.*

Note. The necessity part of the theorem can actually be strengthened: If X is a (T_5) space, then every subspace of X is also a (T_5) space.

Proof. First we suppose that X is a (T_5) space and prove that every subspace Y of X is also a (T_5) space. Let A, B be separated sets in Y, i.e. let $A \cap \bar{B}_Y = \varnothing$ and $\bar{A}_Y \cap B = \varnothing$ where $\bar{A}_Y$ and $\bar{B}_Y$ denote the closure of A and of B in the subspace Y. By Lemma I.10.1 the closure of a set $S \subseteq X$ with respect Y is $\bar{S} \cap Y$ where $\bar{S}$ denotes the closure of S with respect to X. Therefore $A \cap \bar{B}_Y = A \cap \bar{B} \cap Y = A \cap \bar{B}$ and so $A \cap \bar{B} = \varnothing$ and similarly $\bar{A} \cap B = \varnothing$. Hence by axiom (T_5) we can find disjoint sets O_A and O_B open with respect to X and such that $A \subseteq O_A$ and $B \subseteq O_B$. Then the sets $O_A \cap Y$ and $O_B \cap Y$ are disjoint open sets in the subspace Y which contain A and B, respectively.

This shows the necessity in its stronger form as is stated in the foregoing note.

Now we suppose that every subspace Y of a topological space X is a (T_4) space and we prove that X is a (T_5) space. For let A and B denote sets in X satisfying $A \cap \bar{B} = \emptyset$ and $\bar{A} \cap B = \emptyset$. We consider the subspace $Y = c(\bar{A} \cap \bar{B})$. The sets $\bar{A} \cap Y$ and $\bar{B} \cap Y$ are closed sets of Y and they are disjoint because

$$(\bar{A} \cap Y) \cap (\bar{B} \cap Y) = (\bar{A} \cap \bar{B}) \cap Y = (\bar{A} \cap \bar{B}) \cap c(\bar{A} \cap \bar{B}) = \emptyset.$$

Since Y is a (T_4) space there exist disjoint open sets O_A and O_B in the subspace Y such that $\bar{A} \cap Y \subseteq O_A$ and $\bar{B} \cap Y \subseteq O_B$. Since $Y = c(\bar{A} \cap \bar{B})$ is open in X, by Lemma I.10.3 both O_A and O_B are open in X. Moreover, $A \subseteq O_A$ and $B \subseteq O_B$ because for instance

$$\bar{A} \cap Y = \bar{A} \cap c(\bar{A} \cap \bar{B}) = (\bar{A} \cap c\bar{A}) \cup (\bar{A} \cap c\bar{B}) = \bar{A} \cap c\bar{B} \supseteq A \cap c\bar{B} = A.$$

Therefore O_A and O_B are disjoint open sets of X containing the sets A and B. Consequently axiom (T_5) holds in X.

The following result is of great practical interest:

Theorem 2. *Every pseudometric space is a* (T_5) *space and every metric space is completely normal.*

Proof. Let A and B be separated sets in a pseudometric space X. If $a \notin \bar{B}$, then $c\bar{B}$ being open there is an $\epsilon = \epsilon_a > 0$ such that $S_\epsilon[a] \subseteq c\bar{B}$. Therefore $d(a, b) \geqslant \epsilon_a$ for every $a \in A$ and $b \in B$. Consider the union O_A of all balls $S_{\epsilon/2}[a]$ where $a \in A$ and $\epsilon = \epsilon_a$. Then O_A is open and $A \subseteq O_A$. Determine the open set O_B in a similar manner so that $B \subseteq O_B$. The sets O_A and O_B are disjoint because if we had $x \in O_A \cap O_B$, then $d(a, x) < \frac{1}{2}\epsilon_a$ and $d(x, b) < \frac{1}{2}\epsilon_b$ for some $a \in A$ and $b \in B$ whence $d(a, b) < \max(\epsilon_a, \epsilon_b)$ which is a contradiction. This shows that X is a (T_5) space. If X is a metric space, then $d(a, b) > 0$ for any pair of distinct points $a, b \in X$. Thus if we let $\epsilon = \frac{1}{2}d(a, b)$, then $O_a = S_\epsilon[a]$ and $O_b = S_\epsilon[b]$ are disjoint open sets containing a and b, respectively. Therefore every metric space is a Hausdorff space and it is completely normal.

Theorem 3. *If a linearly ordered set has the least upper bound property, then its interval topology is completely normal.*

Proof. Every interval topology satisfies axiom (T_1) and so it is sufficient to show that axiom (T_5) holds in X. Let A and B be separated sets in X. The set $O = c(\bar{A} \cup \bar{B})$ is open and so by Theorem I.5.1 it is the union of disjoint open intervals, say $I_s = (a_s, b_s)$ where $s \in S$. We select a point i_s from every interval I_s which we keep fixed.

If $a \in A$ then there is an open interval in $c\bar{B}$ which contains a. We determine a special open interval I_a in $c\bar{B}$ containing a. Let us consider the following possibilities:

(1) There is an $x_a > a$ such that (a, x_a) is void. Then we let $I_{ar} = [a, x_a)$.

(2) There is an $x_a > a$ such that $[a, x_a] \subseteq \bar{A} \cap c\bar{B}$. In this case we let $I_{ar} = [a, x_a)$.

(3) There is an $x_a \in O$ and an interval (a_s, b_s) such that $(a_s, b_s) \subseteq [a, x_a) \subseteq c\bar{B}$. In this case we choose such an x_a and set $I_{ar} = [a, i_s)$.

If one cannot choose I_{ar} according to (1) or (2) then $a \in \bar{O}$ and so the third choice is possible. Thus an interval I_{ar} is determined for every $a \in A$ such that $I_{ar} \subseteq c\bar{B}$. In a similar fashion we can find a point y_a and construct an interval $I_{al} = (y_a, a]$ or $I_{al} = (i_s, a]$ for every $a \in A$ such that $I_{al} \subseteq c\bar{B}$. We put $I_a = I_{al} \cup I_{ar}$ and $O_A = \{I_a : a \in A\}$. Clearly O_A is open and $A \subseteq O_A$. Using an analog procedure we construct an open set O_B which contains B.

Now we show that O_A and O_B are disjoint sets. First we have $I_{ar} \cap I_{br} = I_{al} \cap I_{bl} = \emptyset$ for every $a \in A$ and $b \in B$ because for instance if $a < b$, then by $I_{ar} \subseteq c\bar{B}$ and $I_{bl} \subseteq c\bar{A}$ we have $b \notin I_{ar}$ and $a \notin I_{bl}$. Next consider the intersections $I_{ar} \cap I_{bl}$ where $a < b$. If I_{ar} comes under case (2), then by $I_{ar} \subseteq \bar{A}$ and $I_{bl} \subseteq c\bar{A}$, we obtain $I_{ar} \cap I_{bl} = \emptyset$ and the same holds if I_{bl} comes under case (2). Let I_{ar} and I_{bl} both come under case (3), say $I_{ar} = [a, i_s)$ and $I_{bl} = (i_t, b]$. Now if $x_a \leqslant y_b$ then clearly $i_s < i_t$, and if $y_b < x_a$ then by $[a, x_a] \subseteq c\bar{B}$ and $[y_b, b] \subseteq c\bar{A}$ we have $[y_b, x_a] \subseteq O$ and so $i_s \leqslant i_t$. In any case we see that I_{ar} and I_{bl} do not intersect. The situation is similar if $b < a$. Therefore O_A and O_B are disjoint and so X is a (T_5) space.

Theorem 4. *A topological space X is a (T_5) space if and only if the following proposition holds: If $S \subseteq X$ and if A is a subset of S which satisfies $A \subseteq S^i$ and $\bar{A} \subseteq S$, then there is an open set O_A such that $A \subseteq O_A \subseteq \bar{O}_A \subseteq S$.*

Proof. The sets A and $B = cS$ are separated because

$$c\bar{B} = B^e = S^i \supseteq A$$

and $c\bar{A} \supseteq cS = B$. According to axiom (T_5) there exist disjoint open sets O_A and O_B such that $A \subseteq O_A$ and $B \subseteq O_B$. Since $O_A \subseteq cO_B \subseteq cB = S$

the set O_A satisfies the requirements. To prove the converse statement we suppose that A and B are separated sets. If we let $S = cB$, then $A \subseteq c\bar{B} = (cB)^i = S^i$ and $\bar{A} \subseteq cB = S$. Hence there is an open set O_A satisfying $A \subseteq O_A \subseteq \bar{O}_A \subseteq S = cB$. If we let $O_B = c\bar{O}_A$, then O_A and O_B are disjoint open neighborhoods of A and of B, respectively.

A topological space X is called *perfectly normal* if it is normal and if every closed set is the intersection of countably many open sets. It can be shown that every perfectly normal space is completely normal but there exist completely normal spaces which are not perfectly normal. Every metric space is perfectly normal. This will be shown in Exercise 6 below.

EXERCISES

1. Show that every regular topology on a countable set is completely normal.

(This follows immediately from Theorem 2.3.)

2. Show that the half-open interval topologies of any linearly ordered set are completely normal.

(Let A and B be sets in the linearly ordered set X such that $A \cap \bar{B} = \bar{A} \cap B = \emptyset$ where $\bar{A}$ and $\bar{B}$ denote the closure of A and of B relative to the half-open interval topology $\mathscr{T}_+$. If $a \in A$, then $a \in c\bar{B}$ and so there is a point $x_a \in X$ such that $[a, x_a) \subseteq c\bar{B}$. We define $O_A = \bigcup\{[a, x_a) : a \in A\}$. Then O_A is open and $A \subseteq O_A$. If we construct O_B in a similar fashion, then O_A and O_B are disjoint because for instance if $a \in A, b \in B$, and $a < b$, then by $[a, x_a) \subseteq c\bar{B}$ we have $b \notin [a\ x_a)$ and so $[a, x_a) \cap [b, x_b)$ is void.)

3. Let X be the set of all ordered pairs (m, n) of integers $m, n = 0, \pm 1, \pm 2, \ldots$. Let $C \in \mathscr{C}$ if $(0, 0) \in C$ or if $C \cap \{(m, n) : n = 0, \pm 1, \pm 2, \ldots\}$ is finite for all but finitely many values of $m = 0, \pm 1, \pm 2, \ldots$. Show that $\mathscr{C}$ satisfies the axioms for closed sets and the topology defined is completely normal.

(Points are closed sets and so X is a (T_1) space. We show that X is a (T_5) space: If A and B are disjoint sets and $(0, 0) \notin A \cup B$, then A and B are both open. If $(0, 0) \in A$, then $(0, 0) \notin B$ and so B is open. Then we can choose $O_A = cB$ and $O_B = B$ because if B were not closed, then by $(0, 0) \notin B$ we would have $\bar{B} = B \cup \{(0, 0)\}$ and by $A \cap \bar{B} = \emptyset$ this is excluded).

4. Prove the conclusion of Theorem 3 without supposing the least upper bound property. Use Theorems 1, 3 and the result of Exercise I.10.4.

[Let $\mathscr{X}$ be the Dedekind completion of the linearly ordered set X.

This consists of intervals $(-\infty, x)$ and the lower classes L without upper bound. If $A, B \in \mathscr{X}$ and $A < B$, then there is an $x \in X$ such that $x \notin A$ but $x \in B$. It follows that $\mathscr{Y} = \{(-\infty, x)\}$ $(x \in X)$ satisfies the conditions given in Exercise 1.10.4. Hence the interval topology of $\{(-\infty, x)\}$ $(x \in X)$ is the subspace topology. Since $\mathscr{X}$ has the least upper bound property its interval topology is completely normal and so by Theorem 1 $\{(-\infty, x)\}$ $(x \in X)$ and X itself are also completely normal spaces.]

5. Let X be the set of all ordered pairs (m, n) of real numbers m, n. Let the topology be defined by specifying the family $\mathscr{C}$ of closed sets similarly as in Exercise 3. Show that X is completely normal but not perfectly normal.

[The point $(0, 0)$ forms a closed set all by itself. This set is not the intersection of countably many open sets. The neighborhood filter of $(0, 0)$ does not have a countable base.]

6. Show that in a metric space every closed set is the intersection of countably many open sets.

(Given C, consider the open sets $O_n = \bigcup\{S_{1/n}[x] : x \in C\}$. Clearly $C \subseteq O_n$ for every $n = 1, 2, \ldots$ and if $y \notin C$, then $y \notin O_n$ for a sufficiently high value of n. For instance, it is sufficient to choose n so large that $1/n < d(y, C)$ where $d(y, C)$ is the distance of y from the closed set C.)

6. Separated Sets

The condition on the sets A and B which occurs in the (T_5) axiom is particularly important and it deserves the special terminology introduced in Section 2. We recall that *if A and B are sets in a topological space and $\bar{A} \cap B = A \cap \bar{B} = \emptyset$, then A and B are called separated.* A number of interesting results hold for separated sets some of which are collected in this section. First we have a trivial remark:

Lemma 1. *If A and B are disjoint and either both open or both closed, then they are separated.*

Proof. If A and B are both closed, then $\bar{A} \cap B = A \cap \bar{B} = A \cap B = \emptyset$. If they are both open, then cA and cB are closed sets which contain B and A, respectively. Hence $\bar{B} \subseteq cA$ and $\bar{A} \subseteq cB$, that is, $A \cap \bar{B} = \bar{A} \cap B = \emptyset$.

The preceding lemma can be generalized to also include intersecting sets:

Lemma 2. *If A and B are either both open or both closed, then the relative complements $c_A(B)$ and $c_B(A)$ are separated.*

Proof. We have in general

$$c_A(B) \cap \overline{c_B(A)} = A \cap cB \cap \overline{B \cap cA} \subseteq A \cap cB \cap \bar{B} \cap \overline{cA}.$$

Therefore the intersection is a subset of both $A \cap \overline{cA}$ and of $cB \cap \bar{B}$. If A is open, then cA is closed and so $A \cap \overline{cA} = A \cap cA = \emptyset$. Similarly, if B is closed, then $cB \cap \bar{B} = cB \cap B = \emptyset$. Hence $c_A(B) \cap \overline{c_B(A)}$ is void in either case. In the same way we can see that $c_B(A) \cap \overline{c_A(B)}$ is void whenever B is open or A is closed.

Lemma 3. *Let A and B be separated. If $A \cup B$ is open, then A and B are open sets, and if $A \cup B$ is closed, then A and B are closed sets.*

Proof. First let $A \cup B$ be a closed set. Then

$$\overline{A \cup B} = \bar{A} \cup \bar{B} = A \cup B$$

and so $\bar{A} \subseteq A \cup B$ and $\bar{B} \subseteq A \cup B$. However, $\bar{A} \cap B = \emptyset$ and $A \cap \bar{B} = \emptyset$, so $\bar{A} \subseteq A$ and $\bar{B} \subseteq B$. This shows that A and B are closed. Now let $A \cup B$ be open. Then $(A \cup B) \cap c\bar{B}$ is an open set, i.e. $(A \cap c\bar{B}) \cup (B \cap c\bar{B}) = A$ is open. Similarly, $(A \cup B) \cap c\bar{A} = B$ is an open set.

The foregoing reasoning also demonstrates the validity of the following statements: If $\bar{A} \cap B = \emptyset$ and $A \cup B$ is closed, then A is closed. If $A \cap \bar{B} = \emptyset$ and $A \cup B$ is open, then A is open. A simple consequence of these statements is: If A and B are disjoint closed sets, $A_1 \subseteq A$ $B_1 \subseteq B$ and if $A_1 \cup B_1$ is closed, then A_1 and B_1 are both closed sets. However, this is a triviality because $A_1 = A \cap (A_1 \cup B_1)$ and $B_1 = B \cap (A_1 \cup B_1)$.

Theorem 1. *If O_1 and O_2 are open sets such that $O_1 \cup O_2 = X$, then the boundary of $O_1 \cap O_2$ is the union of two disjoint closed sets. These are subsets of $(cO_1)^b$ and $(cO_2)^b$, respectively.*

Proof. For the sake of simplicity we put $C_1 = cO_1$ and $C_2 = cO_2$ so that $C_1 \cap C_2 = \emptyset$. We have $(O_1 \cap O_2)^b = (c(O_1 \cap O_2))^b = (C_1 \cup C_2)^b \subseteq C_1{}^b \cup C_2{}^b$. Since C_1 and C_2 are disjoint closed sets, $C_1{}^b$ and $C_2{}^b$ are also disjoint and closed. Therefore $(O_1 \cap O_2)^b$ is the union of the disjoint closed sets $(O_1 \cap O_2)^b \cap C_1{}^b$ and $(O_1 \cap O_2)^b \cap C_2{}^b$.

Theorem 2. *If $A \cup B = X$ and if cA and cB are separated, then for every set S in X we have*

$$\bar{S} = \overline{(S \cap \bar{A} \cap A)} \cup \overline{(S \cap \bar{B} \cap B)}.$$

Proof. The set on the right-hand side is a subset of $\bar{S}$ and so it is sufficient to show that $\bar{S}$ is contained in it. Starting from the relation $S = (S \cap A) \cup (S \cap cA)$ we obtain $\bar{S} = \overline{S \cap A} \cup \overline{S \cap cA}$ and $\bar{S} \cap A \subseteq (\overline{S \cap A} \cap A) \cup \overline{S \cap cA}$. Since cA and cB are separated $\overline{cA} \subseteq B$ and so $S \cap cA \subseteq S \cap B$ and $S \cap cA \subseteq cA$ imply $\overline{S \cap cA} \subseteq \overline{S \cap B} \cap B$. Therefore we proved that $\bar{S} \cap A \subseteq (\overline{S \cap A} \cap A) \cup (\overline{S \cap B} \cap B)$. By symmetry the same inclusion holds for $\bar{S} \cap B$ and hence also for the union of the sets $\bar{S} \cap A$ and $\bar{S} \cap B$. However, $A \cup B = X$ and so this union is $\bar{S}$.

For uniformizable spaces and in particular for pseudometrizable spaces there are some simple sufficient conditions that two sets be separated. The following conditions are sufficient but they are not necessary:

Lemma 4. *If A and B are sets in a uniformizable space X and if there is a uniformity U in some uniform structure $\mathscr{U}$ of X such that $(a, b) \notin U$ for every $a \in A$ and $b \in B$, then A and B are separated sets.*

Proof. We choose a symmetric uniformity $V \in \mathscr{U}$ such that $V \circ V \subseteq U$. For every $a \in A$ we consider the interior $V[a]^i$ and define

$$O_A = \bigcup\{V[a]^i : a \in A\}.$$

Similarly, we let $O_B = \bigcup\{V[b]^i : b \in B\}$. Then O_A and O_B are disjoint open sets because $V \circ V \subseteq U$ and $(a, b) \notin U$ for any $a \in A$ and $b \in B$. Therefore O_A and O_B are separated and so the subsets $A \subseteq O_A$ and $B \subseteq O_B$ are also separated.

If A and B are sets in a metric space, we define the *distance of these sets* as $d(A, B) = glb\{d(a, b): a \in A \text{ and } b \in B\}$. The distance of a point a from a set B is defined as $d(a, B) = d(\{a\}, B)$. If A and B intersect, then $d(A, B) = 0$, but $d(A, B) = 0$ does not imply that A and B have points in common. There are even separated sets A and B whose distance is null; for instance, if X is the real line, then $A = \{x: x < 0\}$ and $B = \{x: x > 0\}$ are such sets.

Lemma 5. *If A and B are sets in a pseudometric space and if $d(A, B) > 0$, then A and B are separated.*

Proof. We can use the previous lemma because by the hypothesis $(a, b) \notin U_\epsilon$ for every $\epsilon < d(A, B)$ and $a \in A$, $b \in B$. The use of uniform structures can, however, be avoided by adapting the proof of Lemma 4 instead of using the lemma itself.

EXERCISES

1. Let X be an infinite set topologized by the topology of finite complements. Find a necessary and sufficient condition that $A, B \subseteq X$ be separated sets.

2. Let X be a metric space and let $\mathscr{P}(X)$ be the power set of the set X. Show that $d(A, B)$ is not a pseudometric for $\mathscr{P}(X)$.

3. Prove Lemma 5 directly.

7. Connected Spaces and Sets

This is perhaps the proper time to discuss in detail the notion of connectedness of a topological space and relevant basic theorems. For connectedness depends on the existence of certain separated sets or equivalently on the existence of certain disjoint open sets and so this concept belongs to the group which we call separation properties of the space. *A topological space X is called connected if the only sets in X which are both open and closed are the improper subsets ø and X.* Other simple ways of characterizing connectedness are given in:

Theorem 1. *The space X is disconnected if and only if any one of the following statements holds*:

(1) *X is the union of two nonvoid separated sets.*

(2) *X is the union of two nonvoid disjoint open sets.*

(3) *X is the union of two nonvoid disjoint closed sets.*

Proof. If X is not connected, then there exists an open-closed set A which is different from ø and X. Then cA is also open-closed and so the sets A, cA satisfy the requirements of (1), (2), and (3). Conversely, suppose that $X = A \cup B$ where A, B are nonvoid separated sets. Then $X = A \cup B$ being open, by Lemma 6.3 both A and B are open sets. By $cA = B$ we see that A is an open-closed set and so X is a disconnected space.

A subset Y of a topological space X is called a connected set in X if the topology induced on Y by the topology of X makes Y a connected space. Briefly, Y is a connected set if Y is a connected subspace of X. At this point the notion of separated sets really enters into play, for we have the following:

Theorem 2. *A set Y is a connected set in the topological space X if and only if it cannot be decomposed into two nonvoid separated sets of X.*

Proof. Let $Y = A \cup B$ where $\bar{A} \cap B = \emptyset$ and $A \cap \bar{B} = \emptyset$. By part (iii) of Lemma I.10.1 we have $\bar{A}_Y = \bar{A} \cap Y = \bar{A} \cap (A \cup B) = \bar{A} \cap A = A$. This shows that A is a closed set relative to the topology of the subspace Y. Similarly, B is a closed set of the subspace Y. By hypothesis we have $Y = A \cup B$, hence if Y is a connected set in X, then either $A = \emptyset$ or $B = \emptyset$. Conversely, if Y is not a connected space, then there exist disjoint closed sets A and B in Y such that $Y = A \cup B$. Since A is closed in Y we have $\bar{A}_Y = A$, i.e., $\bar{A} \cap Y = A$. Therefore $\bar{A} \cap B = \bar{A} \cap Y \cap B = A \cap B = \emptyset$ and similarly $A \cap \bar{B} = \emptyset$. The sets A and B are separated in X.

Lemma 1. *If Y is a connected set in X and if Z is such that $Y \subseteq Z \subseteq \bar{Y}$, then Z is connected.*

Corollary. *If Y is a connected set, then $\bar{Y}$ is also connected.*

Proof. Let $Z = A \cup B$ where A and B are separated sets. Then either $Y \subseteq A$ or $Y \subseteq B$ because together with A and B the sets $A \cap Y$ and $B \cap Y$ are also separated and their union is Y. Suppose that $Y \subseteq A$. Then $\bar{Y} \subseteq \bar{A}$ and $B = Z \cap B \subseteq \bar{Y} \cap B \subseteq \bar{A} \cap B = \emptyset$. In the same way we can show that if $Y \subseteq B$, then $A = \emptyset$. Consequently, A or B is void and Z is connected.

Lemma 2. *Let Y_i $(i \in I)$ be connected sets one of which is not separated from any other Y_i $(i \in I)$. Then $Y = \bigcup Y_i$ is connected.*

Corollary 1. *If $\{Y_i\}$ $(i \in I)$ is a family of connected sets and if $\bigcap Y_i$ is not void, then $\bigcup Y_i$ is connected.*

Corollary 2. *If A, B are connected and not separated, then $A \cup B$ is connected.*

Proof. Let $Y = A \cup B$ where A and B are separated sets. Since Y_i is connected we have $Y_i \subseteq A$ or $Y_i \subseteq B$ for each index $i \in I$. Otherwise $A \cap Y_i$ and $B \cap Y_i$ would give a decomposition of Y_i into nonvoid separated sets. Suppose for instance that $Y_0 \subseteq A$ where Y_0 is not separated from any Y_i $(i \in I)$. Then we have also $Y_i \subseteq A$ for every $i \in I$. Consequently $Y = \bigcup Y_i \subseteq A$ and $B = \emptyset$. This shows that Y is connected.

Lemma 3. *Let $\{Y_i\}$ $(i = 1, 2, \ldots)$ be a sequence of connected sets such that Y_i and Y_{i+1} are not separated for $i = 1, 2, \ldots$. Then $Y = \bigcup Y_i$ is connected.*

Proof. It is possible to construct a direct proof by using similar reasonings as in the proof of Lemma 2 but it is quicker to use the lemma itself: First by induction on i we see that $Y_1 \cup \ldots \cup Y_i$ is connected for every i: In fact, $Y_1 \cup \ldots \cup Y_i$ and Y_{i+1} are not separated because Y_i and Y_{i+1} are not separated and so if $Y_1 \cup \ldots \cup Y_i$ is connected, then by the second corollary of Lemma 2 $(Y_1 \cup \ldots \cup Y_i) \cup Y_{i+1}$ is also connected. The first corollary applies to the sequence $\{Y_1 \cup \ldots \cup Y_i\}$ $(i = 1, 2, \ldots)$ and it shows that $Y = \bigcup Y_i$ is connected.

For uniformizable spaces one can give a simple and useful necessary condition that X be connected. Later we shall see that in certain cases the condition is not only necessary but also sufficient. First we introduce the concept of a U-chain where U is a uniformity of a uniform structure $\mathscr{U}$: *Given a set and a uniform structure $\mathscr{U}$ we say that two points a and b can be connected by a U-chain where U is a uniformity in $\mathscr{U}$ if there exists a finite sequence $x_1, \ldots, x_n \in X$ such that $x_1 = a$, $x_n = b$, and $(x_i, x_{i+1}) \in U$ for* every $i < n$. We shall use this definition only in the special case when U is symmetric.

Theorem 3. *If X is a connected uniform space and $\mathscr{U}$ is any uniform structure for the topology of X, then any pair of points $a, b \in X$ can be connected by a U-chain for every symmetric U in $\mathscr{U}$.*

Proof. Let U be a symmetric uniformity and let S_a denote the set of those points $x \in X$ which can be connected with a by a U-chain. Since $a \in S_a$ our set is not void. If we can prove that its boundary is void, then it is an open-closed set and so by the connectedness of X it must be X. This will show that every point of X can be connected with a by a U-chain. To prove that $(S_a)^b$ is void suppose that $b \in (S_a)^b$. Let $V \in \mathscr{U}$ be symmetric and such that $V \circ V \subseteq U$. The neighborhood $V[b]$ of b meets both S_a and cS_a, say $x \in S_a \cap V[b]$ and $y \in cS_a \cap V[b]$. By the symmetry of V, $(x, y) \in U$. However, in that case any U-chain connecting a with x can be extended to a U-chain connecting a with y. Since $y \in cS_a$ this is a contradiction.

Theorem 4. *Let X be a linearly ordered set with the least upper bound property and let X be topologized by its interval topology. Then every connected set in X is an interval.*

If in addition no open interval with distinct end points is void, that is, if X is densely ordered, then every interval is a connected set in X.

Corollary. *A subset of the real line is connected if and only if it is an interval.*

Proof. Let S be a set in X such that there exist three distinct points $a < x < b$ in X with the property that $a, b \in S$ but $x \notin S$. Then

$S = A \cup B$ where $A = (-\infty, x) \cap S$ and $B = (x, \infty) \cap S$ are disjoint sets. We have $\bar{A} \subseteq (-\infty, x]$ and $\bar{B} \subseteq [x, +\infty)$ so A and B are separated sets and S is not connected. Consequently, if S is a connected set and if $a, b \in S$, then $[a, b] \subseteq S$. As usual let $\alpha = \text{glb } S$ if S is bounded from below and let α be the symbol $-\infty$ if S has no lower bound. Similarly, we let β denote lub S or $+\infty$ according as S is bounded from above or not. The foregoing remark implies that S is (α, β), $[\alpha, \beta)$, $(\alpha, \beta]$, or $[\alpha, \beta]$.

We show that under the additional hypothesis X itself is connected. Then every interval in X is also connected because it has the property required in Exercise I.10.4 and so its interval topology is the subspace topology induced by X. Suppose that O_1 and O_2 are nonvoid open sets in X. By Theorem I.5.1 these are unions of disjoint open intervals. Let (a, b) be such a nonvoid interval in O_1. First suppose that a is not the symbol $-\infty$, i.e., a is a point of X. If $a \notin O_2$, using $a \notin O_1$ we see that $O_1 \cup O_2$ is not a cover of X. If $a \in O_2$, then O_2 being open there is an open interval (α, β) in O_2 which contains a. Since X is densely ordered (α, β) and (a, b) intersect and so O_1 and O_2 are not disjoint sets. Similar reasoning can be applied also in the case when b is not the symbol $+\infty$ but a point of X. Therefore in every case O_1 and O_2 intersect or their union does not cover X. Hence X is not the union of disjoint nonvoid open sets and so it is connected.

Theorem 5. *A product space $\prod X_s$ is connected if and only if every factor X_s is connected.*

Proof. Suppose that every X_s $(s \in S)$ is a connected space. We show that any two nonvoid open sets O and Q in $X = \prod X_s$ whose union is X have a nonvoid intersection. Let $x \in O$ and $y \in Q$. By the definition of the product topology there are open sets O_s $(s \in S)$ and Q_s $(s \in S)$ such that $x \in \prod O_s \subseteq O$ and $y \in \prod Q_s \subseteq Q$ and $O_s = Q_s = X_s$ for all except at most the indices $s_1, \ldots, s_n$. We choose for every $s \neq s_1, \ldots, s_n$ a point $p_s \in X_s$ and define the sets $A_1, \ldots, A_n$ in X as follows: Let

$$A_1 = \{\xi : \xi_{s_i} = x_{s_i} \quad \textit{for} \quad 1 < i \leqslant n \quad \textit{and} \quad \xi_s = p_s \quad \textit{for} \quad s \neq s_1, \ldots, s_n\},$$

for $1 < k < n$ let

$$A_k = \{\xi : \xi_{s_i} = y_{s_i} \quad \textit{for} \quad 1 \leqslant i < k \quad \textit{and} \quad \xi_{s_i} = x_{s_i} \quad \textit{for} \quad k < i \leqslant n \quad \textit{and} \quad \xi_s = p_s \quad \textit{for} \quad s \neq s_1, \ldots, s_n\},$$

and let

$$A_n = \{\xi : \xi_{s_i} = y_{s_i} \quad \textit{for} \quad 1 \leqslant i < n \quad \textit{and} \quad \xi_s = p_s \quad \textit{for} \quad s \neq s_1, \ldots, s_n\}.$$

By Theorem I.11.2 the set A_k considered as a subspace of X is homeomorphic to X_{s_k} and so it is a connected set in X. Moreover, A_k and A_{k+1} intersect for every k $(0 < k < n)$, namely, if ξ is the point such that $\xi_{s_i} = y_{s_i}$ for $i \leqslant k$, $\xi_{s_i} = x_{s_i}$ for $i > k$, and $\xi_s = p_s$ for every other s, then $\xi \in A_k \cap A_{k+1}$. By Lemma 3, $A = A_1 \cup \ldots \cup A_n$ is a connected set in X. Since O and Q are open and $X = O \cup Q$ the sets $A \cap O$ and $A \cap Q$ are open sets in A whose union is A. By our construction $A \cap O$ and $A \cap Q$ are not void, and so by the connectedness of A the nonvoid sets $A \cap O$ and $A \cap Q$ must intersect. Therefore $O \cap Q$ is not void and so X is connected. This shows that the condition given in the theorem is sufficient. Its necessity is obvious, because if a factor X_s is not connected, then there are nonvoid disjoint open sets O_s and Q_s such that $X_s = O_s \cup Q_s$ and so $X = \{\xi: \xi_s \in O_s\} \cup \{\xi: \xi_s \in Q_s\}$ is a decomposition of X into a pair of nonvoid disjoint open sets.

We close with a few remarks on terminology: A compact connected set is usually called a *continuum* and an open connected set is often called a *domain*. Some people prefer to call any closed connected set a continuum. If X is not connected and $X = A \cup B$ where A and B are disjoint open sets, we say that $\{A, B\}$ is a *partition* of X. We can also speak about partitions consisting of more than two parts.

EXERCISES

1. Let X_1 and X_2 be disjoint sets in X such that their union is dense in X. Show that if the connected set S meets both X_1 and X_2, then it meets $X_1{}^b \cup X_2{}^b$.

[Suppose that $S \cap (X_1{}^b \cup X_2{}^b)$ is void. Then by $X = \bar{X}_1 \cup \bar{X}_2$ the set S can be decomposed into the union of $S_1 = S \cap X_1{}^i$ and $S_2 = S \cap X_2{}^i$. The sets S_1 and S_2 are nonvoid open sets in S and so S is not connected.]

2. We say that the sets $S_1, \ldots, S_n$ form a chain if $S_k \cap S_{k+1}$ is nonvoid for every k $(0 < k < n)$. Show that X is connected if and only if given any two nonvoid sets O_{α_1} and O_{α_2} of an open cover $\{O_\alpha\}$ of X, there is a chain of sets $S_k \in \{O_\alpha\}$ such that $S_1 = O_{\alpha_1}$ and $S_n = O_{\alpha_2}$.

[Let O_i $(i = 1, 2)$ be the union of all chains whose first set S_1 is O_{α_i} and let O be the union of those O_α's which do not meet $O_1 \cup O_2$. Since $X = O_1 \cup O_2 \cup O$ and X is connected, O is void and O_1 and O_2 are identical. The converse statement is obvious from the definition of a connected space.]

3. Show that the plane under the usual topology remains connected even after removing a countable set of points from it.

(Let S be countable and let x, y be points not in S. Consider the points z of the perpendicular bisector of the segment xy. There is a z such that neither xz nor zy meets S. The segments xz and zy connect x and y.)

4. Show the following: If X is a normed vector space over the reals whose algebraic dimension is at least two, then X remains connected even after removing a countable set S from X.

[There is a real $\lambda \neq \frac{1}{2}$ such that the connected subsets

$$S_1^\lambda = \{x : x = a_1 + \mu(1 - \lambda)a_1 + \mu\lambda a_2\}$$

and

$$S_2^\lambda = \{x : x = a_2 + \mu\lambda a_1 + \mu(1 - \lambda)a_2\}$$

where $-\infty < \mu < +\infty$ do not meet S. Since, save for a constant factor, S_1^λ and S_2^λ are isometric with the reals and have a common point, $S_1^\lambda \cup S_2^\lambda$ is connected and contains a_1 and a_2.]

5. Show that the union of an increasing family of connected sets is connected.

6. Prove in a simple way that the product of finitely many connected spaces is connected.

[The sets $\{\xi: \xi_1 = x_1\}$ and $\{\xi: \xi_2 = y_2\}$ are connected, have a common point, and their union contains both (x_1, x_2) and (y_1, y_2).]

7. Let X be the set of the ordered pairs $(0, 0)$, $(i, 0)$, and (i, k) for every $i, k = 1, 2, \ldots$. Define a topology on X by specifying the neighborhood filters of its points: Let every (i, k) be an isolated point; let the neighborhoods of $(i, 0)$ be the sets containing a set

$$\{(i, 0)\} \cup \bigcup \{(i, k) : k \geqslant n\}$$

for some $n = 1, 2, \ldots$; and let the neighborhoods of $(0, 0)$ be the sets containing a set $\{(0, 0)\} \cup \bigcup\{(i, k): i, k \geqslant n\}$ for some $n = 1, 2, \ldots$. Verify axioms (Nb. 1) – (Nb. 4) and show that X is a disconnected nonregular Hausdorff space.

8. Let X be the set of all ordered pairs (x, y) of rational numbers such that $y \geqslant 0$. Define a topology on X by specifying the neighborhood filters: A set $N_{(a,b)}$ is a neighborhood of (a, b) if besides (a, b) it also contains a set of the form

$$\left\{(x, 0) : \left|x - \left(a + \frac{b}{\sqrt{3}}\right)\right| < \epsilon \quad or \quad \left|x - \left(a - \frac{b}{\sqrt{3}}\right)\right| < \epsilon\right\}$$

for a sufficiently small value of $\epsilon > 0$. Show that this defines a connected Hausdorff space of denumerably many points.

(Determine the closures of the neighborhoods $N_{(a,b)}$.)

9. Let X be a topological space and let R be an equivalence relation. Show that X/R is connected if X is connected.

10. Show that X is connected if and only if $A^b \neq \emptyset$ for every proper subset A of X.

8. Maximal Connected Subsets

Here we define what is meant by a component or maximal connected subset of a topological space and collect the basic information on components. The definition and the results are very simple: *A set A in a topological space X is called a component of X or a maximal connected set in X if A is connected and not a subset of any other connected set in X.* Therefore if A is a component and if B is a connected set such that $A \subseteq B$, then $A = B$. Distinct components are separated sets because if A and B are connected sets and if A and B are not separated, then by Corollary 2 of Lemma 7.2 $A \cup B$ is a connected set which contains A and B as proper subsets. This separation property implies that if A and B are components in X, then either $A = B$ or $A \cap B = \emptyset$. Each point x of X belongs to some component, namely, to the union of all connected sets containing x. Hence the set of all components of a space X determines an equivalence relation on X.

Components are closed nonvoid sets. In fact, sets consisting of a single point are connected and so $\emptyset$ is not a maximal connected set. Furthermore, if A is a component, then it is connected and so by Lemma 7.1 its closure is also connected. Hence by the maximality property of components, $\bar{A} = A$.

A space X is called *totally disconnected* if every set $\{x\}$ consisting of a single point x is a component of X. Since components are closed sets, by Theorem 1.1 a totally disconnected space is a (T_1) space. However, X can be totally disconnected without being discrete. For instance, 0 and all rationals of the form $1/n$ $(n = 1, 2, \ldots)$ form a metric space X if the distance is defined as $d(x, y) = |x - y|$. The topology induced by this metric d is not the discrete topology because $\{0\}$ is not an open set. Nevertheless, the space is totally disconnected: The open-closed sets $\{1/n\}$ $(n = 1, 2, \ldots)$ are themselves components and the only connected set containing 0 is $\{0\}$.

It is clear that a topological space X is connected if and only if it has only one component. Moreover, if a set S is contained in some connected set of the topological space X, then the union of all connected sets containing S is the unique component of X which contains S.

If X has finitely many components, then every component of X is both open and closed. For every component is closed, so the complement of a component is the union of finitely many closed sets and so it is closed. In the case of infinitely many components this proposition may fail. For example, $\{0\}$ is a component of the space consisting of 0 and the rationals $1/n$ $(n = 1, 2, \ldots)$ but $\{0\}$ is not an open set. This shows that a component and its complement need not be separated sets. The following example shows even more:

Example 1. Let X be a subspace of the plane, namely, let X be the union of the lines $L_{\pm 1} = \{(\xi, \eta): \eta = \pm 1\}$ and of the boundaries of the rectangles whose vertices are $(\pm n, \pm (n-1)/n)$ $(n = 1, 2, \ldots)$. Then L_{+1} and L_{-1} are components of X and neither of them is open. Moreover, if A and B are disjoint closed sets in X whose union is X, then either A or B contains both L_{+1} and L_{-1}. Hence L_{+1} and L_{-1} cannot be separated by a partition of X.

A topological space X is locally connected if for every $x \in X$ each neighborhood N_x contains a connected neighborhood of the point x. In simpler words, X is locally connected if every point of X has arbitrarily small connected neighborhoods. The fact that these neighborhoods can be arbitrarily small is essential because X may have connected neighborhoods without having any connected neighborhood in a prescribed small neighborhood N_x.

Example 2. Let X be the union of the following connected sets in the Euclidean plane: the closed interval $I = [-1, +1]$, the circles $C_n = \{x: d(x, 0) = 1 - (1/n)\}$ $(n = 1, 2, \ldots)$ where $d(x, 0)$ denotes the Euclidean distance of the point x from the center $0 = (0, 0)$, and the circle $C_\infty = \{x: d(x, 0) = 1\}$. Since each circle intersects the interval I, by Lemma 7.2 $X = I \cup (\bigcup C_n) \cup C_\infty$ is connected. There are only two points on the circle C_∞ which have connected neighborhoods of arbitrarily small diameter, namely, the end points of I. Hence X is connected but not locally connected.

The character of the components of open subspaces affects the local connectedness of the space. More precisely we have the following:

Theorem 1. *A space is locally connected if and only if the components of every open subspace are open.*

Note. Since the subspaces considered are open sets in X a set is open in the subspace if and only if it is open in X.

Proof. First let a point x and an open subspace Y containing the point x be given. Then x belongs to some component K_x of Y. By hypothesis K_x is open and connected in Y. Hence it is open and connected in X. Therefore X is locally connected. Now suppose that X is locally connected. Let Y be an open set in X and let K be a component of Y. For every $x \in K$ there is a connected open set O_x in Y which contains x. By

Corollary 2 of Lemma 7.2, $O_x \cup K$ is connected in Y and so K being maximal $O_x \cup K = K$, i.e., $O_x \subseteq K$. Hence $K = \{O_x : x \in K\}$ which shows that K is open in Y, and since Y is an open subspace of X, K is open in X.

Theorem 2. *A product space X is locally connected if and only if all factors are locally connected and all but finitely many factors are connected.*

Proof. The condition is sufficient. For let N_x be a neighborhood of $x \in X$ and let $x \in \prod O_s \subseteq N_x$ where $O_s = X_s$ unless $s = s_1, \ldots, s_n$. We may choose n so large that for every $s \neq s_1, \ldots, s_n$ the spaces X_s are connected. By the local connectedness of X_{s_i} $(i = 1, \ldots, n)$ there is a connected neighborhood N_{s_i} of the projection x_{s_i} of x in X_{s_i} such that $N_{s_i} \subseteq O_{s_i}$. For every other choice of s we set $N_s = X_s$. Then $\prod N_s$ is a neighborhood of x and $\prod N_s \subseteq N_x$. By Theorems 7.5 and I.11.1 $\prod N_s$ is a connected subspace of X. Therefore X is locally connected. The necessity of the condition can be shown by using:

Lemma 1. *If A is a connected set in the product space $\prod X_s$, then its projection into every factor space X_s is connected in X_s.*

Note. This is a special case of Lemma IV.8.1.

Proof. Suppose that $\pi_s(A) = A_s$ is not connected. Then there is a partition of A_s into disjoint open sets, say $A_s = (A_s \cap O_s) \cup (A_s \cap Q_s)$ where O_s and Q_s are open sets in X_s. Hence $A \cap \{\xi : \xi_s \in O_s\}$ and $A \cap \{\xi : \xi_s \in Q_s\}$ are nonvoid open sets of the subspace A and their union is A itself. Consequently A is not connected.

Now we show the necessity of the hypotheses in Theorem 2. Given s, a point $x_s \in X_s$, and a neighborhood N_{x_s} of x_s, let $x \in X$ be a point such that its projection in X_s is $\pi_s(x) = x_s$. Moreover, let N_x be a connected neighborhood of x which is contained in the neighborhood $\{\xi : \xi_s \in N_{x_s}\}$. The projection $\pi_s(N_x)$ is then a connected neighborhood of x_s and it is contained in N_{x_s}. Hence each X_s is locally connected. Next, let A be a connected set in X whose interior is not void. Since X is locally connected such sets A exist. We choose an open set of the type $\prod O_s$ contained in A. Since $O_s = X_s$ except for finitely many values of s, we have $\pi_s(A) \supseteq \pi_s(\prod O_s) = X_s$ and so $\pi_s(A) = X_s$. By the preceding lemma $\pi_s(A)$ is connected, so X_s is connected for all but finitely many values of the index s.

The expression X is *locally connected at a point* x is also used. This means that x has arbitrarily small connected neighborhoods, that is, for every prescribed N_x there is a connected neighborhood of x which is entirely contained in N_x.

EXERCISES

1. Let S_n $(n = 1, 2, ...)$ be the closed segment in the plane which connects the points $(0, 0)$ and $(1, 1/n)$ and let S_∞ be a subset of the interval $0 \leqslant x \leqslant 1$. Show that the union of the sets S_n and S_∞ is a connected subspace of the plane. Show also that by omitting $(0, 0)$ the sets S_1, S_2, ... become components of the remaining set.

2. Determine the components of an open subset of the real line. (Every open set O is the union of disjoint open intervals. These are the components of O.)

3. Show that if X is locally connected, then any open subset of X is a locally connected space.

4. A space X is locally connected if and only if it has a base consisting of connected open sets.
(The necessity is a consequence of Theorem 1 and the sufficiency of the condition is obvious from the definition of local connectedness.)

5. Let the plane set X be the union of the y-axis and of the straight lines $y = nx$ $(n = 1, 2, ...)$. Show that X is connected but not locally connected.

6. Let C be a connected set in the connected space X and let O be an open-closed set in $X - C$. Show that $C \cup O$ is connected.
[Let $C \cup O = Q_1 \cup Q_2$ where Q_1 and Q_2 are disjoint open-closed sets in $C \cup O$. Then C being connected $C \subseteq Q_1$ or $C \subseteq Q_2$ and so $Q_2 \subseteq O$ or $Q_1 \subseteq O$. Suppose that $Q_1 \subseteq O$. Since O is open-closed in $X - C$ and $Q_1 \subseteq O \subseteq cC$ the set Q_1 is open-closed in O and so O being open-closed in $X - C$ we see that Q_1 is open-closed also in $X - C$. By hypothesis Q_1 is open-closed also in $C \cup O$, so it is open-closed in $(X - C) \cup (C \cup O) = X$. However, X is connected and so Q_1 is void.]

7. Determine the components of ΠX_s if the components of every X_s are known.
(Use the theorems on the product of connected spaces and Lemma 7.2 or Lemma 8.1, Theorem 7.5, and Theorem I.11.1 to show that C is a component in ΠX_s if and only if $C = \Pi C_s$ where C_s is a component of X_s for each index s.)

9. (T) Axiom and Complete Regularity

In this section we shall introduce the most important separation axiom and discuss its relations to the remaining separation axioms. The axiom, which we shall call axiom (T) was first mentioned by Urysohn in a paper on the problem of finding a metric for a given topology. In another form it was formally introduced by Tychonoff who discovered its significance for the process of compactifications. The full importance of axiom (T) was realized only after Weil proved that a topological space is uniformizable if and only if axiom (T) holds.

Definition 1. *By a scale of open sets with index set D we mean a family (O_d) $(d \in D)$ of open sets O_d such that the index set D is a dense subset of the interval* $[0, 1]$ *and $\bar{O}_{d_1} \subseteq O_{d_2}$ for every pair $d_1 < d_2$.*

For instance, D can be the set of diadic rationals $d = m/2^n$ where $m = 0, 1, \ldots, 2^n$ and $n = 1, 2, \ldots$. In this case we speak about a *diadic scale* of open sets. As another example we may choose D to be the set of all rationals d $(0 \leqslant d \leqslant 1)$. If X is a metric space and $x \in X$, then $(S_d[x])$ $(d \in D)$ is a scale of open sets because the closure of the ball $S_d[x] = \{\xi\colon d(\xi, x) < d\}$ is $\bar{S}_d[x] \subseteq \{\xi\colon d(\xi, x) \leqslant d\} \subseteq S_{d'}[x]$ whenever $d < d'$.

Definition 2. *A topological space X is called a* (T) *space if for any closed set A and for any $b \notin A$ there exists a scale of open sets (O_d) $(d \in D)$ such that $b \in O_d \subseteq cA$ for every $d \in D$.*

The particular choice of the index set D is of no importance: If the index set Δ is a dense subset of $[0, 1]$ and if a scale of open sets exists for some particular choice of D, then there is a scale also with index set Δ. This follows from the following lemma:

Lemma 1. *If A and B are disjoint sets in a topological space and if there exists a scale of open sets (O_d) $(d \in D)$ such that $B \subseteq O_d \subseteq cA$ for every $d \in D$, then there exists a scale (Q_i) $(i \in I)$ of open sets with index set $I =$ $[0, 1]$ such that $B \subseteq Q_i \subseteq cA$ for every i $(0 \leqslant i \leqslant 1)$.*

Proof. We define Q_i as the union of all sets O_d for which

$$\tfrac{1}{4} \leqslant d \leqslant \tfrac{1}{4} + i/2.$$

Then Q_i is an open set and it is defined for every $i \in I$. Since $B \subseteq O_d \subseteq cA$ we have $B \subseteq Q_i \subseteq cA$ for every $i \in I$. Moreover, given $i_1 < i_2$ we can choose d_1 and d_2 such that

$$\tfrac{1}{4} + i_1/2 < d_1 < d_2 < \tfrac{1}{4} + i_2/2.$$

Hence $Q_{i_1} \subseteq \bar{O}_{d_1} \subseteq O_{d_2} \subseteq Q_{i_2}$. Since $\bar{O}_{d_1}$ is closed we obtain $\bar{Q}_{i_1} \subseteq Q_{i_2}$. This shows that (Q_i) $(i \in I)$ is a scale of open sets.

Hence one can assume that D is fixed, for example, it is no restriction to say that (O_d) $(d \in D)$ is a diadic scale, i.e., $d = m/2^n$ for some $m = 0, 1, \ldots, 2^n$ and $n = 1, 2, \ldots$.

We can easily show that every (T) space is a (T_3) space: For let a closed set A and a point b not belonging to A be given. We choose a scale (O_d) $(d \in D)$ such that $b \in O_d \subseteq cA$ and select d, d' in D. If $d < d'$, then $b \in O_d \subseteq \bar{O}_d \subseteq O_{d'} \subseteq cA$ and so the disjoint open sets $O_A = c\bar{O}_d$ and $O_b = O_d$ separate A and b. There exist topological spaces which are regular but do not satisfy axiom (T). Similarly there are (T) spaces in which axiom (T_4) does not hold. Since every (T) space is a (T_3) space, by Theorem 2.1 we have the following:

Theorem 1. *If a topological space satisfies axiom* (T_0) *and* (T), *then it is a Hausdorff space.*

A Hausdorff space satisfying axiom (T) is called a *completely regular space*. In another terminology our (T) spaces are called completely regular. For Hausdorff spaces satisfying the (T) axiom sometimes the term "Tychonoff space" is used.

A topological space satisfying axioms (T_0) and (T_4) is not necessarily a (T_3) space and so axioms (T_0) and (T_4) together are not strong enough to imply complete regularity. However, (T_3) and (T_4) imply axiom (T) and so we have:

Theorem 2. *Every* $(T_3) + (T_4)$ *space is a* (T) *space. Therefore every normal space is completely regular.*

Proof. We shall use axiom (T_4) in the form stated in Theorem 3.2: If O is open and if A is a closed subset of O, then there is an open set O_A such that $A \subseteq O_A \subseteq \bar{O}_A \subseteq O$. Here it is convenient to choose D to be the set of diadic rationals. Let A be a closed set, let $O = cA$, and let a point $b \in O$ be given. Since X is a (T_3) space there is an open set O_0 such that $b \in O_0 \subseteq \bar{O}_0 \subseteq O$. We define O_1 to be the original set O. This defines the elements of the diadic scale (O_d) $(d \in D)$ for $d = 0$ and $d = 1$. We proceed by induction: Let D_n be the set of those diadic rationals which are of the form $d = m/2^n$ where m is odd and $n > 0$. Using axiom (T_4) we can find an open set $O_{1/2}$ such that

$$\bar{O}_0 \subseteq O_{1/2} \subseteq \bar{O}_{1/2} \subseteq O_1 .$$

Suppose that the sets O_d are already defined for every d in

$$D_1 \cup D_2 \cup \ldots \cup D_n$$

such that $\bar{O}_{d'} \subseteq O_{d''}$ for every $d' < d''$. We extend the definition for every d in $D_1 \cup D_2 \ldots \cup D_n \cup D_{n+1}$: If d' and d'' are adjacent elements of $D_1 \cup D_2 \cup \ldots \cup D_n$ and $d' < d''$, then using (T_4) we choose O_d for $d = \frac{1}{2}(d' + d'')$ such that $\bar{O}_{d'} \subseteq O_d \subseteq \bar{O}_d \subseteq O_{d''}$. In this way we obtain a diadic scale (O_d) $(d \in D)$ with the property that $b \in O_d$ and $O_d \subseteq O = cA$ for every $d \in D$. Hence X is a (T) space.

If we perform a slight modification in the foregoing proof we obtain a characterization of (T_4) spaces in terms of scales of open sets:

Theorem 3. *A topological space X is (T_4) space if and only if for every pair of disjoint closed sets A and B there exists an open scale (O_d) $(d \in D)$ with the property that $B \subseteq O_d \subseteq cA$ for every $d \in D$.*

Proof. Using axiom (T_4) we can find an open set O_0 such that $B \subseteq O_0 \subseteq \bar{O}_0 \subseteq O_A = cA$. We choose O_0 as the smallest set and $O_1 = O_A$ as the largest set of the diadic scale (O_d) $(d \in D)$. The remaining sets O_d for $0 < d < 1$ are constructed in the same fashion as in the proof of Theorem 2. This shows that the condition is necessary. Its sufficiency is obvious.

In Section IV.6 we shall give a similar characterization of a larger class of spaces which includes both (T) spaces and (T_4) spaces.

EXERCISES

1. Show directly that every pseudometric space is a (T) space.

2. Let a commutative group X be topologized such that translates of open sets are open. Show by a simple example that the topology need not satisfy axiom (T).

(The topology of finite complements on an infinite group.)

3. Let the linearly ordered set X be densely ordered. Show that its interval topology is completely regular.

[Let $x \in X$ and an open interval (a, b) containing x be given. Determine for every diadic rational $d = m2^{-n}$ a point pair a_d, b_d such that $a < a_d < x < b_d < b$ and $a_{d'} < a_{d''} < b_{d''} < b_{d'}$ for every $d' < d''$. Such pairs can be constructed by induction on n. The open intervals (a_d, b_d) $(d \in D)$ form a diadic scale and $x \in (a_d, b_d) \subseteq (a, b)$.]

4. Show that the interval topology of every linearly ordered set is a (T) topology.

[Perform the following modifications in the solution of Exercise 3: Let (a, b) be an open interval containing x. If there is a point α in (a, x) such that (α, γ) is void for some $\gamma > \alpha$, then set $a_d = \alpha$ for every $d \in D$.

Similarly, if there is a point $\beta < b$ such that (δ, β) is void for some $\delta < \beta$, then set $b_d = \beta$ for every $d \in D$. If such α or β does not exist, construct the a_d's or b_d's as before.]

5. Show by an example that the following situation can occur: $\mathscr{T}_1 < \mathscr{T}_2 < \mathscr{T}_3$ are topologies on the same set X, $\mathscr{T}_1$ and $\mathscr{T}_3$ are (T) spaces but $\mathscr{T}_2$ is not.

(Let X be an infinite set, let $\mathscr{T}_1$ be the nondiscrete topology, let $\mathscr{T}_2$ be the topology of finite complements, and let $\mathscr{T}_3$ be the discrete topology of X.)

6. Let $\{\mathscr{T}_i\}$ $(i \in I)$ be a family of (T) topologies on the set X. Show that $\text{lub}\{\mathscr{T}_i\}$ also satisfies axiom (T).

[Let $x \in O$ where O is open relative to the least upper bound. Then there exist indices $i_1, \ldots, i_n$ and sets $O^{i_1}, \ldots, O^{i_n}$ open relative to the corresponding topologies $\mathscr{T}_{i_1}, \ldots, \mathscr{T}_{i_n}$ and such that

$$x \in O^{i_1} \cap \ldots \cap O^{i_n} \subseteq O.$$

Choose a scale of open sets $(O_d^{i_k})$ $(d \in D)$ for every $k = 1, \ldots, n$ such that $x \in O_d^{i_k} \subseteq O^{i_k}$. Define O_d as $O_d = O_d^{i_1} \cap \ldots \cap O_d^{i_n}$. Since the closure of $O_d^{i_k}$ relative to $\mathscr{T}_k$ is closed relative to $\text{lub}\{\mathscr{T}_i\}$, the closure $\bar{O}_d$ of O_d relative to $\text{lub}\{\mathscr{T}_i\}$ is contained is the closure of $O_d^{i_k}$ taken relative to $\mathscr{T}_k$. Hence $\bar{O}_{d'} \subseteq O_{d''}$ for every $d' < d''$.]

7. Show that every family of completely regular topologies formed on the same set X has a greatest lower bound among its completely regular lower bounds.

(The greatest lower bound can be strictly coarser than the greatest lower bound of the family concerned. Its existence follows from Exercise 6.)

8. Let X_s be a (T) space for every index $s \in S$. Show that ΠX_s is a (T) space.

[Let $x \in \Pi O_s$ where $O_s = X_s$ for all but finitely many indices. For every O_s there is a diadic scale $(O_s{}^d)$ $(d \in D)$ such that

$$x_s \in O_s{}^d \subseteq O_s.$$

If $O_s = X_s$, we define $O_s{}^d = X_s$ for every $d \in D$. We let $O^d = \Pi O_s{}^d$. Using Lemma I.11.4 we see that (O^d) $(d \in D)$ is a diadic scale and $x \in O^d \subseteq \Pi O_s$.]

10. Uniformization and Axiom (T)

The present section is devoted to the proof of Weil's theorem on the identity of (T) spaces and of uniformizable spaces:

Theorem 1. *A topological space X is uniformizable if and only if axiom* (T) *holds in X.*

Since every normal space is completely regular and every $(T_3) + (T_4)$ space satisfies axiom (T) we have the following:

Corollary. *If a topological space satisfies axioms* (T_3) *and* (T_4), *then it is uniformizable. Therefore every normal space is uniformizable.*

Proof of Theorem 1. *Sufficiency.* Let X be a topological space satisfying axiom (T) and let $\mathscr{T}$ denote the topology of X. Our object is to find a uniform structure $\mathscr{U}$ such that the uniform topology associated with $\mathscr{U}$ is the topology $\mathscr{T}$. We shall construct $\mathscr{U}$ by determining a subbase $\mathscr{U}_S$ for $\mathscr{U}$. We associate with every nonvoid set O and every point $\xi \in O$ a diadic scale (O_d) $(d \in D)$ such that $\xi \in O_d \subseteq O$ for every $d \in D$. It is convenient to denote the set O_d for $d = m/2^n$ $(m = 0, ..., 2^n$ and $n = 1, 2, ...)$ by $O(m, n)$ and define $O(m, n)$ also for $m = -1$ and $m = 2^n + 1$ as $O(-1, n) = \emptyset$ and $O(2^n + 1, n) = X$.

For every $m = 0, ..., 2^n$ we put $O_m{}^n = O(m + 1, n) \cap c\overline{O(m - 1, n)}$. Then $\{O_m{}^n\}$ $(m = 0, ..., 2^n)$ *is a family of open sets which is a cover of X*. For $\overline{O(m - 1, n)} \subseteq O(m, n)$ and so $O_m{}^n \supseteq O(m + 1, n) \cap cO(m, n)$ for $0 < m < 2^n$; moreover, $O_0{}^n = O(1, n)$ and $O^n_{2^n} \supseteq cO(2^n, n)$. Therefore

$$O_0{}^n \cup ... \cup O^n_{2^n} \supseteq O(1, n) \cup ... \cup \{O(m + 1, n) \cap cO(m, n)\} \cup ... \cup cO(2^n, n) = X.$$

Using similar reasonings we can easily show that $O^n_{m_1}$ *and* $O^n_{m_2}$ *are disjoint unless* m_1 *and* m_2 *are adjacent integers*: Indeed, for every m we have $O_m{}^n \subseteq O(m + 1, n)$ and if $k > 1$, then

$$O^n_{m+k} \subseteq cO(m + k - 1, n) \subseteq cO(m + 1, n),$$

whence $O_m{}^n \cap O^n_{m+k} = \emptyset$. A further property of these sets $O_m{}^n$ which we shall use is: *If* m_1 *and* m_2 *are adjacent integers, then there is a suitable integer m such that* $O^{n+1}_{m_1} \cup O^{n+1}_{m_2} \subseteq O_m{}^n$. For let $m_1 = 2\mu$ and let $m_2 = 2\mu + \epsilon$ where $\epsilon = \pm 1$. Then

$$\begin{aligned} O^{n+1}_{2\mu} &= O(2\mu + 1, n + 1) \cap c\overline{O(2\mu - 1, n + 1)} \\ &\subseteq O(\mu + 1, n) \cap c\overline{O(\mu - 1, n)} \end{aligned}$$

because $O(2\mu + 1, n + 1) \subseteq O(\mu + 1, n)$ and

$$O(\mu - 1, n) \subseteq O(2\mu - 1, n + 1).$$

Similarly,

$$O^{n+1}_{2\mu+\epsilon} = O(2\mu + \epsilon + 1, n + 1) \cap c\overline{O(2\mu + \epsilon - 1, n + 1)}$$
$$\subseteq O(\mu + 1, n) \cap c\overline{O(\mu - 1, n)}$$

because

$$O(2\mu + \epsilon + 1, n + 1) \subseteq O(2\mu + 2, n + 1) = O(\mu + 1, n)$$

and

$$O(\mu - 1, n) = O(2\mu - 2, n + 1) \subseteq O(2\mu + \epsilon - 1, n + 1).$$

Now we determine a subbase $\mathscr{U}_S$ of a uniform structure $\mathscr{U}$ for X; namely, we define a set $S(O, \xi, n) \in \mathscr{U}_S$ for every choice of O, $\xi \in O$, and n as follows:

$$S(O, \xi, n) = \{(x, y): x, y \in O_m{}^n \textit{ for some } m\}.$$

Axiom (Us. 4) is clearly satisfied because $\{O_m{}^n\}$ $(m = 0, 1, ..., 2^n)$ is a cover of X. Axiom (Us. 5) also holds because every $S(O, \xi, n)$ is symmetric. In order to verify axiom (Us. 6) it is sufficient to show that $S(O, \xi, n + 1) \circ S(O, \xi, n + 1) \subseteq S(O, \xi, n)$ for every O, ξ, and n. For let $(x, y) \in S(O, \xi, n + 1)$ and $(y, z) \in S(O, \xi, n + 1)$ be given. Then $x, y \in O^{n+1}_{m_1}$ for some m_1 and also $y, z \in O^{n+1}_{m_2}$ for some m_2. Since y belongs to both $O^{n+1}_{m_1}$ and $O^{n+1}_{m_2}$ the integers m_1 and m_2 must be adjacent and so $O^{n+1}_{m_1} \cup O^{n+1}_{m_2} \subseteq O_m{}^n$ for a suitable value of m. Consequently, $x \in O_m{}^n$ and $z \in O_m{}^n$ which implies that $(x, z) \in S(O, \xi, n)$. Hence axiom (Us.6) holds and $\mathscr{U}_S$ is a subbase for a uniform structure $\mathscr{U}$.

We prove that the uniform topology associated with $\mathscr{U}$ is the topology $\mathscr{T}$. First let Q be open relative to $\mathscr{U}$. Then for every $x \in Q$ there is a $U \in \mathscr{U}$ such the $U[x] \subseteq Q$. The set U contains the intersection of finitely many sets $S(O, \xi, n)$. Since $\{O_m{}^n\}$ $(m = 0, 1, ..., 2^n)$ is a cover of X we have $x \in O_m{}^n$ for some index m and so $x \in O_m{}^n \subseteq S(O, \xi, n)[x]$. Consequently, Q contains with each point x the intersection of finitely many sets $O_m{}^n$. These sets are open in $\mathscr{T}$ and so Q is open relative to the topology $\mathscr{T}$.

Now let Q be open relative to $\mathscr{T}$ and let $\xi \in Q$. There is a diadic scale $\{O_d\}$ $(d \in D)$ associated with ξ and Q, which determines the sets $S(Q, \xi, n) \in \mathscr{U}_S$. Since $\xi \in O_0 = O(0, n)$ and $O(0, n) \subseteq O(m - 1, n)$ for $m > 0$ we have $\xi \in O_0{}^n$ and $\xi \notin O_m{}^n$ for $m > 0$. This implies that $S(Q, \xi, n)[\xi] = O_0{}^n = O(1, n) \subseteq Q$. Hence for every $\xi \in Q$ there is a $U \in \mathscr{U}$, namely, $U = S(Q, \xi, n)$, such that $U[\xi] \subseteq Q$. Therefore Q is open relative to the uniform topology. This completes the proof of the sufficiency in Theorem 1.

Proof of Theorem 1. *Necessity.* Here we prove that every uniform topology satisfies axiom (T). Let an open set O and a point x in O be given. The object is to construct a scale $(O_d)\,(d \in D)$ such that $x \in O_d \subseteq O$ for every $d \in D$. Since O is an open set and $x \in O$, there is a $U_1 \in \mathscr{U}$ satisfying $x \in U_1[x] \subseteq O$. By Lemma I.8.1 we can select a sequence $\{U_{2^{-n}}\}$ $(n = 0, 1, \ldots)$ of symmetric uniformities such that $U_{2^{-(n+1)}} \circ U_{2^{-(n+1)}} \subseteq U_{2^{-n}}$ for every n. For every positive diadic rational $d = 2^{-n_1} + 2^{-n_2} + \ldots + 2^{-n_k}$ we define

$$U_d = U_{2^{-n_1}} \circ \ldots \circ U_{2^{-n_k}}.$$

It is simple to show that $U_{d_1} \circ U_{d_2} \subseteq U_{d_1+d_2}$ for any pair of positive diadic rationals d_1, d_2 satisfying $d_1 + d_2 < 1$. Since the same reasoning will be used again in the future we state this fact in the form of a lemma:

Lemma 1. *Let $\mathscr{U}$ be a uniform structure and let $(U_{2^{-n}})$ $(n = 1, 2, \ldots)$ be a sequence of symmetric uniformities in $\mathscr{U}$ such that*

$$U_{2^{-(n+1)}} \circ U_{2^{-(n+1)}} \subseteq U_{2^{-n}}$$

for every $n = 1, 2, \ldots$. Define U_d for every diadic rational

$$d = 2^{-n_1} + \ldots + 2^{-n_k}$$

where $0 < n_1 < \ldots < n_k$ as follows:

$$U_d = U_{2^{-n_1}} \circ \ldots \circ U_{2^{-n_k}}.$$

Then for every pair of positive diadic rationals d_1, d_2 satisfying $d_1 + d_2 < 1$ we have $U_{d_1} \circ U_{d_2} \subseteq U_{d_1+d_2}$.

Proof. We show that $U_d \circ U_{2^{-n}} \subseteq U_{d+2^{-n}}$ for every diadic rational $d = 2^{-n_1} + \ldots + 2^{-n_k}$ and for every $n \geqslant 1$. This special case can be proved by induction on k because if $n_i \neq n$ for every i, then by definition $U_d \circ U_{2^{-n}} = U_{d+2^{-n}}$, and if $n_i = n$ for some i, then

$$U_d \circ U_{2^{-n}} = U_{d-2^{-n}} \circ U_{2^{-n}} \circ U_{2^{-n}} \subseteq U_{d-2^{-n}} \circ U_{2^{-(n-1)}}$$

where $d - 2^{-n}$ is a diadic rational of $k - 1$ terms. The lemma follows.

Now let $0 < d_1 < d_2$ be given. From what we just proved we see that $U_{d_1} \circ U_{d_2-d_1} \subseteq U_{d_2}$ and so by Lemma I.9.7 we have $\overline{U_{d_1}[x]} \subseteq U_{d_2}[x]^i$. Hence if we define O_d $(0 < d < 1)$ as $O_d = U_d[x]^i$, then $\bar{O}_{d_1} \subseteq O_{d_2}$ for any pair of rationals $0 < d_1 < d_2 < 1$. Since $U_1[x] \subseteq O$ we have $x \in O_d \subseteq O$ for every $d < 1$. Therefore (O_d) $(d \in D)$ is a scale of open sets satisfying the requirements of axiom (T). This shows that axiom (T) holds in every uniform space.

EXERCISES

1. Show that the uniform topology associated with a uniform structure $\mathscr{U}$ is a (T_2) topology if and only if $\bigcap\{U: U \in \mathscr{U}\} = I$.

[If X is a (T_0) space under the uniform topology, then for every distinct pair $x, y \in X$ there is a $U \in \mathscr{U}$ such that $(x, y) \notin U$. Hence if X is a (T_0) space, then $\bigcap\{U: U \in \mathscr{U}\} = I$. Conversely, if this intersection is I, then given $x \neq y$ there is a U such that $(x, y) \notin U$. Let $V \in \mathscr{U}$ be symmetric and let $V \circ V \subseteq U$. Then $V[x]$ and $V[y]$ are disjoint neighborhoods of x and y, respectively.]

2. Find a necessary and sufficient condition that the uniform topology be a Hausdorff topology when the uniform structure is given by a subbase.

(The same as in Exercise 1: $\bigcap\{U: U \in \mathscr{U}_S\} = I$.)

3. Show that a topological space X is a (T_2) space if and only if the diagonal I is a closed set in the product space $X \times X$.

[Let X be a (T_2) space and let O_x and O_y be disjoint open neighborhoods of the distinct points x and y, respectively. Then $O_x \times O_y$ is an open neighborhood of (x, y) not intersecting I. Hence cI is open. Conversely, if I is closed, then for every $x \neq y$ there are open sets O_x and O_y such that $(x, y) \in O_x \times O_y \subseteq cI$. The sets O_x and O_y are disjoint open neighborhoods of x and y, respectively.]

4. Let X be a metric space with a nonbounded metric d. Show that the uniform structure constructed in the proof of Theorem 1 is not the metric structure $\mathscr{U}_d$.

(If $U \in \mathscr{U}$, then there are finitely many points $x_1, \ldots, x_n \in X$ such that $U[x_1] \cup \ldots \cup U[x_n] = X$. If $\epsilon > 0$, then $U_\epsilon \in \mathscr{U}_d$ does not have this property. We call $\mathscr{U}$ a precompact structure.)

5. Let $\mathscr{U}$ be the structure constructed in the proof of Theorem 1 and let d be a metric for the topology of X satisfying $\mathscr{U}_d \leqslant \mathscr{U}$. Show that d is a bounded metric.

11. Axioms of Separation in Product Spaces

Here we discuss some of the necessary and sufficient conditions under which a particular separation axiom holds in a product space. First of all we have the following:

Theorem 1. *The product space* $X = \prod X_s$ *satisfies the* (T_i) *axiom* $(i = 0, 1, \text{ or } 2)$ *if and only if every factor is a* (T_i) *space.*

Proof. First we prove that the conditions are necessary. Let us suppose that X_s is not a (T_i) space for some $s \in S$ and $i \leqslant 2$. Then there is a pair of distinct points a_s and b_s in X_s for which the statement of the (T_i) axiom fails to hold. For every index $\sigma \neq s$ we choose a point $x_\sigma = a_\sigma = b_\sigma$ in X_σ arbitrarily. Then the statement of axiom (T_i) fails for the point pair $a = \{a_\sigma\}$, $b = \{b_\sigma\}$ $(\sigma \in S)$.

For instance, if $i = 2$ and if $a \in \Pi O_\sigma$ and $b \in \Pi Q_\sigma$, then ΠO_σ and ΠQ_σ intersect. Indeed, O_s and Q_s have a common point x_s, and by our construction $x_\sigma \in O_\sigma \cap Q_\sigma$ for every $\sigma \neq s$. Thus $x = \{x_\sigma\}$ $(\sigma \in S)$ belongs to both ΠO_σ and ΠQ_σ. If $i = 0$ or $i = 1$, the reasoning is similar.

Next we prove that the conditions given in Theorem 1 are sufficient. We suppose that X_s is a (T_i) space for every $s \in S$. Let a and b be distinct points in X. Then there is at least one index $s \in S$ such that $a_s \neq b_s$. If X_s is a (T_2) space we can find disjoint open sets O_{a_s} and O_{b_s} in X_s such that $a_s \in O_{a_s}$ and $b_s \in O_{b_s}$. The cylinders $O_a = Z(s, O_{a_s})$ and $O_b = Z(s, O_{b_s})$ are disjoint open sets in X; moreover, $a \in O_a$ and $b \in O_b$. Hence X is a Hausdorff space. If axiom (T_0) or (T_1) holds in X_s for every $s \in S$, then a similar reasoning shows that X is a (T_0) space or a (T_1) space, respectively.

Theorem 2. *The product space $X = \Pi X_s$ satisfies the (T_3) axiom if and only if every factor is a (T_3) space.*

Corollary. *The product $X = \Pi X_s$ is regular if and only if every factor X_s is regular.*

Proof. First we suppose that X_s is not a (T_3) space for some index $s \in S$ and prove that X is not either. If X_s is not a (T_3) space, then there is a point $x_s \in X_s$ and a closed set A_s not containing x_s such that if O_{x_s} and O_{A_s} are open sets and if $x_s \in O_{x_s}$ and $A_s \subseteq O_{A_s}$, then

$$O_{x_s} \cap O_{A_s} \neq \emptyset .$$

With x_s given, we define x in X by choosing x_σ for every $\sigma \neq s$ arbitrarily and let A denote the closed cylinder $Z(s, A_s)$. Let $x \in \Pi O_{x_\sigma}$ and let O_A be an open set containing A. Then

$$O_{A_s} = \{\xi_s : \xi = \{\xi_\sigma\} \in O_A \quad \textit{and} \quad \xi_\sigma = x_\sigma \quad \textit{for every} \quad \sigma \neq s\}$$

is an open set in X_s. It is clear that $A_s \subseteq O_{A_s}$ and so O_{x_s} and O_{A_s} intersect. Therefore the open sets ΠO_{x_σ} and O_A intersect and so x and A cannot be separated by open sets.

Next we prove that the condition is sufficient. Let us suppose that X_s is a (T_3) space for every $s \in S$. Let O be an open set in X and let $x \in O$. By the definition of the product topology we have $x \in \Pi O_s \subseteq O$ where O_s is open in X_s and $O_s = X_s$ for all but finitely many indices $s \in S$. If $O_s = X_s$ we define $Q_s = X_s$. If $O_s \subset X_s$, then using axiom (T_3) we can select an open set Q_s in X_s such that $x_s \in Q_s \subseteq \bar{Q}_s \subseteq O_s$. Since $Q_s = X_s$ for all but finitely many choices of s, the set $Q = \Pi Q_s$ is open in X; moreover, by Lemma I.11.3, $\Pi \bar{Q}_s$ is a closed set and $Q \subseteq \Pi \bar{Q}_s \subseteq O$. Therefore $x \in Q \subseteq \bar{Q} \subseteq O$ which shows that X is a (T_3) space.

The product of an arbitrary family of (T) spaces is a (T) space. This was proved in the solution of Exercise 9.8. It can be proved also by using Theorem 10.1 and Theorem I.12.2. In particular, the product of completely regular spaces is completely regular.

The product of normal topological spaces need not be normal even if the number of factors is finite. Actually, the product of completely normal spaces need not be normal even in the case of two equal factors. This is shown by the following example:

Example 1. Let X be the set of reals topologized by the right half-open interval topology. It is proved in the solution of Exercise 5.2 that X is a completely normal space. Now we show that $X \times X$ is not normal. First of all, any subset A of the diagonal line $D = \{(x, -x) : x \in X\}$ is a closed set in $X \times X$. In fact, if $(x, y) \notin D$, then (x, y) can be enclosed in a disk which does not intersect D, and if $(x, -x)$ is a point in cA, then the set $\{(\xi, \eta) : \xi \geqslant x \textit{ and } \eta \geqslant -x\}$ is an open neighborhood of $(x, -x)$ which is disjoint from A. Now let $A = \{(x, -x) : x \textit{ is rational}\}$ and let $B = \{(x, -x) : x \textit{ is irrational}\}$. These are disjoint closed sets in $X \times X$. Let O_A and O_B be open sets in $X \times X$ containing A and B, respectively. To every $(x, -x) \in B$ there corresponds an $\epsilon_x > 0$ such that the square

$$\{(\xi, \eta) : x \leqslant \xi < x + \epsilon_x \quad \text{and} \quad -x \leqslant \eta < -x + \epsilon_x\}$$

belongs entirely to O_B. Let B_n be the set of those points in B for which $\epsilon_x \geqslant 1/n$. Using the same reasoning as in Example 3.1 we can find a positive integer n and an interval I on the diagonal line D such that every point of I is an accumulation point of B_n in the Euclidean sense. Hence B_n has an accumulation point $(x, -x)$ for a rational value of x. To $(x, -x)$ there corresponds a rectangle which lies entirely in O_A. If a point of B_n is sufficiently close to $(x, -x)$, then the corresponding rectangles will intersect. Hence O_A and O_B are not disjoint and so $X \times X$ is not a (T_4) space.

EXERCISES

1. Let $X = \{\infty\} \cup S$ where S is an arbitrary set. Let $C \subseteq X$ be closed if $\infty \in C$ or if C is finite. Show that the topology defined in this way is completely normal.

2. Define a topology on $X = \{\infty\} \cup S$ where S is a noncountable set by specifying its closed sets: $C \subseteq X$ is closed if $\infty \in C$ or if C is countable. Show that the topology of X is completely normal.

(If $\infty \notin O$, then O is open. Hence if A and B are separated sets and $\infty \notin A \cup B$, then we can set $A = O_A$ and $B = O_B$. If $\infty \in A$ and if A and B are separated, then B is open-closed: For $\infty \notin B$ and $\bar{B} \subseteq \{\infty\} \cup B$ and so $\bar{B} \subseteq B$. Hence we can set $O_A = cB$ and $O_B = B$.)

3. Let X be the topological space defined in Exercise 1. Show that the sets $A = \{(x,\infty): x \in S\}$ and $B = \{(\infty, x): x \in S\}$ are separated sets in $X \times X$.

[We have $\bar{A} \subseteq \{(x,\infty): x \in X\}$ and $\bar{B} \subseteq \{(\infty, x): x \in X\}$. Equality holds if X is infinite.]

4. Using the separated sets A and B introduced in the preceding exercise show that $X \times X$ is not a (T_5) space when S is not countable.

[Let O_A and O_B be open sets containing A and B, respectively. If $x \in S$, then $\{x\}$ is open in X, and since $(x,\infty) \in O_A$, there is an open set $O_x{}^A$ in X which contains ∞ and is such that $(x,\infty) \in \{x\} \times O_x{}^A \subseteq O_A$. Hence $cO_x{}^A$ is finite, say $n(x)$ is the number of elements of $cO_x{}^A$. There is a noncountable subset S_0 of S and a positive integer n such that $n(x) \leqslant n$ for every $x \in S_0$. Choose $n + 1$ points in B, say $(\infty, x_1), \ldots, (\infty, x_{n+1})$. Since $cO_x{}^B$ is finite there is an $x \in S_0$ such that $x \notin cO^B_{x_k}$ for every $k = 1, \ldots, n + 1$. This implies that $(x, x_k) \in O^B_{x_k} \times \{x_k\}$ for every k. The set $cO_x{}^A$ contains at most n points and so at least one of the points (x, x_k) must belong to $\{x\} \times O_x{}^A$. Hence O_A and O_B intersect.]

5. Suppose that $X = \Pi X_s$ is a (T_5) space. Show that every X_s $(s \in S)$ is a (T_5) space.

(Choose an $x_\sigma \in X_\sigma$ for every $\sigma \neq s$. By Theorem I.11.2 the subspace $\{x : \pi_\sigma(x) = x_\sigma \textit{ for every } \sigma \neq s\}$ is homeomorphic to X_s. Since X is a (T_5) space, by Theorem 5.1 this subspace is a (T_5) space.)

12. Separable Spaces and Countability Axioms

The axioms introduced in this section are of a very restrictive nature and so it is very pleasant to deal with spaces which satisfy anyone of these axioms. When we are faced with a new problem it is always tempting to solve it first in an easy case and relax the hypothesis afterward. Nevertheless, one should not acquire the habit of trying to solve problems first for separable spaces or for spaces satisfying one of the countability axioms. The reasonings will often not generalize or if they do then the force of these axioms was not used in the first place.

If A is a set in a topological space X and if $\bar{A} = X$, then we say that A is *dense* in X. If the space X contains a countable dense set A, then X is called *separable*. The simplest nontrival example of separable spaces is furnished by the Euclidean spaces. If the space is finite dimensional, then the points all of whose coordinates are rational form a denumerable dense set in X. This can be seen immediately by considering the open base consisting of the cubical neighborhoods

$$\{x : | x_k - a_k | < \epsilon \quad \text{for} \quad k = 1, ..., n\}$$

where each a_k is rational. The "infinite-dimensional Euclidean space" or in other words the "separable Hilbert space" is also separable. A denumerable dense set is formed by those points $a = (a_1, ..., a_n, 0, ...)$ whose coordinates are rationals and only finitely many of these are different from zero. An example of a nonseparable space is obtained by considering the discrete topology of a noncountable set. It follows that nonseparable spaces exist already among metrizable spaces.

A topological space X is said to satisfy the second countability axiom if the family of its open sets has a countable base. A space satisfying the second countability axiom is sometimes called a *rational space.*

Lemma 1. *If X has a countable base, then it is separable.*

Proof. Let $\mathscr{B} = \{B_n\}$ $(n = 1, 2 ...)$ be a base for the open sets of the rational space X. Choose a point a_n from each B_n. Then $A = \{a_n\}$ $(n = 1, 2, ...)$ is a countable dense subset of X because every nonvoid open set meets A.

Lemma 2. *Every separable pseudometric space has a countable base.*

Proof. Let $A = \{a_n\}$ $(n = 1, 2, ...)$ be a dense subset of X. For every $m = 1, 2, ...$ we consider the open balls $S_{1/m}[a_n]$ $(n = 1, 2, ...)$. These form a countable collection of open sets. If O is open and $x \in O$, then $S_{2/m}[x] \subseteq O$ for a sufficiently large value of m, and since A is dense in X we have $a_n \in S_{1/m}[x]$ for some $a_n \in A$. It follows that

$$x \in S_{1/m}[a_n] \subseteq S_{2/m}[x] \subseteq O$$

and so the balls $S_{1/m}[a_n]$ $(m, n = 1, 2, ...)$ form a countable base.

Theorem 1. *A pseudometric space is separable if and only if it satisfies the second axiom of countability.*

Proof. This is an immediate consequence of the preceding two lemmas.

The equivalence of separability and of rationality holds only for pseudometric spaces. For instance, the right half-open interval topology of the reals leads to a separable space which does not satisfy the second countability axiom.

Lemma 3. *The product of countably many separable spaces is separable.*

Proof. Let $X = \Pi X_n$ and let $A_n = \{a_{mn}\}$ $(m = 1, 2, ...)$ be a dense subset of X_n for every $n = 1, 2, ...$. Let A be the set of those points $x \in X$ which have the following properties: (1) $x_n \in A_n$ for every $n = 1, 2, ...$; and (2) there is an index $n(x)$ and a positive integer m depending on x but not on n such that $x_n = a_{mn}$ for every $n > n(x)$. This set A is countable and it is dense in the product space X. Indeed, if ΠO_k is a nonvoid open set, then there is an index n such that $O_k = X_k$ for every $k > n$, and since A_k is dense in X_k we can choose an $a_{i_k k}$ in O_k for each $k \leqslant n$. We let $x_k = a_{i_k k}$ for $k \leqslant n$ and $x_k = a_{1k}$ for $k > n$. Then $x = (x_1, x_2, ...)$ belongs to ΠO_n.

The second countability axiom has many interesting consequences. Earlier, in Exercise 1.9, we proved for instance that if X is a (T_0) space which satisfies the second countability axiom, then the cardinality of its family of open sets is at most that of the continuum. Later we shall prove that all regular spaces satisfying the second countability axiom are metrizable. Here we discuss another remarkable consequence of this axiom which is known under the name "*Lindelöf property*."

If a set $Y \subseteq X$ is covered by a family of open sets O_i $(i \in I)$ and if $\mathscr{B} = \{B\}$ is a base for $\mathscr{O}$, then Y can be covered by a subfamily $\{B_{ix}\}$ $(i \in I$ and $x \in O_i)$ of $\mathscr{B}$ where B_{ix} is chosen such that $x \in B_{ix} \subseteq O_i$. If $\mathscr{B}$ is a countable set, then $\{B_{ix}\}$ is a countable collection and so choosing for each B_{ix} one O_j containing it we can replace the original cover $\{O_i\}$ $(i \in I)$ of Y by a subcover consisting only of countably many sets. Hence we have:

Lemma 4. *If X satisfies the second countability axiom and if Y is a subset of X which is covered by a collection of open sets, then there is a countable subcollection which covers Y.*

This property of rational spaces is called the *hereditary Lindelöf property*. If open covers of the whole space X can be replaced by countable subcovers, we say that X has the *Lindelöf property*. Hence if X has the hereditary Lindelöf property, then so does every subspace Y of X.

Conversely, if every open set of X has the Lindelöf property, then X has the hereditary Lindelöf property. Indeed, if Y is not open and $Y \subseteq \bigcup O_i$, then $\bigcup O_i$ being open we can select a countable subfamily of $\{O_i\}$ $(i \in I)$ which covers $\bigcup O_i$. Of course this will also cover Y.

A topological space X is said to satisfy the first countability axiom if the neighborhood filter $\mathcal{N}(x)$ *of each point* $x \in X$ *has a countable base* $\mathcal{B}(x) = \{B_x{}^n\}$ $(n = 1, 2, ...)$. Clearly, if X satisfies the second countability axiom, then it will also satisfy the first one but not conversely. Every pseudometric space satisfies the first countability axiom. If the first countability axiom is satisfied in a (T_1) space, then by Theorem 1.1, for every x there are denumerably many neighborhoods $N_x{}^1$, $N_x{}^2$, ... such that $\bigcap N_x{}^n = \{x\}$. Moreover, if X is a regular space, then we may assume that these neighborhoods $N_x{}^n$ are closed sets. It is remarkable that this property can hold for every $x \in X$ even if X does not satisfy the first axiom of countability. An example of such a space is given in Exercise 5.3.

Theorem 3. *If X is a* (T_3) *space and satisfies the second countability axiom, then it is a* (T_5) *space.*

Note. A similar result is stated in Theorem 2.3. The two theorems are independent even in the special case of Hausdorff topologies: The cardinality of the reals is not countable and they form a rational space while the space described in Exercise 5.3 is formed on a countable set but it does not have a countable base.

Proof. Let A and B be separated sets in X. Since X has a countable base there exist countable families $\{O_n\}$ and $\{Q_n\}$ of open sets covering A and B, respectively. Since A and B are separated and X is a (T_3) space we can assume that $\bar{O}_n \subseteq c\bar{B}$ and $\bar{Q}_n \subseteq c\bar{A}$. Let O be the union of the open sets $O_n \cap c\bar{Q}_1 \cap \dots \cap c\bar{Q}_n$ where $n = 1, 2, ...$ and let Q be defined similarly. Since $A \subseteq \bar{A} \subseteq c\bar{Q}_n$ for every $n = 1, 2, ...$ and $A \subseteq \bigcup O_n$, we see that $A \subseteq O$. Similarly, we see that $B \subseteq Q$. Hence O and Q are disjoint open sets covering the separated sets A and B, respectively.

EXERCISES

1. Show that every subspace of a separable pseudometric space is separable.

(It is sufficient to show that Y has a countable base. By Lemma 2 the original space X has a countable base and so the same holds for Y.)

2. Give a simple proof that the product of finitely many separable spaces is separable.

3. Give an example of a nonmetrizable space which satisfies the second countability axiom.

[The topology of finite complements on a denumerable set has countably many open sets but it is not metrizable. For the space is a (T_1) space but not a Hausdorff space.]

4. Show that the half-open interval topology $\mathscr{T}_+$ of the set of reals does not have a countable base.

5. Show by an example that Lemma 3 cannot be extended to the case of noncountably many factors.

6. Show that the space described in Exercise 5.3 does not satisfy the first axiom of countability.

[Let $O_1, O_2, \ldots$ be a sequence of open sets containing $(0, 0)$. By the definition of the closed sets, for each n there are integers $x_n, y_n \geqslant n$ such that $(x_n, y_n) \in O_n$ for every n. The set $N = X \cap c\{(x_1, y_1), (x_2, y_2), \ldots\}$ is a neighborhood of $(0, 0)$ and none of the O_n's is a subset of N.)

7. Let A^* denote the set of those points x of the topological space X for which $O_x \cap A$ is an uncountable set for every choice of the open neighborhood O_x containing x. Show that if A is a noncountable set in a Lindelöf space X, then A^* is not void.

[Suppose that every $x \in X$ has an open neighborhood such that $O_x \cap A$ is countable. The open cover $\{O_x\}$ $(x \in X)$ of X can be replaced by countable subcover, say by $\{O_{x_n}\}$ $(n = 1, 2, \ldots)$. Since $O_{x_n} \cap A$ is countable it follows that A is at most denumerable.]

8. Show that A^* is always closed.

9. Show that the half-open interval topologies of the reals have the Lindelöf property.

(The intervals $[a, b)$ form a base for the topology $\mathscr{T}_+$. Given any open cover of X by such half-open intervals, consider for every integer $m = 0, \pm 1, \ldots$ families of half-open intervals $[\alpha, \beta)$ such that each of these intervals is a subinterval of one of the given intervals, they are mutually disjoint, and their union is a half-open interval of X whose left end point is m. Zorn's lemma can be applied and the resulting maximal family is a cover of $[m, \infty)$. Since the elements of this family are disjoint intervals for every $n = 1, 2, \ldots$, there are at most denumerably many among these intervals whose length is at least $1/n$. Hence the family is countable.)

10. Show that a pseudometric space X has the Lindelöf property if and only if there is a countable dense set in X.

[In other words, show that X is a Lindelöf space if and only if it is separable. For every $n = 1, 2, \ldots$ consider the open cover $S_{1/n}[x]$ $(x \in X.)$ If X is a Lindelöf space there is a countable subcover with centers, say at $x_1^n, x_2^n, \ldots$. The family $\{x_m^n\}$ $(m, n = 1, 2, \ldots)$ is a countable dense subset of X. Conversely, if X has a countable dense subset $\{x_m\}$ $(m = 1, 2, \ldots)$, then the open balls $S_{1/n}[x_m]$ $(m, n = 1, 2, \ldots)$ form a countable base for the open sets of X.]

11. Show the following proposition: If a pseudometric space X has the Lindelöf property, then so does every subspace of X.

(This is an immediate consequence of Exercise 1 and the preceding exercise.)

NOTES

The (T_3) axiom was introduced by Vietoris [1]. A space is called (T_3) at a point x if for every O_x there is a Q_x such that

$$x \in Q_x \subseteq \bar{Q}_x \subseteq O_x .$$

Theorem 2.3 was first proved by Tietze in his paper on separation in topological spaces [2]. A second proof of the theorem can be found in one of Urysohn's articles [3]. Semiregular spaces were discussed by Stone [4] and in Hewitt's thesis [5]. Hereditary semiregular or completely semiregular spaces were also studied by Katetov [6]. The result stated in Exercise 2.8 occurs in a note by Norris [7].

Theorem 4.2 can be found in Nöbeling's book [8], in Lefschetz's book [9], and in Dieudonné's article on paracompactness [10]. Results related to similarity of ordered n-tuples of sets first occur in an addendum by Hurewicz to a joint paper of Hurewicz and Menger [11]. A further generalization of this result was given by Kuratowski [12].

Perfectly normal spaces were introduced by Alexandroff and Urysohn [13]. The implication (perfectly normal) $\rightarrow$ (completely normal) is proved in Urysohn's article [3] (see footnote 41 on p. 286). Perfectly normal spaces were studied by Katetov [14] who also considered products of such spaces.

Once Fréchet raised the problem of characterizing the Hausdorff separation property by means of the derivation operator. The solution was given by Monteiro [15]: *X is a Hausdorff space if and only if for every $A \subseteq X$ and points $a_1, a_2 \in A'$ there is a decomposition $A = A_1 \cup A_2$ with $A_1 \cap A_2 = \emptyset$ such that* $a_2 \notin A_1'$ and $a_1 \notin A_2'$. Later, similar results were obtained on regular, normal, and completely normal spaces by Ribeiro [16].

A closer study of connectivity belongs to algebraic topology. For instance the number of components has a simple meaning in homology theory. However, there are other results which were obtained by different means. See for instance Borsuk [17]. The following is proved: *If A and B are homeomorphic subspaces of the n-dimensional Euclidean space* E_n, *then* $E_n - A$ *and* $E_n - B$ *have the same number of components.* See also a more recent article by Granas [18].

Exercise 7.7 is due to Urysohn and it was published in the paper which was mentioned previously [3]. Exercise 7.8 is a more recent contribution by Bing [19].

We gave no example for a regular not completely regular space in Section 9. The product space given in Example 11.1 is completely regular because the factors are completely regular but it is not normal. Another example can be found for instance in Tychonoff's article [20] where the notion of complete regularity is introduced. Here one finds also an example of a (T) space which is not a (T_4) space. Other examples of such spaces are discussed by Hewitt [21]. A further example is discussed in the text.

The first proof of the sufficiency in Theorem 10.1 is due to A. Weil who discovered the result. A second proof can be found in Kelley's book [22] and the proof given in Section 10 is a variant of a proof which can be found in Nöbeling's book [8]. The necessity was proved also by Weil but for special types of uniformizable spaces, e.g., topological groups, the result was already known.

The article by Katetov [14] contains some simple but interesting results on separation in product spaces. For instance we find the following: *If* $X \times Y$ *is completely normal then either every countable subset of Y is closed or X is perfectly normal.*

Separability was introduced by Fréchet in his thesis [23]. Lindelöf's theorem for the m-dimensional Euclidean space can be found in [24]. The first and second countability axioms were introduced by Hausdorff. Theorem 12.3 was discovered by Tychonoff [25]. Vijayaraghavan [26] gave an example of a connected linearly ordered space formed on a set of cardinality $\mathfrak{c}$ which is not separable. The *character of a space* is defined to mean the least cardinal number of a basis for the open sets of the space. Rational spaces have character at most ω.

References

1. L. Vietoris, Stetige Mengen. *Monatsh. Math.* **31**, 173–204 (1921).
2. H. Tietze, Beiträge zur allgemeinen Topologie. I. Axioms für verschiedene Fassungen des Umgebungs Begriff. *Math. Ann.* **88**, 290–312 (1923).
3. P. Urysohn, Über die Mächtigkeit der zusammenhangenden Mengen. *Math. Ann.* **94**, 262–295 (1925).

4. M. Stone, Applications of the theory of Boolean rings to general topology. *Trans. Amer. Math. Soc.* **41**, 375–481 (1937); see Section III.1.
5. E. Hewitt, A problem of set-theoretic topology. *Duke Math. J.* **10**, 309–333 (1943).
6. M. Katetov, Remarque sur les espaces topologiques dénombrables. *Ann. Soc. Polon. Math.* **21**, 120–122 (1948).
7. M. J. Norris, A note on regular and completely regular topological spaces. *Proc. Amer. Math. Soc.* **1**, 754–755 (1950).
8. G. Nöbeling, "Grundlagen der analytischen Topologie," pp. 84 and 85. Springer-Verlag, Berlin, 1954.
9. S. Lefschetz, "Algebraic Topology." Amer. Math. Soc. Publications, Vol. **27**, New York, 1942.
10. J. Dieudonné, Une généralisation des espaces compacts. *J. Math. Pures Appl.* [9] **23**, 65–76 (1944).
11. W. Hurewicz and K. Menger, Ein Additionssatz. *Math. Ann.* **100**, 627–633 (1928).
12. C. Kuratowski, Sur le prolongement des fonctions continues et les transformations en polytopes. *Fund. Math.* **24**, 259–268 (1935).
13. P. Alexandroff and P. Urysohn, On compact topological spaces. *Trudy Math. Inst. Steklov* **31**, 95 (1950); this is a revised second edition of the original article.
14. M. Katetov, Complete normality of Cartesian products. *Fund. Math.* **35**, 271–274 (1948).
15. A. Monteiro, Caractérisation des espaces de Hausdorff au moyen de l'opération de dérivation. *Portugal. Math.* **1**, 333–339 (1940).
16. H. Ribeiro, Caractérisation des espaces réguliers normaux et complètement normaux au moyen de l'opération de dérivation. *Portugal. Math.* **2**, 13–19 (1941).
17. K. Borsuk, Set-theoretic approach to the disconnection theory of the Euclidean space. *Fund. Math.* **37**, 217–241 (1950).
18. A. Granas, On local disconnection of Euclidean spaces. *Fund. Math.* **41**, 42–48 (1954).
19. R. Bing, A connected countable Hausdorff space. *Proc. Amer. Math. Soc.* **4**, 474 (1953).
20. A. Tychonoff, Über die topologische Erweiterung von Raumen. *Math. Ann.* **102**, 544-561 (1929).
21. E. Hewitt, Rings of real-valued continuous functions, I. *Trans. Amer. Math. Soc.* **64**, 45–99 (1948).
22. J. Kelley, "General Topology." Van Nostrand, Princeton, New Jersey, 1955.
23. M. Fréchet, Sur quelques points du calcul fonctionnel. *Rend. Circ. Mat. Palermo* **22**, 1–74 (1906).
24. E. Lindelöf, Sur quelques points de la théorie des ensembles. *C. R. Acad. Sci. Paris* **137**, 697–700 (1903).
25. A. Tychonoff, Über einen Metrisationsatz von P. Urysohn. *Math. Ann.* **95**, 139–142 (1926).
26. T. Vijayaraghavan, On two problems relating to linear connected topological spaces. *J. Indian Math. Soc.* [N.S.] **11**, 28–30 (1947).

CHAPTER III

Compactness and Uniformization

1. Compactness

A topological space X is called *compact* if from every open cover of X one can select a finite subcover. Thus if X is compact and $X = \bigcup O_i$, then there are finitely many sets $O_{i_1}, \ldots, O_{i_m}$ among the O_i's such that $X = O_{i_1} \cup \ldots \cup O_{i_m}$. Notice that X can be compact without being a Hausdorff space.

There are very simple examples of compact spaces: If X is finite or if more generally the family of open sets of X is finite, the space is compact. The topology of finite complements is compact no matter how large the cardinality of X may be. Nontrivial examples of compact spaces are the subspaces formed by closed bounded sets of the real line: We shall see that the topology induced on any closed bounded set of a finite-dimensional Euclidean space is compact.

The notion of compactness was first introduced in 1923 by Alexandroff and Urysohn who used the expression "bicompact." The word "compact" was already reserved to designate another property of topological spaces which we now call sequential compactness or compactness in the Fréchet sense. We shall see that for metric spaces these two notions are equivalent.

A family $\{S_i\}$ $(i \in I)$ of subsets of a set X is said to have the *finite intersection property* if every finite subfamily of $\{S_i\}$ $(i \in I)$ has a nonvoid intersection.

Theorem 1. *A topological space X is compact if and only if every family of closed sets having the finite intersection property has a nonvoid intersection.*

Proof. X is compact if and only if from every open cover of X one can select a finite subcover. If we replace the open sets by their complements we see that X is compact if and only if given any family of closed sets $\{C_i\}$ $(i \in I)$ satisfying $\bigcap C_i = \emptyset$ there is a finite subfamily $C_{i_1}, \ldots, C_{i_m}$ such that $C_{i_1} \cap \ldots \cap C_{i_m} = \emptyset$. This is equivalent to the condition stated in the theorem.

A space X is called *countably compact* if from every countable covering of X by open sets one can select a finite subcovering. Hence the notion of compactness is more restrictive than countable compactness. A point x of a topological space X will be called a 1-*accumulation point* of a set S in X if $N_x \cap S$ is infinite for every neighborhood of x. This notion was first introduced by Hausdorff who used the expression β-accumulation point.

Theorem 2. *A topological space X is countably compact if and only if every infinite set $S \subseteq X$ has a* 1-*accumulation point in X.*

Proof. If X contains an infinite set without 1-accumulation points, then it contains a denumerable set S having no 1-accumulation point. Then for every $x \in X$ we can determine an open set O_x such that $O_x \cap S$ is finite. For every finite subset F of S, including the void set $\varnothing$, let $O_F = \bigcup\{O_x : O_x \cap S = F\}$. The family of these sets is a countable open cover of X. Since each O_F meets the infinite set S in only finitely many points, the countable cover $\{O_F\}$ admits no finite subcover. Therefore X is not countably compact.

The sufficiency can be proved by assuming that X is not countably compact and constructing a denumerable set $S = \{x_1, x_2, \ldots\}$ without 1-accumulation point. Indeed if (O_n) $(n = 1, 2, \ldots)$ is a family of open sets covering X and no finite subfamily covers X, then we can choose a sequence of distinct points $x_1, x_2, \ldots$ such that $x_n \notin O_1 \cup \ldots \cup O_n$ for every n. No point x of X is a 1-accumulation point of $S = \{x_1, x_2, \ldots\}$ because if $x \in O_n$, then $O_n \cap S$ is finite.

Let (x_n) $(n = 1, 2, \ldots)$ be a sequence of points in a topological space X. A point x of X is called a *limit point* of (x_n) if each neighborhood N_x contains all but finitely many points of the sequence. If (x_n) has a limit point, then it is called a *convergent sequence*. In a general topological space a convergent sequence might have more than one limit point but this is of no importance right now. If a topological space X is such that from every sequence (x_n) of points of X one can select a convergent subsequence, then X is called *sequentially compact*. Compactness and sequential compactness are independent notions: There exist compact spaces which are not sequentially compact and vice versa.

A point x of a topological space X is called an *accumulation point of an infinite sequence* (x_n) $(n = 1, 2, \ldots)$ if for every neighborhood of x we have $x_n \in N_x$ for infinitely many indices n. For instance, if the point x occurs infinitely often among the terms of (x_n), then x is an accumulation point. Thus a point x might be an accumulation point of a sequence (x_n) while it is not an accumulation point of the set of points $x_1, x_2, \ldots$.

A space X is said to have the *Bolzano-Weierstrass property* if every infinite sequence has an accumulation point in X. It is clear that sequential compactness implies the Bolzano-Weierstrass property. Theorem 2 shows that countable compactness also implies this property. We are going to prove that conversely the Bolzano-Weierstrass property implies countable compactness. This will prove the following:

Theorem 3. *A topological space X is countably compact if and only if it has the Bolzano-Weierstrass property.*

Proof. Suppose X is not countably compact. Then there is a denumerable family of open sets O_n $(n = 1, 2, ...)$ whose union covers X but which admits no finite subcover. Hence we can select a sequence of points x_n $(n = 1, 2, ...)$ such that $x_n \notin O_1 \cup ... \cup O_n$. The sequence (x_n) has no accumulation points in X. For, given a point x we have $x \in O_n$ for some index n and so at most $x_1, ..., x_{n-1}$ belong to O_n.

Theorem 4. *If the first axiom of countability holds in X and if X is countably compact, then it is sequentially compact.*

Proof. Let (x_n) $(n = 1, 2, ...)$ be an infinite sequence of points. By Theorem 3 this sequence has an accumulation point x. By hypothesis, the neighborhood filter of x has a countable base. We may assume that the elements of the base are open sets and they form a decreasing sequence: $O_1 \supseteq O_2 \supseteq ...$. Since x is an accumulation point of (x_n), for every index k we can choose an element x_{n_k} in O_k. The subsequence $x_{n_1}, x_{n_2}, ...$ is convergent to x: Indeed, given N_x there is a k such that $O_k \subseteq N_x$ and so $x_{n_k}, x_{n_{k+1}}, ...$ all belong to N_x.

We conclude this section with a theorem of Haar and König which gives for instance a complete characterization of the compact subspaces of the real line:

Theorem 5. *A linearly ordered set X is compact in its interval topology if and only if it is order complete, i.e., if and only if every nonvoid subset has a least upper bound and a greatest lower bound.*

Proof. First let us suppose that X is order complete; then X has a smallest element a and a largest element b. We prove that X is compact: For let an open cover $\{O_i\}$ $(i \in I)$ be given. Since the set of all proper and improper open intervals is a base for the interval topology we may assume that the sets O_i are intervals. We denote by S the set of those points $x \in X$ for which the interval $[a, x]$ can be covered by finitely many intervals of the family $\{O_i\}$. The set S is not empty because a is covered by some interval $O_i = (-\infty, b_i)$ and so O_i is a nonvoid subset of S. Let $s = \text{lub } S$; then s being a point of X, it is covered by an interval of the family $\{O_i\}$ $(i \in I)$, say by (a_j, b_j) where b_j may be the symbol $+\infty$. Since $a_j < s$ the interval $(-\infty, b_j)$ can also be covered by finitely many intervals of $\{O_i\}$ $(i \in I)$. Thus if we can prove that b_j is not an element of X but the symbol $+\infty$, then it will follow that $(-\infty, +\infty)$ can be covered by finitely many sets of the family $\{O_i\}$. Now suppose that b_j is a point of X. Then there is an interval (c_k, d_k) in $\{O_i\}$ which contains b_j and so since $c_k < b_j$, $(-\infty, d_k)$ has a finite cover. However, $(-\infty, b_j] \subseteq (-\infty, d_k)$ and b_j exceeds the least upper bound s, so we have a contradiction.

Now we suppose that X is not order complete and show that the interval topology of X is not compact: If X contains a nonvoid set S without a least upper bound in X, then an open cover of X is given by the following family of open intervals: $(-\infty, s)$ for every $s \in S$, also $(u, +\infty)$ for every upper bound u of S. This cover cannot contain any finite subcover since then the largest s or the smallest u occurring in the subcover would be the least upper bound of S. If X contains a nonvoid set S without a greatest lower bound, a similar reasoning shows that X is not compact.

A simple application of Theorem 5 leads to a necessary and sufficient condition for the compactness of a well-ordered topological space:

Theorem 6. *A well-ordered set X is compact in its interval topology if and only if it contains a maximal element.*

Proof. The necessity of the condition is an immediate consequence of Theorem 5. The sufficiency follows from Theorem 5 and the concept of well ordering.

Similarly, the Haar-König theorem shows that *a subspace of the real line is compact if and only if it is a closed and bounded subset.*

EXERCISES

1. Let C be a closed set in the compact space X. Show that C is a compact subspace.

(Every open cover of C can be extended to an open cover of X by joining the set cC. By hypothesis the extended cover admits a finite subcover. Discard cC from this finite subcover to obtain a finite subcover of the original open cover of C.)

2. Let $\mathcal{T}_1$ and $\mathcal{T}_2$ be topologies on the set X. Show that if $\mathcal{T}_1$ is compact and $\mathcal{T}_1 \geqslant \mathcal{T}_2$, then $\mathcal{T}_2$ is compact.

3. Show that the quotient space X/R is compact if X is compact. (X/R under $\mathscr{T}/R$ and X under $\mathscr{T}_R$ are topologically isomorphic and by the preceding exercise the topology $\mathscr{T}_R$ is compact.)

4. A set S is called an $\mathscr{F}_\sigma$ set of a topological space if it is the union of countably many closed sets. Show that if X is compact, then every $\mathscr{F}_\sigma$ set is a Lindelöf space.

5. Let X be a topological space and let $\mathscr{O}_c$ be the family consisting of $\varnothing$ and of those open sets $O \in \mathscr{O}$ whose complement is compact. Show that $\mathscr{O}_c$ satisfies the axioms for open sets of a topology $\mathscr{T}_c$ and show that $\mathscr{T}_c$ is compact.

[The union of two compact subspaces of X is compact. This implies axiom (O.2). Axiom (O.3) follows from Exercise 1.]

6. Show that the right half-open interval topology of a linearly ordered set X can be compact only if every increasing sequence of points $x_n \in X$ contains not more than finitely many distinct points.

[Suppose there is an infinite strictly increasing sequence of distinct points, say $x_1 < \ldots < x_n < \ldots$. Consider the family of the following open sets: $(-\infty, x_1)$, $[x_n, x_{n+1})$ for every $n = 1, 2, \ldots$, and $[x, \infty)$ for every x such that $x_n < x$ for $n = 1, 2, \ldots$. This is an open cover of X and contains no finite subcover.]

7. Show the existence of a compact Hausdorff topology on any nonvoid set X.

(This follows from the axiom of choice: Well order X such that it has a maximal element. This is always possible: If X is well ordered and x is its first element, modify the ordering by specifying x to be the last element of X. The interval topology of X relative to such an ordering is compact.)

8. Let S be the set of all ordinals less than or equal to the first uncountable ordinal ω_1 . Show that the interval topology of S is compact but S is not a hereditary Lindelöf space.

(The compactness follows either from Theorem 5 or from the fact that a decreasing sequence of ordinals is necessarily finite: Let ω_1 be the first uncountable ordinal and let O_1 be a set of a cover by open intervals such that $\omega_1 \in O_1$. Let ω_2 be the first element of O_1 if this element is not ω_1 and otherwise let ω_2 be the first element less than ω_1 not in O_1 . If $\omega_1 \in O_1, \ldots, \omega_n \in O_n$ are determined such that $\omega_1 > \omega_2 > \ldots > \omega_n$, let ω_{n+1} be the first element of $O_1 \cup \ldots \cup O_n$ except if this element is ω_n in which case let ω_{n+1} be the first element less than ω_n not in $O_1 \cup \ldots \cup O_n$. After a finite number of steps this procedure must stop, that is, $O_1 \cup \ldots \cup O_n = S$ for a sufficiently high value of n. Since there are noncountably many ordinals less than the first uncountable ordinal which have predecessors, there is a noncountable set in S such that its elements form open sets by themselves. S therefore does not have the Lindelöf property.)

2. Compact Metric Spaces

The main purpose of this section is to prove that for pseudometric spaces compactness, sequential compactness, countable compactness, and the Bolzano-Weierstrass property are equivalent notions. We already know that sequential compactness, countable compactness, and the Bolzano-Weierstrass property are equivalent in any space satisfying the first axiom of countability. Since compactness implies countable compactness it is sufficient to prove that if a pseudometric space is countably compact, then it is compact. This is an immediate consequence of the following:

Lemma 1. *A countably compact pseudometric space X has a countable base.*

Proof. First we show that for every $\epsilon > 0$ there are finitely many points $x_1, ..., x_m$ in X such that the balls $S_\epsilon[x_k]$ $(k = 1, ..., m)$ cover X. For if such finite system of points does not exist we can find an infinite sequence (x_m) $(m = 1, 2, ...)$ of distinct points such that $S_\epsilon[x_m]$ contains no points of (x_m) besides the center x_m itself. Using this sequence we construct a countable covering of X by open sets which contains no finite subcovering: For we consider the ball $S_{\epsilon/2}[x]$ about each point x which is not covered by the union of the balls $S_\epsilon[x_m]$. Then the open set which is the union of all these balls $S_{\epsilon/2}[x]$ together with the family of the balls $S_\epsilon[x_m]$ $(m = 1, 2, ...)$ is an open covering of X. This covering does not contain any finite subcovering because x_m is not covered by any $S_\epsilon[x_n]$ $(m \neq n)$ nor by any of the balls $S_{\epsilon/2}[x]$ which are included in the cover. Hence if the proposition stated in the beginning of the proof does not hold, then X is not countably compact.

Now we determine for every $\epsilon = 1/n$ $(n = 1, 2, ...)$ finitely many points $x_1^n, ..., x_{m_n}^n$ such that the balls $S_{1/n}[x_k^n]$ $(k = 1, ..., m_n)$ cover X. These balls form a countable family of open sets. We show that it is a base: Given an open set O and a point $x \in O$ there is an $\epsilon > 0$ such that $x \in S_\epsilon[x] \subseteq O$ and for any $n = 1, 2, ...$ there is a point x_k^n with the property that $x \in S_{1/n}[x_k^n]$. If n is so large that $2/n < \epsilon$, then $S_{1/n}[x_k^n] \subseteq S_\epsilon[x]$ and so x is covered by the base element $S_{1/n}[x_k^n]$ which is entirely contained in O.

We proved the following:

Theorem 1. *For pseudometric spaces compactness, sequential compactness, countable compactness, and the Bolzano-Weierstrass property are equivalent notions.*

Corollary. *Every compact subspace of a metric space is a closed bounded subset.*

(This can be obtained also without Theorem 1: Since every metric space is a Hausdorff space by Theorem 3.4 only closed subsets can be compact. The boundedness of compact subsets can be proved directly from the definition of compactness.)

The boundedness of the compact subspaces is not a statement about the open sets but it concerns the metric. Nevertheless, it is a topological property because it is a statement on *all metrics* compatible with the topology of the space. In particular, a compact space X has a finite diameter $d(X)$ relative to every metric d generating its topology. This is useful information and it shows for instance that finite-dimensional Euclidean spaces are not compact and that the only possible compact subspaces are closed bounded subsets. One of the classical results, called the Heine-Borel theorem, states that actually every such subset is a compact subspace:

Theorem 2. *A subspace of the m-dimensional Euclidean space E_m is compact if and only if it is closed and bounded.*

The sufficiency follows from the so called Bolzano-Weierstrass theorem:

Theorem 3. *Every bounded sequence (x_n) $(n = 1, 2, \ldots)$ of points x_n in the m-dimensional Euclidean space E_m contains a convergent subsequence.*

Proof. First we show the result when $m = 1$. In this case the proof proceeds according to an old but useful principle. We consider the intervals $-M \leqslant x \leqslant 0$ and $0 \leqslant x \leqslant M$ where $M > 0$ is chosen such that $| x_n | \leqslant M$ for every $n = 1, 2, \ldots$. At least one of these intervals contains an infinity of terms of the sequence (x_n). If $[-M, 0]$ contains infinitely many x_n's, we put $a_0 = -M$ and $b_0 = 0$. If it contains only a finite number of terms, then $[0, M]$ contains infinitely many x_n's. In this case we put $a_0 = 0$ and $b_0 = M$. In either case we define the interval $[a_0, b_0]$ such that $b_0 - a_0 = M$ and it contains infinitely many terms of (x_n).

Now we assume that the interval $[a_k, b_k]$ is constructed already such that $a_{k-1} \leqslant a_k < b_k \leqslant b_{k-1}$; $b_k - a_k = M \cdot 2^{-k}$ and it contains infinitely many terms x_n. Then we choose

$$a_{k+1} = a_k \quad \text{and} \quad b_{k+1} = \tfrac{1}{2}(a_k + b_k)$$

or

$$a_{k+1} = \tfrac{1}{2}(a_k + b_k) \quad \text{and} \quad b_{k+1} = b_k$$

according as $a_k \leqslant x \leqslant \frac{1}{2}(a_k + b_k)$ contains an infinity of x_n's or not. Hence in either case $a_k \leqslant a_{k+1} < b_{k+1} \leqslant b_k$; $b_{k+1} - a_{k+1} = M \cdot 2^{-(k+1)}$ and $[a_{k+1}, b_{k+1}]$ contains infinitely many x_n's. In this way we can define two monotone sequences $a_0 \leqslant a_1 \leqslant a_2 \leqslant \ldots$ and $b_0 \geqslant b_1 \geqslant b_2 \ldots$ such that $b_k - a_k = M \cdot 2^{-k}$ for every $k = 0, 1, 2, \ldots$ and $[a_k, b_k]$ contains an infinity of x_n's. Since $a_0 \leqslant a_k < b_k \leqslant b_0$ both sequences are bounded and hence convergent. We have $(b_k - a_k) \to 0$ and so $\lim a_k = \lim b_k$.

We select a term x_{n_1} from the interval $[a_1, b_1]$, and in general a term x_{n_k} of (x_n) $(n = 1, 2, \ldots)$ with $n_k > n_{k-1}$ from the interval $[a_k, b_k]$. Since each interval contains infinitely many terms this is possible. We obtain a subsequence (x_{n_k}) of (x_n) such that $a_k \leqslant x_{n_k} \leqslant b_k$ for every $k = 1, 2, \ldots$. Consequently the convergence of (a_k) and (b_k) to the same limit implies that (x_{n_k}) is a convergent subsequence of (x_n).

Now we generalize the result to E_m: Given a bounded sequence of points $x_n = (x_n{}^1, \ldots, x_n{}^m)$ in E_m the sequence of first coordinates $x_n{}^1$ is bounded and so we can select a subsequence of (x_n) such that the sequence of first coordinates is convergent. Similarly, starting from this subsequence we can select a second subsequence of (x_n) such that the sequence of the first and also the sequence of the second coordinates are convergent. In m steps we obtain a subsequence of the original sequence (x_n) such that the sequence of its kth coordinates is convergent for $k = 1, 2, \ldots, m$. Therefore (x_n) contains a convergent subsequence. This completes the proof of the Bolzano-Weierstrass theorem.

The method of subdividing an interval in a diadic fashion which is used in the beginning of the preceding proof is useful also elsewhere; for instance, in the proof of the intermediate value theorem. At present, however, we could have easily avoided this method by using instead the following simple result:

Lemma 2. *Every infinite sequence (x_n) $(n = 1, 2, \ldots)$ of elements x_n of a linearly ordered set X contains a monotonic subsequence.*

Proof. Suppose (x_n) $(n = 1, 2, \ldots)$ contains no infinite increasing subsequence. Then there is a first index k such that $x_{k+l} < x_k$ for every $l \geqslant 1$. We let $x_{n_1} = x_k$ and consider the terms $x_{n_1+1}, \ldots$. Again, since

(x_n) contains no increasing subsequence there is a first $k \geqslant 1$ such that $x_{n_1+k+l} \leqslant x_{n_1+k}$ for every $l \geqslant 1$. We let $x_{n_2} = x_{n_1+k}$ and consider the terms $x_{n_2+1}, \ldots$ to determine x_{n_3}, etc. The infinite subsequence $(x_{n_1}, x_{n_2}, \ldots)$ is decreasing.

EXERCISES

1. Show the following: If X is compact and pseudometrizable, then its diameter $d(X)$ is finite for every pseudometric d compatible with its topology.

[The space is covered by the union of the open balls $S_1[x]$ $(x \in X)$. Since X is compact it can be covered by a finite subcollection. Hence for any $x, y \in X$ we have $d(x, y) < 2 + \max\{d(x_i, x_j)\}$ where $x_1, \ldots, x_n$ denote the centers of the finitely many balls covering X.]

2. Let the real-valued function f be continuous in the closed interval $[a, b]$. Suppose that for every x in $[a, b)$ there is a y in $(x, b]$ such that $f(x) \leqslant f(y)$. Show that $f(a) \leqslant f(b)$.

[First assume that a y can be found for every x such that the strict inequality $f(x) < f(y)$ takes place. For every $x \in [a, b]$ we determine a y_x such that $f(x) < f(y_x)$. Since f is continuous there is an open subinterval I_x containing x and not containing b such that $f(\xi) < f(y_x)$ for every $\xi \in I_x$. Here open means open relative to the subspace $[a, b]$ so that I_a is a right half-open interval. We determine I_b such that $f(\xi) < f(b) + \epsilon$ for every $\xi \in I_b$. Since $[a, b]$ is compact there are finitely many intervals $I_a, I_{x_1}, \ldots, I_{x_n}, I_b$ covering $[a, b]$. Now determine a subsequence of these intervals such that

$$y_a \in I_{x_{k_1}}, \ldots, y_{x_{k_i}} \in I_{x_{k_{i+1}}}, \ldots, y_{x_{k_m}} \in I_b .$$

We see that

$$f(a) < f(y_a), \ldots, f(y_{x_{k_i}}) < f(y_{x_{k_{i+1}}}), \ldots, f(y_{x_{k_m}}) < f(b) + \epsilon.$$

Hence $f(a) \leqslant f(b)$.

The general case can be reduced to the present one by considering instead of f the function g_ϵ: $g_\epsilon(x) = f(x) + \epsilon(x - a)$.]

3. Prove: A set S on the real line is a compact subspace of the reals under the usual topology if and only if every real-valued function defined and continuous on S has a maximum.

[First let S be compact. Determine an interval I_x for every $x \in S$ such that $f(\xi) < f(x) + 1$ for every $\xi \in I_x \cap S$. Since the open cover $\{I_x\}$ $(x \in S)$ contains a finite subcover it follows that f is bounded. Choose the points x_1, x_2, ... such that $f(x_n) > (\text{lub } f) - (1/n)$. If x is an accumulation point of (x_n) $(n = 1, 2, ...)$, then $f(x)$ is the maximum value of f. Next suppose that S is not compact. If S is not bounded, define f by $f(x) = x$ and f will not have a maximum. If S is bounded, then it cannot be closed. Let x_0 be a boundary point of S which does not belong to S. Define f by $f(x) = 1/(x - x_0)$ for every $x \in S$. Then f is continuous but not bounded.]

4. Show that a set in a finite-dimensional real or complex normed vector space is compact if and only if it is closed and bounded.

(The necessity follows from Exercise 1 and Theorem 1. The sufficiency argument is similar to the proof of the sufficiency in Theorem 3.)

5. The Hilbert cube is the set $H = \{x\colon x = (x_n)$ *and* $|x_n| \leqslant 1/n$ *for every* $n = 1, 2, ...\}$ metrized by $d(x, y) = \Sigma |x_n - y_n|^2$. Show that H is a compact space.

[We use Theorem 1: Let (x^n) be a sequence of points in H. The sequence of first coordinates $(x_1{}^n)$ has an accumulation point x_1. We select a subsequence $({}^1x^n)$ of (x^n) such that $|{}^1x_1{}^n - x_1| < 1/n$ for every $n = 1, 2, ...$. The sequence of second coordinates $({}^1x_2{}^n)$ of $({}^1x^n)$ has an accumulation point x_2. Determine the subsequence $({}^2x^n)$ of $({}^1x^n)$ such that $|{}^2x_2{}^n - x_2| < 1/n$ for every $n = 1, 2, ...$. Using induction determine the subsequence $({}^mx^n)$ and the real numbers x_k such that $|{}^mx_k{}^n - x_k| < 1/n$ for every $n = 1, 2, ...$ and for every $k = 1, ..., m$. The point $x = (x_k)$ belongs to H and is an accumulation point of the sequence (x^n).]

6. Let X be a hereditary Lindelöf space and let S be a noncountable closed set in X. Show that S can be decomposed as $S = P \cup Q$ where P is a *perfect set*, i.e., $P = P'$, and Q is countable. Characterize further the points of P.

(Call a point $x \in X$ on ω-accumulation point of S if $N_x \cap S$ is uncountable for every neighborhood N_x. Let P be the set of ω-accumulation points of S and let $Q = S - P$. By the hereditary Lindelöf property the set Q is countable. Every $p \in P$ is an ω-accumulation point of S and so also of P because Q is countable. Hence $P \subseteq P'$. We show that P is closed: If $x \in \bar{P}$, then $O_x \cap P$ is not void; say $p \in O_x \cap P$. Then O_x is a neighborhood of p and as such it contains noncountably many elements of P. Hence x is an ω-accumulation point of S and so $x \in P$ and $\bar{P} \subseteq P$.)

7. There are uniformizable Lindelöf spaces X such that not every subspace of X is a Lindelöf space.

[Let S be the set of ordered pairs $p = (x, y)$ of real numbers $0 \leqslant x, y \leqslant 1$. We order S by the ordering relation $p = (x, y) < q = (u, v)$ if $x < u$ or if $x = u$ and $y < v$. Under this ordering S becomes a densely and linearly ordered set. It has the least upper bound property and contains a smallest and a largest element. By Theorem 1.5 the interval topology of S is compact. By Exercise II.9.4 and Theorem II.10.1 S is uniformizable. However, the space is not a hereditary Lindelöf space. For instance, the open intervals $((x, 0), (x, 1))$ $(0 \leqslant x \leqslant 1)$ form a noncountable family of mutually disjoint open sets of S.]

8. Use the Heine-Borel theorem to show that the space described in the preceding exercise is compact.

[Suppose that S is covered by a family of open intervals. For every x $(0 \leqslant x \leqslant 1)$ there is an interval in the family which covers $(x, 0)$ and there is another which covers $(x, 1)$. The union of these contains for a sufficiently small positive value of ϵ the set

$$((x - \epsilon, 0), (x + \epsilon, 1)) \cap c[(x, \epsilon), (x, 1 - \epsilon)].$$

The set $[(x, \epsilon), (x, 1 - \epsilon)]$ is homeomorphic to the closed interval $[\epsilon, 1 - \epsilon]$ and so it is covered by finitely many open intervals of the given family. If $x = 0$ or $x = 1$, slight modifications are needed. Hence for every x $(0 \leqslant x \leqslant 1)$ there is an $\epsilon > 0$ and finitely many open intervals such that their union covers $((x - \epsilon, 0), (x + \epsilon, 1))$. By the compactness of the closed interval $[0, 1]$ or by the compactness of S under the usual topology of the plane, it follows that finitely many of these intervals give already a cover of S.]

3. Subspaces and Separation Properties of Compact Spaces

The first half of this section contains the basic results on subspaces of compact topological spaces and more generally results on compact subspaces of arbitrary topological spaces. In the second part we shall study the validity of separation axioms in compact spaces and give some criteria on connectedness.

Theorem 1. *If X is a compact topological space and Y is a closed set in X, then Y is a compact subspace of X.*

Note. A similar proposition holds for countably compact spaces and for Lindelöf spaces.

Proof. Let $\{O_i\}$ $(i \in I)$ be an open covering of Y. The set $O_0 = cY$ is open and so O_0 and the sets O_i $(i \in I)$ together form an open covering of X. By hypothesis there is a suitable finite subfamily $\{O_{i_j}\}$ $(j \in J)$ covering X. Since $O_0 \cap Y = \emptyset$ those open sets O_{i_j} for which $i_j \neq 0$ form a finite open cover of Y.

Theorem 2. *If X is a topological space such that every open subset of X is a Lindelöf space, then every subset is a Lindelöf space.*

Proof. Let S be an arbitrary subset of X and let $\{O_i\}$ $(i \in I)$ be a cover of S. Then the family $\{O_i\}$ $(i \in I)$ is an open cover of the open set $\bigcup\{O_i: i \in I\}$. Hence by hypothesis there is a countable subfamily $\{O_{i_j}\}$ $(j \in J)$ which covers $\bigcup\{O_i: i \in I\}$. This subfamily is also a cover of the set S.

Theorem 3. *Let X be a topological space and let $\{Y_s\}$ be a finite family of subsets of X. If every Y_s is compact, then $\bigcup Y_s$ is a compact subspace of X.*

Note. A similar proposition holds for countable families of Lindelöf spaces.

Proof. If $\{O_i\}$ $(i \in I)$ is an open cover of $Y = \bigcup Y_s$, then it is an open cover of Y_s for every $s \in S$. Hence if Y_s is compact, then $\{O_i\}$ contains a finite subfamily which is a cover of Y_s. The union of these families is a finite subcover of Y.

Theorem 4. *A subset of a compact Hausdorff space is a compact subspace if and only if it is closed.*

Proof. The sufficiency is a consequence of Theorem 1. To prove the necessity let Y be a compact subspace of X and let $x \in cY$. We show the existence of an open set O_x such that $x \in O_x \subseteq cY$. This will prove that cY is open and Y is closed. Since X is a (T_2) space for every $y \in Y$ there exist disjoint open sets $O_x{}^y$ and O_y containing x and y, respectively. By varying y in Y we obtain an open cover $\{O_y\}$ $(y \in Y)$ of the compact space Y. Let the sets $O_{y_1}, \ldots, O_{y_n}$ form a finite subcover. An open set $O_x^{y_k}$ corresponds to each y_k such that $O_x^{y_k}$ and O_{y_k} are disjoint. Therefore $O_x = O_x^{y_1} \cap \ldots \cap O_x^{y_n}$ is open, it contains x and

$$O_x \cap Y \subseteq O_x \cap \bigcup O_{y_k} = \emptyset .$$

This shows that cY is open and so Y is closed.

Theorem 5. *If X is a compact (T_3) space, then axiom (T_4) also holds in X.*

Note. Notice that no other separation axiom is assumed.

Corollary. *Every compact (T_3) space is uniformizable.*

Proof. Let A and B be disjoint closed sets in X. The object is to find disjoint open sets O_A and O_B such that $A \subseteq O_A$ and $B \subseteq O_B$. By axiom (T_3), for each $b \in B$ there exist disjoint open sets O_A and O_b such that $A \subseteq O_A$ and $b \in O_b$. The family $\{O_b\}$ $(b \in B)$ is an open cover of B which by Theorem 1 is a compact subspace of X. Hence the open cover $\{O_b\}$ $(b \in B)$ contains a finite subcover, say $\{O_{b_1}, \ldots, O_{b_n}\}$. Let $O_A{}^k$ denote the open set which corresponds to O_{b_k} $(k = 1, \ldots, n)$. We define $O_A = O_A{}^1 \cap \ldots \cap O_A{}^n$ and $O_B = O_{b_1} \cup \ldots \cup O_{b_n}$. These are disjoint open sets, $A \subseteq O_A$ and $B \subseteq O_B$, and so X is a (T_4) space. The corollary follows from the corollary of Theorem II.10.1.

Theorem 6. *Every compact Hausdorff space is normal.*

Corollary. *Every compact Hausdorff space is uniformizable.*

Proof. In view of the preceding theorem it is sufficient to show that every compact Hausdorff space X is a (T_3) space. The proof of this fact is identical with the proof of Theorem 5 with the set B replaced by a point b.

The reasoning used in the proof of Theorem 5 can be improved so that it leads to a much stronger result which was first obtained by Tychonoff:

Theorem 7. *If the Lindelöf space X is a* (T_3) *space, then axiom* (T_4) *also holds in X.*

Note. This result can be still further strengthened: X is a *fully* (T_4) *space.* (See Theorems 6.2 and 7.3.)

Corollary. *Every* (T_3) *Lindelöf space is uniformizable.*

Proof. Let A and B be disjoint closed sets in X. By Theorem 1, A and B are also Lindelöf spaces. Since X is a (T_3) space we can find for each $b \in B$ an open set O_b such that $\bar{O}_b \cap A = \emptyset$. The family $\{O_b\}$ $(b \in B)$ is an open cover of B and so there is a countable subcover, say $\{O_{b_j}\}$ $(j = 1, 2, \ldots)$. We can determine in a similar way a countable open cover $\{O_{a_i}\}$ $(i = 1, 2, \ldots)$ of A such that $\bar{O}_{a_i} \cap B = \emptyset$ for every i. We define for every positive integer k the open sets $O_A{}^k = O_{a_k} \cap c(\bar{O}_{b_1} \cup \ldots \cup \bar{O}_{b_k})$ and $O_B{}^k = O_{b_k} \cap c(\bar{O}_{a_1} \cup \ldots \cup \bar{O}_{a_k})$. Then $O_A{}^k$ and $O_B{}^l$ are disjoint for every choice of the indices k and l because for instance if $k \leqslant l$ then $O_A{}^k \cap O_B{}^l \subseteq O_{a_k} \cap c(\bar{O}_{a_1} \cup \ldots \cup \bar{O}_{a_l}) \subseteq O_{a_k} \cap c(O_{a_k}) = \emptyset$. Therefore the open sets $O_A = \bigcup O_A{}^k$ and $O_B = \bigcup O_B{}^l$ are disjoint; moreover, $A \subseteq O_A$ and $B \subseteq O_B$ because $c(\bar{O}_{b_1} \cup \ldots \cup \bar{O}_{b_k}) \supseteq A$ and similarly $c(\bar{O}_{a_1} \cup \ldots \cup \bar{O}_{a_k}) \supseteq B$. This shows that X is a (T_4) space.

The hypothesis of Theorem 5 can be relaxed by replacing compactness by a weaker notion called paracompactness. Since Lindelöf spaces satisfying the (T_3) axiom are paracompact, Theorem 7 is a corollary of this extended form of Theorem 5. (See Theorem 6.2.)

The next theorem concerns the corollary of Theorem 5:

Theorem 8. *There is at most one uniform structure which is compatible with a compact topology.*

Note. By Theorem 6 *every compact Hausdorff space is uniformizable.* However, if X is only a compact (T_0) or (T_1) space, then it cannot be uniformized.

Proof. We prove that if such a structure exists, then it consists of the neighborhoods N_I of the diagonal I of $X \times X$. Suppose that $\mathcal{U}$ is a uniform structure for the compact space X. First given $U \in \mathcal{U}$ we can determine a symmetric $V \in \mathcal{U}$ such that $V \circ V \subseteq U$. We set $O_x = V[x]^i$ so that $O_x \times O_x \subseteq V[x] \times V[x] \subseteq V \circ V \subseteq U$ for every $x \in X$. Clearly $\bigcup(O_x \times O_x)$ is open in $X \times X$ and of course $I \subseteq \bigcup(O_x \times O_x) \subseteq U$. Hence every $U \in \mathcal{U}$ majorizes some neighborhood of I.

Next let a neighborhood N_I be given. For every $x \in X$ there is a U_x in $\mathcal{U}$ with the property that $U_x[x] \times U_x[x] \subseteq N_I$. If we choose a symmetric $V_x \in \mathcal{U}$ such that $V_x \circ V_x \subseteq U_x$, then $\{V_x[x]^i\}$ $(x \in X)$ is an open cover of X and so there is a finite subcover

$$\{V_{x_1}[x_1]^i, \dots, V_{x_n}[x_n]^i\},$$

say. Let $V = V_{x_1} \cap \dots \cap V_{x_n}$. If $(x, y) \in V$, then $(x_k, x) \in V_{x_k}$ for some index k and so by $V \subseteq V_{x_k}$ we have $(x_k, y) \in V_{x_k} \circ V_{x_k} \subseteq U_{x_k}$. Since $V_{x_k} \subseteq U_{x_k}$ we also have $(x_k, x) \in U_{x_k}$. Therefore

$$(x, y) \in U_{x_k}[x_k] \times U_{x_k}[x_k] \subseteq N_I$$

and so $V \subseteq N_I$. This shows that every neighborhood N_I majorizes some set $V \in \mathcal{U}$.

There are also noncompact spaces such that only one uniform structure is compatible with the topology of the space.

Lemma 1. *Let X be a uniform space with uniform structure $\mathcal{U}$. If A is compact and B is closed and disjoint from A, then there is a symmetric uniformity U in $\mathcal{U}$ such that $(a, b) \notin U$ for every $a \in A$ and $b \in B$.*

Proof. Since cB is open and $A \subseteq cB$, for every $a \in A$ there is a uniformity $U_a \in \mathcal{U}$ such that $U_a[a] \subseteq cB$. We choose a symmetric $V_a \in \mathcal{U}$ such that $V_a \circ V_a \subseteq U_a$. The open sets $V_a[a]^i$ $(a \in A)$ cover the compact set A and so there is a finite subcover, say $V_{a_1}[a_1]^i, \ldots, V_{a_n}[a_n]^i$. If we let $U = V_{a_1} \cap \ldots \cap V_{a_n}$, then $(a, b) \notin U$ for every $a \in A$ and $b \in B$: For $a \in V_{a_k}[a_k]^i$ for some k and so $(a_k, a) \in V_{a_k}$. This shows that $(a, b) \notin V_{a_k}$ for every $b \in B$ because $V_{a_k} \circ V_{a_k} \subseteq U_{a_k}$ and

$$U_{a_k}[a_k] \cap B = \emptyset .$$

Since $(a, b) \notin V_{a_k}$ of course $(a, b) \notin U$ and so U satisfies the requirements.

Lemma 2. *Let X be a pseudometric space with pseudometric d. If $A \subseteq X$ is compact and B is closed and disjoint from A, then $d(A, B) > 0$.*

Proof. We can either use the foregoing lemma or give a direct proof based on Theorem 2.1: Suppose that there are points $a_k \in A$ and $b_k \in B$ for every $k = 1, 2, \ldots$ so that $d(a_k, b_k) < 1/k$. Since A is compact the infinite sequence (a_k) $(k = 1, 2, \ldots)$ has an accumulation point $a \in A$. Hence for every $\epsilon > 0$ there are infinitely many indices k such that $d(a, a_k) < \epsilon/2$ and so by the triangle inequality

$$d(a, b_k) < (\epsilon/2) + (1/k) \leqslant \epsilon$$

for k sufficiently large. Therefore a is an accumulation point of B, so either B is not closed or A and B intersect.

If both A and B are compact, then the lemma can be further sharpened:

Lemma 3. *If X is a metric space with metric d and A, B are disjoint compact sets, then there are points $a \in A$ and $b \in B$ such that*

$$d(a, b) = d(A, B).$$

Note. Since $A \cap B = \emptyset$ and d is a proper metric $d(a, b) > 0$, so $d(A, B) > 0$.

Proof. There are points $a_k \in A$ and $b_k \in B$ for every $k = 1, 2, \ldots$ such that $d(a_k, b_k) < d(A, B) + (1/k)$. Using the same reasoning as in the proof of Lemma 2 we can show the existence of a point $a \in A$ such that for every $\epsilon > 0$ there is some point b_ϵ satisfying $d(a, b_\epsilon) < d(A, B) + \epsilon$. Choosing $\epsilon = 1/k$ for $k = 1, 2, \ldots$ we can find another sequence of points $b_k \in B$ such that $d(a, b_k) < d(A, B) + (1/k)$. The sequence (b_k) has an accumulation point $b \in B$ and for this point we have $d(a, b) \leqslant d(A, B)$. Since $d(a, b) \geqslant d(A, B)$ always holds, we have equality.

Theorem 9. *A compact uniform space X is connected if and only if any pair of points $a, b \in X$ can be connected by a U-chain for every symmetric U in a uniform structure $\mathscr{U}$ for X.*

Proof. The necessity of this condition was proved earlier and it is stated in Theorem II.7.3. To see the sufficiency suppose that $X = A \cup B$ where A and B are disjoint closed sets. Since X is compact so are the closed sets A and B. Hence by Lemma 1 there is a $U \in \mathscr{U}$ such that $(a, b) \notin U$ for every $a \in A$ and $b \in B$. It follows that a sequence $x_1 = a$, $x_2, ..., x_n$ satisfying $(x_i, x_{i+1}) \in U$ cannot leave the set A and so $a \in A$ cannot be connected by a U-chain with points of B.

Corollary. *A compact pseudometric space is connected if and only if any pair of points can be connected by a finite ϵ-chain for every $\epsilon > 0$.*

The *finite ϵ-chain condition* can be used to show that certain sets which often occur in practice are connected. For instance, it is plain that a finite closed interval of the real line satisfies the finite ϵ-chain condition and so it is connected. More generally a *closed n-dimensional interval* or *closed n-cell* is connected. For the points $(x_1, ..., x_i, ..., x_n)$ and $(x_1, ..., y_i, ..., x_n)$ can be connected by the ϵ-chain of points

$$(x_1, ..., \xi_i^k, ..., x_n) \qquad (k = 1, ..., m)$$

where $\xi_i^k = x_i + (k/m)(y_i - x_i)$ and $m > (1/\epsilon)(y_i - x_i)$. Hence any pair of points $x = (x_1, ..., x_n)$ and $y = (y_1, ..., y_n)$ of an n-cell $\{x : a_i \leqslant x_i \leqslant b_i \text{ for } i = 1, ..., n\}$ can be connected by an ϵ-chain. Similarly, we can easily show that every closed ball in the n-dimensional Euclidean space is connected.

EXERCISES

1. Give a simple example of a compact (T_1) space which is not a Hausdorff space.

2. Prove Theorem 5 by using axiom (T_4) in the form as is stated in Theorem II.3.2.

(Let A be a closed subset of the open set O. For every $a \in A$ determine O_a such that $a \in O_a \subseteq \bar{O}_a \subseteq O$. Let $O_{a_1}, ..., O_{a_n}$ be a cover of A and let $O_A = O_{a_1} \cup ... \cup O_{a_n}$. Then $A \subseteq O_A$ and

$$\bar{O}_A = \bar{O}_{a_1} \cup ... \cup \bar{O}_{a_n} \subseteq O.)$$

3. Show that if all subsets of a Hausdorff space are compact subspaces, then its topology is discrete.

4. Let $\mathcal{T}_1$ be a compact Hausdorff topology on a set X. Show that (i) if $\mathcal{T}_2$ is a compact topology on X and $\mathcal{T}_1 \leqslant \mathcal{T}_2$, then $\mathcal{T}_1 = \mathcal{T}_2$ and (ii) if $\mathcal{T}_2$ is a Hausdorff topology on X and $\mathcal{T}_2 \leqslant \mathcal{T}_1$, then $\mathcal{T}_1 = \mathcal{T}_2$.

[Let C_2 be closed relative $\mathcal{T}_2$. By Theorem 1 it is compact relative to $\mathcal{T}_2$. By Exercise 1.2 it is compact relative to the Hausdorff topology $\mathcal{T}_1$ and so by Theorem 4 it is closed relative to $\mathcal{T}_1$. This proves (i). If C_1 is closed relative to $\mathcal{T}_1$, then it is compact relative $\mathcal{T}_1$ and so by Exercise 1.2 it is compact also relative to $\mathcal{T}_2$. Hence by Theorem 4 the set C_1 is closed relative to $\mathcal{T}_2$.]

5. Prove that every compact pseudometric space satisfies the second countability axiom.

(Consider for every $n = 1, 2, \ldots$ a finite cover of X by balls $S_{1/n}[x_{mn}]$. These form a denumerable family $\mathcal{B}$ of open sets which is a base.)

6. Let X be a (T_4) space and let $\{O_1, \ldots, O_n\}$ be a finite open cover of X. Show the existence of a finite open cover $\{Q\}$ having the following property: For every $p \in X$ there is an O_i such that

$$p^* = \bigcup\{Q : p \in Q\} \subseteq O_i .$$

(The set p^* is called the *star* of the point p. By Theorem II.4.1 there is an open cover $\{Q_1, \ldots, Q_n\}$ such that $\bar{Q}_i \subseteq O_i$. The family

$$\{O_{i_1} \cap \ldots \cap O_{i_m} \cap c\bar{Q}_{i_{m+1}} \cap \ldots \cap c\bar{Q}_{i_n}\}$$

is a finite open covering of X. Let $p \in X$. Then $p \in \bar{Q}_i$ for some i. If $p \in Q = O_{i_1} \cap \ldots \cap O_{i_m} \cap c\bar{Q}_{i_{m+1}} \cap \ldots \cap c\bar{Q}_{i_n}$, then $Q \subseteq O_i$.)

7. Let X be a compact (T_3) space and let for every finite open cover $\{O_i\}$ of X the set $S(\{O_i\}) \subseteq X \times X$ be defined as

$$S(\{O_i\}) = \{(x, y) : x, y \in O_i \quad \textit{for some} \quad O_i \in \{O_i\}\}.$$

Using the result stated in the preceding exercise prove that

(i) the family $\mathcal{U}_B$ of all sets $S(\{O_i\})$ is a structure base for a uniform structure, and

(ii) the topology associated with $\mathcal{U}_B$ is the given topology of X.

[Axiom (Ub.3) follows from the relation

$$S(\{O_i\}) \cap S(\{Q_j\}) = S(\{O_i \cap Q_j\}) .$$

Axioms (Ub.4) and (Ub.5) obviously hold. To show axiom (Ub.6) determine for a given $\{O_i\}$ a finite open cover $\{Q_j\}$ such that $x^* \subseteq O_i$ for every $x \in X$ where the star of x is taken relative to the finite cover and a suitable O_i $\{Q_j\}$. Then $S(\{Q_j\}) \circ S(\{Q_j\}) \subseteq S(\{O_i\})$. The rest follows from axiom (T_3).]

4. The Product of Compact Topological Spaces

The object of this section is to find a condition which assures the compactness of a product of topological spaces in the product topology. The answer to this problem is given by a theorem of Tychonoff which gives a necessary and sufficient condition that a product space be compact. First we prove the necessity of Tychonoff's condition:

Lemma 1. *If $X = \Pi X_s$ is compact, then every factor X_s is compact.*

Proof. Let $\{O_i\}$ $(i \in I)$ be a collection of open sets in the space X_s. Then the collection of cylinders $Z(s, O_i) = \{x \colon x \in X \text{ and } x_s \in O_i\}$ is a family of open sets in X, and if $\{O_i\}$ $(i \in I)$ is a cover of X_s, then $\{Z(s, O_i)\}$ covers X. Hence if X is compact, then there is a finite subfamily $\{Z(s, O_{i_j})\}$ $(j \in J)$ which covers X. Therefore $\bigcup O_{i_j} = X_s$ and so X_s is compact.

Now we are going to prove that if the number of factor spaces X_s is finite, then the compactness of each X_s implies the compactness of the product. It will be sufficient to consider the case of two factors:

Theorem 1. *The product of two compact spaces X, Y is compact.*

Proof. Let $\{O_i\}$ $(i \in I)$ be an open cover of $X \times Y$. For every point $(x, y) \in X \times Y$ there are open sets $O_x{}^y \subseteq X$ and $O_y{}^x \subseteq Y$ such that $(x, y) \in O_x{}^y \times O_y{}^x \subseteq O_i$ for some O_i containing (x, y). Hence it is sufficient to show that the family of rectangles $O_x{}^y \times O_y{}^x$ ($x \in X$ and $y \in Y$) contains a finite subfamily which is a cover of $X \times Y$. For any fixed $y \in Y$ the sets $O_x{}^y$ $(x \in X)$ form an open cover of the compact space X, so there is a finite subcover consisting, say, of the sets $O^y_{x_1}, \ldots, O^y_{x_m}$. Let us consider the sets $O^{x_1}_y, \ldots, O^{x_m}_y$. Since $y \in O^{x_k}_y$ for every index k, the intersection $Q_y = O^{x_1}_y \cap \ldots \cap O^{x_m}_y$ is a nonvoid open set in Y which contains y. Therefore the finite family $\{O^y_{x_k} \times Q_y\}$ $(k = 1, \ldots, m)$ covers all points of the set $\{(\xi, y) : \xi \in X\}$.

Now we let y vary over Y and consider the family $\{Q_y\}$ $(y \in Y)$ which is an open cover of the compact space Y. There is a finite subfamily $\{Q_{y_l}\}$ $(l = 1, \ldots, n)$ which is a cover of Y. The union of the sets $O^{y_l}_{x_k} \times Q_{y_l}$ (l fixed and $k = 1, \ldots, m(y_l)$) covers $X \times Q_{y_l}$ so that the family $\{O^{y_l}_{x_k} \times Q_{y_l}\}$ $(k = 1, \ldots, m(y_l)$ and $l = 1, \ldots, n)$ is a finite covering of

$X \times Y$. Since $Q_y \subseteq O_{y^k}^{x_k}$, the family $\{O_x{}^y \times O_y{}^x\}$ has a finite subcover of $X \times Y$.

The preceding theorem implies that the product of finitely many compact topological spaces is compact. This is a special case of Tychonoff's theorem. Similar reasoning shows that the product of a compact space and of a Lindelöf space is always a Lindelöf space.

Example. There are Lindelöf spaces X such that $X \times X$ is not a Lindelöf space. As a matter of fact, there are examples in which X is a hereditary Lindelöf space: For instance, let X be the set of reals ordered as usual and let the topology on X be the right half-open interval topology $\mathscr{T}_+$. First we show that X is a hereditary Lindelöf space: Given any family $\{O_i\}$ $(i \in I)$ of open sets we can choose a half-open interval $I_x : x \leqslant \xi < x + \delta_x$ for every x in $O = \bigcup O_i$ such that $I_x \subseteq O_i$ for some $i \in I$. The family $\{I_x\}$ $(x \in O)$ is an open cover of O so it is sufficient to show that there exists a countable subfamily $\{I_{x_j}\}$ in $\{I_x\}$ such that $\bigcup I_{x_j} = \bigcup I_x$.

We fix a positive integer n and consider the subfamily $\{I_x{}^n\}$ of those intervals $I_x{}^n \in \{I_x\}$ for which $\delta_x > 1/n$. If $\xi \in I^n = \bigcup I_x{}^n$ but is not an interior point of any $I_x{}^n$ in the usual sense, then there is an interval of length at least $1/n$ situated at the left of ξ and containing no points x with $\delta_x > 1/n$. Hence the set S of such points ξ is countable and can be covered by the countable family $\{I_\xi{}^n\}$ $(\xi \in S)$. If we omit the left end point of all remaining intervals $I_x{}^n$ we obtain a family of intervals $J_x{}^n$ which are open in the usual sense. Since the real line is a hereditary Lindelöf space relative to its usual topology the open set $J^n = \bigcup J_x{}^n$ can be covered by a countable subfamily of $\{J_x{}^n\}$. Thus I^n has a countable covering by sets of the family $\{I_x{}^n\}$. But $\bigcup I_x = \bigcup I^n$ and so $O = \bigcup I_x$ is also covered by a countable subfamily of $\{I_x\}$ and hence by a countable subfamily of $\{O_i\}$ $(i \in I)$.

Now we consider the set $X \times X$ in the product topology. In order to show that $X \times X$ is not a Lindelöf space, by Theorem 3.1 it is sufficient to find a closed subset S of $X \times X$ which is not Lindelöf. The diagonal set $S = \{(x, -x) : x \in X\}$ is such a closed set: The sets $\{(\xi, \eta) : \xi \geqslant x$ *and* $\eta \geqslant -x\}$ for $x \in X$ form an open cover of S which has no countable subcover since each open set of the cover contains only one point of the uncountable set S. Therefore $X \times X$ is not a Lindelöf space.

The following theorem on the compactness of an unrestricted product of compact spaces is due to Tychonoff. All proofs of the sufficiency of Tychonoff's condition use the axiom of choice and it is known that the theorem can hold only in those set theories in which the axiom of choice is valid.

Theorem 2. *The product* $X = \Pi X_s$ *of an arbitrary family* $\{X_s\}$ $(s \in S)$ *of compact topological spaces is compact.*

There are several proofs; the one which is given below is based on a very remarkable compactness criterion which was discovered by Alexander. It is both a necessary and sufficient condition but it is the sufficiency which is important:

Theorem 3. *If a topological space* X *has a subbase* $\mathscr{S}$ *such that from every covering of* X *by elements of* $\mathscr{S}$ *one can select a finite subcovering, then* X *is a compact.*

Proof. Let $\mathscr{A}$ be a family of open sets such that no finite subfamily of $\mathscr{A}$ is a cover of X. The object is to show that $\mathscr{A}$ itself is not a cover of X. We consider families $\mathscr{Q}$ of open sets Q having the property that $\mathscr{A} \subseteq \mathscr{Q}$ and no finite subfamily of $\mathscr{Q}$ covers the space X. Let $\mathfrak{A}$ be the set of all such open families $\mathscr{Q}$. Since $\mathscr{A} \in \mathfrak{A}$, the set $\mathfrak{A}$ is not void.

If we order $\mathfrak{A}$ by inclusion, then the hypotheses of Zorn's lemma are satisfied: Indeed, if $\mathfrak{L}$ is a linearly ordered subset of $\mathfrak{A}$, then the family $\mathscr{U}$ consisting of all open sets $Q \in \mathscr{Q} \in \mathfrak{L}$ is an upper bound of $\mathfrak{L}$ and no finite subfamily of $\mathscr{U}$ covers the space X. Therefore by Zorn's lemma there exists a maximal family $\mathscr{M}$ in $\mathfrak{A}$. This maximal family has the following property:

If $M \in \mathscr{M}$ *and if* $Q_1, \ldots, Q_n$ *are open sets such that* $Q_1 \cap \ldots \cap Q_n \subseteq M$, *then at least one of the sets* Q_k *belongs to* $\mathscr{M}$.

For if Q_k is an open set and $Q_k \notin \mathscr{M}$, then $\mathscr{M}$ being maximal there exist suitable elements $M_1{}^k, \ldots, M_{m_k}^k$ in $\mathscr{M}$ such that

$$Q_k \cup M_1{}^k \cup \ldots \cup M_{m_k}^k = X .$$

Hence if $Q_1, \ldots, Q_n \notin \mathscr{M}$, then

$$(Q_1 \cap \ldots \cap Q_n) \cup (\bigcup_{i,k} M_i{}^k) = X .$$

Therefore by $Q_1 \cap \ldots \cap Q_n \subseteq M$, also $M \notin \mathscr{M}$. This proves the foregoing statement.

Now we consider the family $\mathscr{S} \cap \mathscr{M}$ where $\mathscr{S}$ is the specific subbase mentioned in the theorem. If a point x is covered by a set $M \in \mathscr{M}$, then $\mathscr{S}$ being a subbase there are sets $S_1, \ldots, S_n \in \mathscr{S}$ such that

$$x \in S_1 \cap \ldots \cap S_n \subseteq M .$$

Since $S_1, \ldots, S_n$ are open sets and $S_1 \cap \ldots \cap S_n \subseteq M \in \mathscr{M}$, by the preceding statement we have $S_i \in \mathscr{M}$ for some index i and so

$$x \in S_i \in \mathscr{M} .$$

Hence if a point x is covered by the family $\mathcal{M}$, then it is also covered by $\mathcal{S} \cap \mathcal{M}$. However, $\mathcal{M}$ does not contain any finite subcover and so $\mathcal{S} \cap \mathcal{M}$ cannot contain any finite subcover. Since $\mathcal{S} \cap \mathcal{M}$ is a subfamily of $\mathcal{S}$ and every cover by elements of $\mathcal{S}$ contains a finite subcover we see that $\mathcal{S} \cap \mathcal{M}$ is not a cover of X. Therefore $\mathcal{M}$ is not a cover and finally by $\mathcal{A} \subseteq \mathcal{M}$ we see that $\mathcal{A}$ is not a cover of X. This finishes the proof.

Proof of Theorem 2. It is sufficient to find a subbase $\mathcal{S}$ such that every cover of X by elements of $\mathcal{S}$ contains a finite subcover. Let

$$\mathcal{S} = \{Z(s, O_s)\} \qquad (s \in S \text{ and } O_s \in \mathcal{O}_s),$$

i.e., let $\mathcal{S}$ be the family of all cylinders in X with open base in X_s ($s \in S$). Suppose that $\mathcal{Z} \subseteq \mathcal{S}$ is a family such that $\mathcal{Z}$ has no finite subfamily covering the entire space X. We prove that $\mathcal{Z}$ itself is not a cover of X. For we fix an $s \in S$ and consider the sets O_s such that $Z(s, O_s) \in \mathcal{Z}$. These sets form a family $\{O_s\}$ which contains no finite subcover of X_s, since otherwise the corresponding cylinders $Z(s, O_s)$ would form a finite subcover of X. Consequently, by the compactness of X_s this family $\{O_s\}$ is not a cover of X_s and so we can choose a point $x_s \in X_s$ such that x_s is not covered by $\{O_s\}$. If we put $x = (x_s)$ ($s \in S$) where x_s is choosen for every $s \in S$ according to the preceding principle, then $x \in X$ is not covered by any set $Z(s, O_s)$ of $\mathcal{Z}$. Hence $\mathcal{Z}$ is not a cover of X and so by Alexander's theorem X is compact.

EXERCISES

1. Show that the torus is compact.
(Use Theorem 1.)

2. Modify the proof of Theorem 1 to obtain the following: *If X is compact and Y is countably compact, then $X \times Y$ is countably compact.*

3. Using the same reasoning as in the proof of Theorem 1 prove the following: *If A and B are compact sets in the topological spaces X and Y, respectively, and if O is an open set in $X \times Y$ containing $A \times B$, then there are open sets $O_A \subseteq X$ and $O_B \subseteq Y$ such that $A \subseteq O_A$, $B \subseteq O_B$, and $O_A \times O_B \subseteq O$.*

[For every $(x, y) \in A \times B$ there are open sets $O_x{}^y$ and $O_y{}^x$ such that $(x, y) \in O_x{}^y \times O_y{}^x \subseteq O$. Fix $y \in B$ and vary x in A. Select a finite subcover $O_{x_1}^y, \ldots, O_{x_m}^y$ of A. Define $Q_y = O_y^{x_1} \cap \ldots \cap O_y^{x_m}$ for every $y \in B$ and select a finite subcover of B. Define O_B to be the union of these sets $Q_{y_1}, \ldots, Q_{y_n}$. Put $O_A = \bigcap_k \bigcup_i O_{x_i}^{y_k}$.]

4. Show that the product of countably many countably compact spaces is countably compact.

[Let S be an infinite set of points $x \in X = \Pi X_n$ where $n = 1, 2, ...$ and let x_n denote the projection of x into X_n. Since X_n is countably compact there is a point $\xi_n \in X_n$ whose open neighborhoods O_{ξ_n} contain the projection x_n of infinitely many points $x \in S$. Let $\xi = (\xi_1, \xi_2, ...)$. Then ξ is 1-accumulation point of S.]

5. Let $X_n = \{1, 2, ...\}$ for every $n = 1, 2, ...$ and let X_n be topologized by the discrete topology. Determine all compact subspaces of $X = \Pi X_n$.

[A set $Y \subseteq X$ is compact if and only if it is closed and bounded in the following sense: There is a bound $m(n) > 0$ for every $n = 1, 2, ...$ such that $y_n \leqslant m(n)$ for every $y \in Y$. The sufficiency of this condition follows from Theorems I.11.1 and 2. The necessity can be seen by assuming that the set $\{y_n\}$ $(y \in Y)$ is not bounded for some index n and constructing a sequence of points $y \in Y$ having no accumulation point in X.]

5. Locally Compact Spaces

A topological space is called locally compact if every point has a neighborhood which is a compact subspace. It is clear that every compact space is locally compact. Less trivial examples of locally compact spaces are the finite-dimensional Euclidean spaces. For every closed bounded subset of E_n is compact and so for example the closed balls $\overline{S_\epsilon(x)}$ are compact neighborhoods of x. A locally compact space need not be a Hausdorff space.

Theorem 1. *If a Hausdorff space is locally compact, then it is regular.*

Proof. Let O be an open set in X and let $x \in O$. We show the existence of an open set Q_x such that $x \in Q_x \subseteq \bar{Q}_x \subseteq O$. Then by part (1) of Theorem II.2.2 it will follow that the space X is a (T_3) space and so it is regular. By the local compactness there is an open set O_x and a compact subspace C_x of X such that $x \in O_x \subseteq C_x$. Since X is a Hausdorff space the compact subspace C_x is a closed set of X. The set $O \cap O_x$ is an open subset of the compact Hausdorff space C_x. By Theorem 3.6 the space C_x is normal and therefore regular. Thus there is a set Q_x which is open in the subspace C_x and is such that $x \in Q_x \subseteq \bar{Q}_x \subseteq O \cap O_x$, where $\bar{Q}_x$ denotes the closure of Q_x relative to the topology of C_x. The set Q_x is open also in the space X because $Q_x = Q \cap C_x$ for some set Q which is open in X and so $Q_x = Q \cap (O \cap O_x)$. Hence it is sufficient to show that the closure of Q_x relative to X is a subset of O. However, C_x is a closed set in X and so the closure of Q_x relative to C_x and relative to X is the same set $\bar{Q}_x$.

Theorem 2. *Every closed subset of a locally compact topological space X is a locally compact subspace. If in addition X is a Hausdorff space, then every open set is also locally compact.*

Proof. Let C be a closed set in X and let $x \in C$. There exist an open set O_x and a compact set C_x with the property that $x \in O_x \subseteq C_x$. Then $C \cap C_x$ is a closed set in C_x and so by Theorem 3.1 it is a compact subspace of C_x. Consider $C \cap O_x$; this is an open set in C and

$$x \in C \cap O_x \subseteq C \cap C_x .$$

Hence C is locally compact.

Now let X be a locally compact Hausdorff space and let O be a nonvoid open set in X. According to Theorem 1 the space X is regular and so for every $x \in O$ there is an open set Q_x such that $x \in Q_x \subseteq \bar{Q}_x \subseteq O$. The closed set $\bar{Q}_x$ is a locally compact subspace and so there are sets O_x and C_x in $\bar{Q}_x$ such that $x \in O_x \subseteq C_x$. The set O_x is open in $\bar{Q}_x$ and C_x is a compact subspace of $\bar{Q}_x$ and so also a compact subspace of O. The set $O_x \cap Q_x$ is open in X because Q_x is an open set in X, and O_x being open in $\bar{Q}_x$ we have $O_x = Q \cap \bar{Q}_x$ for some open set Q of X. Hence $x \in O_x \cap Q_x \subseteq C_x$ where $O_x \cap Q_x$ is open in O and C_x is a compact subspace of O. This shows that O is locally compact.

Theorem 3. *Every locally compact Hausdorff space is completely regular and so it is uniformizable.*

Proof. The uniformizability follows immediately from Theorem II.10.1 and so it is sufficient to deal with the (T) property. In order to show that X is a (T) space let O be an open set in X and let $x \in O$. Since X is locally compact there exist an open set O_x and a compact set C_x such that $x \in O_x \subseteq C_x$. Moreover, X is a (T_3) space and so there is an open set Q_x such that $x \in Q_x \subseteq \bar{Q}_x \subseteq O$. Obviously we have $x \in O_x \cap Q_x \subseteq C_x \cap \bar{Q}_x \subseteq O$. The set $C_x \cap \bar{Q}_x$ is a closed set of the compact subspace C_x of X, thus by Theorem 3.1 it is compact. Since $C_x \cap \bar{Q}_x$ is a subspace of the (T_3) space X it is a (T_3) space and therefore by Theorems 3.5 and II.9.2 it is a (T) space.

The set $O_x \cap Q_x$ is open in X and so it is an open set of the subspace $C_x \cap \bar{Q}_x$. Since this space $C_x \cap \bar{Q}_x$ is a (T) space there is a diadic scale

(O_d) $(d \in D)$ such that $x \in O_d \subseteq O_x \cap Q_x$ for every $d \in D$. The sets O_d are open in the subspace $C_x \cap \bar{Q}_x$ but $O_x \cap Q_x$ being open in X we see that O_d is open in X. We can also show that the closure of O_d in $C_x \cap \bar{Q}_x$ is the same as its closure $\bar{O}_d$ in X. For C_x is a compact subspace of the (T_2) space X and so by Theorem 3.4 it is a closed set. Hence by Lemma I.10.2 the closure of O_d with respect to $C_x \cap \bar{Q}_x$ is $\bar{O}_d$. We proved that (O_d) $(d \in D)$ is a diadic scale not only in the subspace $C_x \cap \bar{Q}_x$ but also in X and $x \in O_d \subseteq O$. Hence X is completely regular.

The following lemma is often used:

Lemma 1. *If A is a compact set in a locally compact Hausdorff space, then there is an open set O such that its closure is compact and $A \subseteq O$.*

Note. If X is a uniform space with uniform structure $\mathscr{U}$, then given any $U \in \mathscr{U}$ the set O can be chosen such that $O \subseteq U[A]$. In particular, if X is a metric space, then there is an O contained in the ϵ-neighborhood of A.

Proof. We cover A by a family of open sets $\{O_x\}$ $(x \in A)$ such that $\bar{O}_x$ is compact for every $x \in A$. By the local compactness of X such a cover exists. Since A is compact there is a finite subcover, say, the union of the sets O_{x_i} $(i = 1, ..., n)$ covers A. Then $A \subseteq O = \bigcup O_{x_i}$ and $\bar{O} = \bigcup \bar{O}_{x_i}$ is compact by Theorem 3.3. If X is a uniform space and U is a uniformity, then the open sets O_x $(x \in A)$ can be all chosen such that $O_x \subseteq U[x]$. Then $\bigcup O_{x_i} \subseteq U[A]$ follows.

Theorem 4. *A product space is locally compact if and only if all but finitely many factors are compact and the remaining ones are locally compact.*

Corollary. *A finite product is locally compact if and only if every factor is locally compact.*

Proof. First we prove the sufficiency. Given $x \in X = \prod X_s$ we can find a compact neighborhood N_s of every projection x_s of x and we can assume that $N_s = X_s$ for all but finitely many indices. Then $\prod N_s$ is a neighborhood of x in X and by Theorems 4.2 and I.11.1 it is a compact subspace of X. Hence X is locally compact. The necessity of the conditions follows equally easily: Given $x_s \in X_s$ let x be a point in X such that its projection into X_s is x_s. If N is a compact neighborhood of x in X, then its projection N_s into X_s is a compact neighborhood of x_s in X_s. Hence every factor of X is locally compact. To see the other condition let $x \in X$ be arbitrary and let N be a compact neighborhood of x. Then the projection N_s is compact and by the definition of the product topology $N_s = X_s$ except for at most finitely many choices of s.

If X is locally compact, then it is the union of a family of compact subspaces whose interiors are nonvoid and cover X. If X is the union of countably many such sets, then X is called *σ-locally compact.* A space is called *σ-compact* if it is the union of countably many compact subspaces. For instance, the finite-dimensional Euclidean spaces are both σ-locally compact and σ-compact. A σ-compact space is of course always a Lindelöf space. Sometimes the notion of rim-compactness comes up: X is called *rim-compact* if each point of X has arbitrarily small neighborhoods whose boundaries are compact.

EXERCISES

1. Show that an infinite-dimensional Hilbert space is not locally compact.

2. Show that a normed vector space over the reals or over the complex field is locally compact if and only if its unit sphere is compact.

(First show by considering the product of the unit sphere and the closed interval [0, 1] that the unit ball is compact if the unit sphere is compact: For this purpose one needs Theorems 3.1 and 4.1. Next identify all points of the product whose second coordinate is 0. Use Exercise 1.3 and show that the quotient space is homeomorphic to the unit ball of X. The "only if" part follows from Theorem 3.1.)

3. Show the following: A normed vector space over the reals or over the complex numbers is locally compact if and only if it is finite dimensional.

[In view of the preceding exercise the necessity follows by proving that in the infinite-dimensional case the unit sphere is not compact. This can be seen by constructing a sequence (x_m) $(m = 1, 2, ...)$ of points such that $\| x_m \| = 1$ and $\| x_m - x_n \| > \frac{1}{2}$ for every $m < n$. The construction of such a sequence can be based on the following proposition: *If X is a normed vector space and Y is a proper complete linear subspace of X, then for every $\epsilon > 0$ there is an $x \in X$ such that $\| x \| = 1$ and $\| x - y \| > 1 - \epsilon$ for every $y \in Y$.* For let $z \notin Y$ and let $d = \text{glb}\{\| z - y \| : y \in Y\}$. Using the completeness of Y one can easily see that $d > 0$. If we choose $y \in Y$ such that $\| z - y \| < (1 + \epsilon)\, d$, then for every $\eta \in Y$ we have $\| z - (y + \eta) \| \geqslant d > \| z - y \| - \epsilon d$. Hence $x = (z - y)/\| z - y \|$ satisfies the requirements. The sufficiency follows from the compactness of the product space: If $x_1, ..., x_n$ is a basis of X, then the points $\lambda_1 x_1 + ... + \lambda_n x_n$ with $|\lambda_1| \leqslant 1, ..., |\lambda_n| \leqslant 1$ form a compact neighborhood of the origin.]

4. Every locally compact subspace of an infinite-dimensional normed vector space X over the field of real or complex numbers is nowhere dense in X.

(If Y is a compact set in X, then by the previous exercise Y contains finitely many linearly independent elements. Hence $\bar{Y}$ is a finite-dimensional subspace of X.)

5. Show that the convex hull of a finite set of points of a real normed vector space is compact.

[The convex hull of the set $\{x_1, ..., x_n\}$ is the set of all points $\lambda_1 x_1 + ... + \lambda_n x_n$ where $\lambda_1 + ... + \lambda_n = 1$ and $0 \leqslant \lambda_i \leqslant 1$ $(i = 1, ..., n)$. It follows that the convex hull is a closed bounded set of the finite-dimensional subspace generated by the elements $x_1, ..., x_n$.]

6. Give an example of a locally compact space X and an equivalence relation such that the quotient space is not locally compact.

(Let X be a locally compact Hausdorff space and let $x_1, ..., x_n, ...$ be a sequence of points without accumulation point in X. Identify all points of the sequence. For instance, X can be the space of reals and we may identify all integers.)

7. Show that a Hausdorff space is locally compact if and only if there is an open cover $\{O_i\}$ $(i \in I)$ of X such that $\bar{O}_i$ is compact for every $i \in I$.

8. Show that the product of finitely many σ-compact spaces is σ-compact.

[Let $X = \bigcup A_n$ and $Y = \bigcup B_n$ where A_n and B_n are compact for every $n = 1, 2, ...$. By Theorems 4.1 and I.11.1 the sets $A_m \times B_n$ are compact subspaces of $X \times Y$ and $X \times Y = \bigcup (A_m \times B_n)$.]

9. Show by an example that the product of denumerably many σ-compact spaces need not be σ-compact.

[Let $X_n = \{1, 2, ...\}$ for every $n = 1, 2, ...$ and let X_n be topologized by the discrete topology. The product $X = \prod X_n$ is not σ-compact. For by Exercise 4.5 a set $Y \subseteq X$ is compact if and only if it is closed and bounded. Let Y^k $(k = 1, 2, ...)$ be compact subspaces of X and let $m(n, k)$ be the bound of the nth coordinate of the points in Y^k. The point whose nth coordinate is $m(n, n) + 1$ for every $n = 1, 2, ...$ is not in $\bigcup Y^k$ and so $\bigcup Y^k$ is a proper subset of X.]

10. The unit sphere S of a normed real vector space X is called *strictly convex* if $\frac{1}{2}\| x + y \| < 1$ for every pair of distinct points $x, y \in S$. It is called *uniformly convex* if for every $\epsilon > 0$ there is a $\delta > 0$ such

that $x, y \in S$ and $\| x - y \| > \epsilon$ imply $\frac{1}{2}\| x + y \| < 1 - \delta$. Show that if X is locally compact and strictly convex, then it is uniformly convex.

[For every pair $x, y \in S$ satisfying $\| x - y \| \geqslant \epsilon$ there are open sets O_x and O_y in X and a positive $\delta(x, y)$ such that $\frac{1}{2}\| \xi + \eta \| < 1 - \delta(x, y)$ for every $\xi \in O_x$ and for every $\eta \in O_y$. The open sets $(O_x \times O_y) \cap (S \times S)$ and $\{(x, y) : x, y \in S \text{ and } \| x - y \| < \epsilon\}$ form an open cover of $S \times S$. By the compactness of the unit sphere S there is a finite subcover corresponding, say, to the points $(x_1, y_1), \ldots, (x_n, y_n)$. Let δ be the smallest of the positive numbers $\delta(x_1, y_1), \ldots, \delta(x_n, y_n)$. Now if $\xi, \eta \in S$ and $\| \xi - \eta \| > \epsilon$, then $(\xi, \eta) \in O_{x_i} \times O_{y_i} \cap S \times S$ for some $i\,(1 \leqslant i \leqslant n)$ and so $\frac{1}{2}\| \xi + \eta \| < 1 - \delta$.]

6. Paracompactness and Full-Normality

A system of sets $S_i\,(i \in I)$ of a topological space X is called a *locally finite system* if every point $x \in X$ has a neighborhood N_x which intersects only a finite number of the sets $S_i\,(i \in I)$. If $\bigcup S_i = X$ we speak about a *locally finite cover of* X. Usually one is interested only in locally finite covers consisting of open sets. The system $\{S_i\}\,(i \in I)$ is called *star-finite* if every S_i is intersected only by a finite number of sets of the system. It is called a *point-finite system* if every $x \in X$ belongs to at most finitely many of the sets $S_i\,(i \in I)$. A cover $\{Q_j\}\,(j \in J)$ of X is called a *refinement* of the cover $\{O_i\}\,(i \in I)$ if each Q_j is contained entirely in some O_i.

Definition 1. *A topological space X is called paracompact if every open cover of X has a locally finite open refinement.*

It is clear that every compact space is paracompact and it is easy to see that every closed subspace of a paracompact space is paracompact. Local compactness does not necessarily imply paracompactness. In the next section we shall prove a theorem on the equivalence of paracompactness and full-normality. This will immediately imply that every metric space is paracompact.

Here we prove some of the basic properties of paracompact spaces. The proofs depend on the following observation:

Lemma 1. *If $\{S_i\}\,(i \in I)$ is a locally finite system, then $\overline{\bigcup S_i} = \bigcup \bar{S}_i$.*

Proof. It is sufficient to show that $\overline{\bigcup S_i} \subseteq \bigcup \bar{S}_i$. If $x \in \overline{\bigcup S_i}$, then by the local-finiteness there is an open set O_x such that $O_x \cap S_i$ is not void only for finitely many indices, say, for $i = i_1, \ldots, i_n$. Since $x \in \overline{\bigcup S_i}$, every neighborhood N_x of x intersects $\bigcup S_i$. Therefore every N_x intersects $S_{i_1} \cup \dots \cup S_{i_n}$ and so

$$x \in \overline{S_{i_1} \cup \dots \cup S_{i_n}} = \bar{S}_{i_1} \cup \dots \cup \bar{S}_{i_n} \subseteq \bigcup \bar{S}_i .$$

Lemma 2. *Every paracompact Hausdorff space is regular.*

Proof. Let A be a closed set in X and let $x \notin A$. For every y $(y \neq x)$ determine a pair of disjoint open sets $O_x{}^y$ and O_y such that $x \in O_x{}^y$ and $y \in O_y$. Since A is closed $O_x{}^y$ can be chosen such that $O_x{}^y \subseteq cA$. The family $\{O_x{}^y, O_y\}$ $(y \in X$ and $y \neq x)$ is an open cover of X. Let the cover $\{Q_j\}$ $(j \in J)$ be a locally finite refinement of this open cover. If $Q_j \cap A \neq \emptyset$, then $Q_j \subseteq O_y$ for some $y \neq x$ because $O_x{}^y \subseteq cA$ for every $y \neq x$. Hence by $x \in O_x{}^y \subseteq c\bar{O}_y$ we get $x \notin \bar{Q}_j$. If we set $O_A = \bigcup\{Q_j : Q_j \cap A \neq \emptyset\}$, then $A \subseteq O_A$ and $x \notin \bar{O}_A$ because by Lemma 1 we have $\bar{O}_A = \bigcup\{\bar{Q}_j : Q_j \cap A \neq \emptyset\}$. The set A and the point x are separated by the open sets O_A and $O_x = c\bar{O}_A$.

Theorem 1. *Every paracompact* (T_3) *space satisfies axiom* (T_4).

Corollary. *Every paracompact Hausdorff space is normal.*

Proof. Let A be a closed set in an open set O. Using axiom (T_3) we determine for each $a \in A$ an open neighborhood O_a such that

$$a \in O_a \subseteq \bar{O}_a \subseteq O .$$

The sets O_a $(a \in A)$ and cA form an open cover of X. Let $\{Q_j\}$ $(j \in J)$ be a locally finite refinement. If $A \cap Q_j \neq \emptyset$, then $Q_j \subseteq O_a$ for some $a \in A$ and so $\bar{Q}_j \subseteq \bar{O}_a \subseteq O$. Hence if we define $O_A = \bigcup\{Q_j : Q_j \cap A \neq \emptyset\}$, then by Lemma 1

$$A \subseteq O_A \subseteq \bar{O}_A = \bigcup\{\bar{Q}_j : Q_j \cap A \neq \emptyset\} \subseteq O.$$

By Theorem II.3.2, X is a (T_4) space.

Theorem 2. *Every Lindelöf* (T_3) *space is paracompact.*

Proof. Let $\{O_i\}$ $(i \in I)$ be an open cover of X. For each x there is an O_i containing x and since X is a (T_3) space we can determine two open sets O_x and Q_x such that

$$x \in Q_x \subseteq \bar{Q}_x \subseteq O_x \subseteq \bar{O}_x \subseteq O_i .$$

The sets $\{Q_x\}$ $(x \in X)$ form an open cover of the Lindelöf space X and so there is a countable subcover consisting, say, of the sets Q_{x_n} ($n = 1, 2, \ldots$). We put $S_1 = O_{x_1}$ and in general we define

$$S_n = O_{x_n} \cap c\bar{Q}_{x_1} \cap \ldots \cap c\bar{Q}_{x_{n-1}}$$

so that S_n is open and $S_n \subseteq O_i$ for some $i \in I$. The family $\{S_n\}$ ($n = 1, 2, \ldots$) is a cover of X: In fact, if $x \notin S_1 = O_{x_1}$, then there is a first index $n > 1$ such that $x \in O_{x_n}$. By the choice of n we have

$$x \notin O_{x_1} \cup \ldots \cup O_{x_{n-1}} \quad \text{so} \quad x \notin \bar{Q}_{x_1} \cup \ldots \cup \bar{Q}_{x_{n-1}} \quad \text{and} \quad x \in S_n .$$

The open cover $\{S_n\}$ ($n = 1, 2, \ldots$) is locally finite. For, if $x \in Q_{x_n}$, then $N_x = \bar{Q}_{x_n}$ is a neighborhood of x which intersects at most $S_1, \ldots, S_n$. Hence $\{S_n\}$ ($n = 1, 2, \ldots$) is a locally finite refinement of the given open cover $\{O_i\}$ ($i \in I$).

Now we turn to the definition of full-normality. If $\{S_i\}$ ($i \in I$) is a system of sets in a set X, then the *star* of an element $x \in X$ is defined to be the set $x^* = \bigcup\{S_i : x \in S_i\}$. More generally we define the star of any subset $A \subseteq X$ as $A^* = \bigcup\{x^* : x \in A\}$ or equivalently as $A^* = \bigcup\{S_i : S_i \cap A \neq \emptyset\}$. A system $\{Q_j\}$ ($j \in J$) of sets $Q_j \subseteq X$ is called a (*strong*) *star refinement* of another system $\{O_i\}$ ($i \in I$) if $\bigcup Q_j = \bigcup O_i$ and if for (every Q_j) every x there is an O_i such (that $Q_j^* \subseteq O_i$) that $x^* \subseteq O_i$.

Definition 2. *A topological space X is called a fully* (T_4) *space if every open cover of X has a star refinement.*

The existence of a large class of fully (T_4) spaces is assured by the following:

Theorem 3. *If X is pseudomatrizable then every open cover of X admits strong star refinements.*

Corollary. *Every pseudometric space is a fully* (T_3) *space.*

Proof. Let $\{O_i\}$ ($i \in I$) be an open cover of the pseudometric space X and let d be any pseudometric which is compatible with the topology of X. For every $x \in X$ there is an O_i containing x and there is an ϵ_x ($0 < \epsilon_x < 1$) such that $S_{9\epsilon_x}[x] \subseteq O_i$. Let such an ϵ_x and O_i be fixed for every $x \in X$. For each x in X we define

$$\epsilon = \epsilon(x) = \text{lub}\{\epsilon_z : d(x, y) < \epsilon_x \;\; \textit{and} \;\; d(y, z) < \epsilon_z \;\; \textit{for some} \;\; y \in X\}.$$

Clearly the condition on z means that $S_{\epsilon_z}[z]$ meets $S_{\epsilon_x}[x]$. We show that the sets $S_{\epsilon_x}[x]$ ($x \in X$) form a star refinement of the given cover $\{O_i\}$ ($i \in I$). Given x choose a u such that $\epsilon_u > \epsilon/2$, and $d(x, y) < \epsilon_x$

and $d(y, u) < \epsilon_u$ for some $y \in X$. We prove that the star of $S_{\epsilon_x}[x]$ is contained in $S_{9\epsilon_u}[u]$. The latter is known to be a subset of some O_i $(i \in I)$. In fact, if v belongs to the star of $S_{\epsilon_x}[x]$, then there are points w and z such that $d(x, w) < \epsilon_x$ and $d(w, z) < \epsilon_z$ and $d(v, z) < \epsilon_z$. Consequently, $d(x, v) < \epsilon_x + 2\epsilon_z$. By the definition of u we have $d(x, y) < \epsilon_x$ and $d(y, u) < \epsilon_u$ for some $y \in X$. Thus $d(x, u) < \epsilon_x + \epsilon_u$ and so $d(u, v) < 2\epsilon_x + 2\epsilon_z + \epsilon_u \leqslant 4\epsilon + \epsilon_u \leqslant 9\epsilon_u$. This proves that v belongs to $S_{9\epsilon_u}[u]$.

Lemma 3. *Every fully* (T_4) *space is a* (T_4) *space.*

Proof. Let A and B be disjoint closed sets in the fully (T_4) space X. The sets cA and cB form an open cover of X. Let $\{Q_j\}$ $(j \in J)$ be a star refinement. We define $O_A = A^*$ and $O_B = B^*$ where the stars are formed relative to the open cover $\{Q_j\}$ $(j \in J)$. The sets O_A and O_B are disjoint open neighborhoods of A and B, respectively: For suppose that $x \in Q_{j_1} \cap Q_{j_2}$ where $Q_{j_1} \cap A \neq \emptyset$ and $Q_{j_2} \cap B \neq \emptyset$. Then $x^* \nsubseteq cA$ and also $x^* \nsubseteq cB$ so that $\{Q_j\}$ $(j \in J)$ is not a star refinement of the open cover $\{cA, cB\}$.

Definition 3. *A fully* (T_4) *space satisfying axiom* (T_1) *is called fully normal.*

A topological space X is called *countably paracompact* if every countable open cover of X admits a locally finite refinement.

EXERCISES

1. Show by an example that Lemma 1 cannot be extended to point-finite systems even if the sets S_i $(i \in I)$ are all open.

2. Let C be a closed set in the paracompact space X. Show that C is paracompact.

[Let $\{O_i \cap C\}$ $(i \in I)$ be an open cover of C. Join cC to the family $\{O_i\}$ $(i \in I)$ to obtain an open cover of X. If $\{Q_j\}$ $(j \in J)$ is a locally finite open refinement of this cover, then $\{Q_j \cap C\}$ $(j \in J)$ is a locally finite cover of C which is a refinement of $\{O_i \cap C\}$ $(i \in I)$.]

3. Show that if every open subspace of a topological space is paracompact, then every subspace is paracompact.

[Let $\{O_i \cap S\}$ $(i \in I)$ be an open cover of a subspace S. Then $\{O_i\}$ $(i \in I)$ is an open cover of the open subspace $O = \bigcup O_i$. If $\{Q_j\}$ $(j \in J)$

is a locally finite refinement of $\{O_i\}$ $(i \in I)$, then $\{Q_j \cap S\}$ $(j \in J)$ is a locally finite refinement of the given cover of S.]

4. A space X is called *metacompact* if every open cover of X admits a point-finite refinement. Extend the results stated in the preceding two exercises to metacompact spaces.

5. Using Example II.3.1 and Exercise II.3.5 show that a locally compact Hausdorff space need not be paracompact.

[The spaces in question are not (T_4) spaces and therefore not paracompact. Every point has a compact neighborhood: If $x = (x_1, x_2)$ where $x_2 > 0$, then for a sufficiently small value of $\epsilon > 0$ the closed disk $C_\epsilon[x]$, and if $x_2 = 0$, then the closed sector $\Sigma_\epsilon[x]$ is a compact neighborhood of x.]

6. A space is called *hypocompact* if every open cover has a star-finite refinement. Prove that the right half-open interval topology of any linearly ordered set with a first element and having the least upper bound property is hypocompact.

(Every open cover of X admits a refinement consisting of right half-open intervals. Using the same reasoning as in the solution of Exercise II.12.9 we can show the existence of a subfamily of every such refinement which consists of mutually disjoint sets. The existence of a first element is not essential and can be replaced by weaker requirements.)

7. Find a paracompact space X such that $X \times X$ is not paracompact.

(Consider the set of nonnegative real numbers under the right half-open interval topology. According to Example II.11.1 $X \times X$ is not normal, hence it is not paracompact.)

8. Let X be compact and let Y be paracompact (hypocompact or metacompact). Using the same reasoning as in the proof of Theorem 4.1 show that $X \times Y$ is paracompact (hypocompact or metacompact).

9. Using the same reasoning as in the solution of Exercise 4.2 show the following: If X is compact and Y is countably paracompact, then $X \times Y$ is countably paracompact.

10. A family of sets S_i $(i \in I)$ is called a *discrete family* if every $x \in X$ has a neighborhood N_x which meets at most one of the sets S_i $(i \in I)$. Show that if $\{S_i\}$ $(i \in I)$ is a discrete family, then so is $\{\bar{S}_i\}$ $(i \in I)$.

[Let O_x be an open neighborhood of x which meets at most one of the sets S_i $(i \in I)$. If O_x meets $\bar{S}_i$, then O_x is a neighborhood of the common points of O_x and $\bar{S}_i$, hence O_x must intersect S_i. It follows that O_x can intersect at most one of the sets $\bar{S}_i$ $(i \in I)$.]

11. A topological space X is called a *collection-wise* (T_4) *space* if for every discrete family of (closed) sets S_i $(i \in I)$ there exists a family of

disjoint open sets O_i $(i \in I)$ such that $S_i \subseteq O_i$ for every $i \in I$. Show that every fully (T_4) space is a collection-wise (T_4) space.

[Let $\{S_i\}$ $(i \in I)$ be a discrete collection of (closed) sets. For every $x \in X$ there is an open set O_x such that O_x intersects at most one of the sets S_i. Since the space is a fully (T_4) space the open cover $\{O_x\}$ $(x \in X)$ has a star refinement $\{Q_j\}$ $(j \in J)$. Define O_i for every $i \in I$ as the star of S_i relative to this refinement. The reasoning used in the proof of Lemma 3 shows that the open sets O_i $(i \in I)$ are disjoint: For suppose that $z \in Q_{j_1} \cap Q_{j_2}$ where $Q_{j_1} \cap S_{i_1} \neq \emptyset$ and $Q_{j_2} \cap S_{i_2} \neq \emptyset$. Then $z^* \cap S_{i_1} \neq \emptyset$ and $z^* \cap S_{i_2} \neq \emptyset$. Since $z^* \in Q_{j_1} \subseteq O_x$ for some O_x which meets only one of the S_i's it follows that $\{Q_j\}$ $(j \in J)$ is not a star refinement.]

12. Show that the following properties of a (T_4) space are equivalent: (i) the space X is countably paracompact; (ii) every countable open covering of X has a point-finite refinement; (iii) every countable open covering $\{O_n\}$ $(n = 1, 2, ...)$ has an open refinement $\{Q_n\}$ $(n = 1, 2, ...)$ such that $\bar{Q}_n \subseteq O_n$ for every $n = 1, 2, ...$.

[The implication (i) $\to$ (ii) is obvious because every locally finite cover is point-finite. Next show that (ii) $\to$ (iii): Let $\{Q_j\}$ $(j \in J)$ be a point-finite refinement of the countable open cover $\{O_n\}$ $(n = 1, 2, ...)$. For every $j \in J$ let the positive integer $n(j)$ be the first index such that $Q_j \subseteq O_{n(j)}$ and let $R_n = \bigcup\{Q_j : n(j) = n\}$. The sets R_n $(n = 1, 2, ...)$ form a point-finite cover of X and $R_n \subseteq O_n$ for every $n = 1, 2, ...$. By Theorem II.4.2 there is an open cover $\{S_n\}$ $(n = 1, 2, ...)$ such that $\bar{S}_n \subseteq R_n \subseteq O_n$. The equivalence of the three properties follows by showing that (iii) $\to$ (i): Let $\{O_n\}$ $(n = 1, 2, ...)$ be given. First construct an open cover $\{Q_n\}$ $(n = 1, 2, ...)$ such that $\bar{Q}_n \subseteq O_n$ for every $n = 1, 2, ...$. Then use that same construction as in the proof of Theorem 2 to obtain a locally finite refinement of $\{O_n\}$ $(n = 1, 2, ...)$.]

13. Show that the space described in Example II.2.1 is not paracompact.

(This is a Hausdorff space but it is not regular.)

14. The space given in Example II.2.1 is not countably paracompact.

[Let $O_n = (X - A) \cup \left\{1, \frac{1}{2}, ..., \frac{1}{n-1}, \frac{1}{n}\right\}$ for every $n = 1, 2, ...$. This is a countable open cover of X which admits no locally finite refinement. For suppose that there is an open neighborhood O_0 of 0 which intersects only finitely many sets of the refinement $\{Q_j\}$ $(j \in J)$. Then on the one hand each Q_j covers only finitely many points of the form $1/n$ $(n = 1, 2, ...)$ and on the other hand for n sufficiently large $1/n$ is covered by Q_j only if Q_j intersects O_0. Hence $\{Q_j\}$ $(j \in J)$ is not a cover of X.]

7. The Equivalence of Paracompactness and Full-Normality

The main object of this section is to prove that for uniformizable spaces the proposition of being a fully (T_4) space is equivalent to paracompactness. The essential part of this result is the implication expressed in the theorem of A. H. Stone:

Theorem 1. *Every fully* (T_4) *space satisfying axiom* (T_3) *is paracompact.*

Since every $(T_1) + (T_4)$ space is regular we obtain the following:

Corollary. *Every fully normal space is paracompact.*

The importance of Stone's theorem can be realized by applying it to pseudometric spaces. Every pseudometric space satisfies axiom (T_3) and by Theorem 6.3 it is a fully (T_4) space. Hence we have:

Theorem 2. *Every pseudometric space is paracompact.*

First we sketch an independent proof of a weaker statement, namely, we show that every pseudometric space is countably paracompact. The proof of Theorem 1 is based on the same ideas but the axiom of choice is needed in the form of the well-ordering theorem. Stone's theorem is proved in the same way. There is only one new point which requires some discussion before entering into the technical details of the proof.

Let (O_i) $(i = 1, 2, ...)$ be a countable open cover of the pseudometric space X. Our object is to construct a locally finite refinement. We choose any pseudometric d which is compatible with the topology of X. Given any set $A \subseteq X$ we define $U^n[A] = \{x : d(x, A) < 2^{-n}\}$ and

$$I^n[A] = \{x : S_{2^{-n}}[x] \subseteq A\}.$$

The first is an open neighborhood of A and the second is a closed subset of A because $I^n[A] = cU^n[cA]$. These sets are commonly used, for instance, in the theory of convex bodies and are called the outer and the inner "parallel domains" of distance 2^{-n}.

First we replace the open cover (O_i) $(i = 1, 2, ...)$ by the refinement $\{Q_i^n\}$ $(i = 1, 2, ...$ and $n = 1, 2, ...)$ where $Q_i^1 = I^m[O_i] \neq \emptyset$ and $Q_i^n = U^n[Q_i^{n-1}]$ for every $n > 1$. The sets Q_i^n $(n > 1)$ form an expanding sequence of open sets whose union Q_i is an open subset of O_i. The process of replacing O_i by Q_i is a well-known geometric operation. For every $n = 1, 2, ...$ we define

$$C_1^n = I^n[Q_1] \qquad \text{and} \qquad C_i^n = I^n[Q_i - \bigcup\{C_j^n : j < i\}].$$

The family of open sets $U^{n+3}[C_i^n]$ $(i = 1, 2, ...)$ is a discrete system and so by Lemma 6.1 the union of the closures of these sets is a closed set which we denote by C^n. This assures that the following sets are open:

$$R_1^n = U^{n+2}[C_1^n] \quad \text{and} \quad R_i^n = U^{n+2}[C_i^n] \cap cC^1 \cap \dots \cap cC^{n-1}.$$

These form again a discrete system but the union of these systems for $n = 1, 2, \dots$ is an open cover of X. One can easily show that we are dealing with a refinement of (O_i). Its local finiteness follows by showing that $U^{m+3}[x]$ for a sufficiently high value of m can meet at most m of the sets R_i^n ($n = 1, 2, \dots$ and $i = 1, 2, \dots$).

The preceding reasoning works also when the index set I of (O_i) $(i \in I)$ is not countable. However, the axiom of choice is needed here to show that the index set I of any open cover $\{O_i\}$ $(i \in I)$ can be well-ordered. If X is not pseudometrizable we must change the definition of the sets $U^n[A]$ and $I^n[A]$. The latter can be defined as $I^n[A] = cU^n[cA]$ provided $U^n[A]$ is meaningful for every $A \subseteq X$. Since X is a fully (T_4) space satisfying axiom (T_3) by the Corollary of Theorem II.10.1 it is uniformizable. We introduce a suitable uniform structure $\mathcal{U}$ and select a sequence of uniformities U^n ($n = 1, 2, \dots$) such that $U^{n+1} \circ U^{n+1} \subseteq U^n$. The choice of these uniformities will depend on the open cover $\{O_i\}$ $(i \in I)$. Then as usual we define $U^n[A]$ to be the uniformity U^n evaluated on the set A. As far as the actual proof is concerned it is not necessary to deal with the whole of $\mathcal{U}$ but it is sufficient to introduce only the uniformities U^n ($n = 1, 2, \dots$) which are needed in the proof. These sets U^n form a subbase for a uniform structure whose topology is weaker than the original topology of the set X. In the next section we shall prove that this weaker uniform structure is always a pseudometric structure.

Proof of Theorem 1. Let $\{O_i\}$ $(i \in I)$ be an open cover of the fully (T_4) space X. First we construct a sequence of open covers $\mathcal{O}^1, \mathcal{O}^2, \dots$ such that $\mathcal{O}^1$ is a star refinement of $\{O_i\}$ $(i \in I)$ and $\mathcal{O}^{n+1}$ is a star refinement of $\mathcal{O}^n$ for every $n \geqslant 1$. We construct a uniform structure for X by introducing

$$U^n = \{(x, y) : x, y \in O^n \quad \textit{for} \quad \textit{some} \quad O^n \in \mathcal{O}^n\}.$$

The family $\{U^n\}$ ($n = 1, 2, \dots$) satisfies the axioms of a subbase for a uniform structure: It is plain that U^n is symmetric and contains I. Moreover, $U^n[x] = (x)_n{}^*$ where $(x)_n{}^*$ denotes the star of the point x relative to the cover $\mathcal{O}^n$. Axiom (Us. 6) follows from the fact that $\mathcal{O}^{n+1}$ is a star refinement of $\mathcal{O}^n$ and so $U^{n+1} \circ U^{n+1} \subseteq U^n$: Indeed, if

$$(x, y) \in U^{n+1}$$

and $(y, z) \in U^{n+1}$, then $x, z \in (y)^*_{n+1}$ and $(y)^*_{n+1} \subseteq O^n$ for some $O^n \in \mathcal{O}^n$. Hence $(x, z) \in U^n$.

We define for every $i \in I$ the sets Q_i^n $(n = 1, 2, \ldots)$ as follows:

$$Q_i^1 = cU^1[cO_i] \qquad \text{and} \qquad Q_i^n = U^n[Q_i^{n-1}] \qquad \text{for} \quad n > 1\,.$$

Since $U^n[x] = (x)_n{}^*$ is open for every $x \in X$, the sets Q_i^n are open for every $n > 1$ and Q_i^1 is closed. If $x \in U^1[Q_i^1]$, then $x, y \in O^1$ for some $O^1 \in \mathcal{O}^1$ where $y \notin U^1[cO_i]$. Hence $y \in U^1[x]$ and so $x \notin cO_i$. This shows that $Q_i^1 \subseteq U^1[Q_i^1] \subseteq O_i$. Suppose that $U^n[Q_i^n] \subseteq O_i$ for some $n \geqslant 1$. Then

$$U^{n+1}[Q_i^{n+1}] = U^{n+1}[U^{n+1}[Q_i^n]] = (U^{n+1} \circ U^{n+1})[Q_i^n] \subseteq U^n[Q_i^n] \subseteq O_i.$$

Thus we see that $Q_i^n \subseteq O_i$ for every $n = 1, 2, \ldots$ and so $Q_i = \bigcup Q_i^n$ is an open subset of O_i.

The family $\{Q_i\}$ $(i \in I)$ is an open cover of X: Given $x \in X$ there is an index $i \in I$ such that $U^1[x] = (x)_1{}^* \subseteq O_i$. For this index i we have $x \in Q_i^1 \subseteq Q_i$ because if $x \in O^1 \in \mathcal{O}^1$, then $O^1 \subseteq O_i$ and so

$$x \in cU^1[cO_i] = Q_i^1.$$

We well-order the index set I and define C_i^n for every $i \in I$ by transfinite induction as

$$C_1^n = cU^n[cQ_1] \qquad \text{and} \qquad C_i^n = cU^n[cQ_i \cup \bigcup\{C_j^n : j < i\}].$$

As $C_i^n \subseteq cU^n[cQ_i]$, by the definition of U^n we have $U^n[C_i^n] \subseteq Q_i$. We show that $\{C_i^n\}$ $(n = 1, 2, \ldots$ and $i \in I)$ is a cover of X: Given $x \in X$, let i be the first index such that $x \in Q_i$. Then we have $x \in Q_i^{n-1}$ for a sufficiently high value of n. Clearly $U^n[x] \subseteq U^n[Q_i^{n-1}] = Q_i^n \subseteq Q_i$. We prove that $x \in C_i^n$: If $x \notin C_i^n$ for some index $i \in I$, then there is a y in cQ_i or in $\bigcup\{C_j^n : j < i\}$ such that $x, y \in O^n$, i.e., $y \in U^n[x]$ for some $O^n \in \mathcal{O}^n$. However, $U^n[x] \subseteq Q_i$ and so y must belong to $\bigcup\{C_j^n : j < i\}$. Then $y \in C_j^n$ for some $j < i$ and so by $x, y \in O^n$ we have $x \in U^n[C_j^n] \subseteq Q_j$ where $j < i$. Therefore i is not the first index with the property that x belongs Q_i.

For each fixed $n = 1, 2, \ldots$ every $O^n \in \mathcal{O}^n$ intersects at most one of the sets C_i^n: Indeed, if $i > j$ and $x \in O^n \cap C_i^n$, then

$$x \notin U^n[cQ_i \cup \bigcup\{C_j^n : j < i\}]$$

and so by $x \in O^n$ we have $O^n \subseteq Q_i \cap \bigcap\{cC_j^n : j < i\} \subseteq cC_j^n$. Since $\mathcal{O}^{n+2}$ is a star refinement of $\mathcal{O}^n$ it follows that every $O^{n+2} \in \mathcal{O}^{n+2}$ intersects at most one of the sets $U^{n+2}[C_i^n]$ $(i \in I)$ and so $\mathcal{O}^{n+2}$ being an open cover these sets form a discrete system. The family of open sets $U^{n+3}[C_i^n]$ $(i \in I)$ is *a fortiori* discrete and so by Lemma 6.1 $C^n = \bigcup \overline{U^{n+3}[C_i^n]}$ is a closed set. For each $n = 1, 2, \ldots$ we define by transfinite induction

$$R_1^n = U^{n+2}[C_i^n] \quad \text{and} \quad R_i^n = U^{n+2}[C_i^n] \cap cC^1 \cap \ldots \cap cC^{n-1}.$$

Since C^n is closed for every n these are open sets in X.

We prove that $\{R_i^n\}$ $(n = 1, 2, \ldots$ and $i \in I)$ is a locally finite refinement of $\{O_i\}$ $(i \in I)$. We first show that every $x \in X$ belongs to some R_i^n: Since $\{C_i^n\}$ $(n = 1, 2, \ldots$ and $i \in I)$ is a cover of X there are sets C_i^n containing x. Let n be the smallest integer such that $\overline{U^{n+3}[C_i^n]}$ contains x for some $i \in I$. Then by Lemma I.9.7, $x \in U^{n+2}[C_i^n]$ and $x \in cC^1 \cap \ldots \cap cC^{n-1}$ so that $x \in R_i^n$. This cover is a refinement of $\{O_i\}$ $(i \in I)$ because

$$R_i^n \subseteq U^{n+2}[C_i^n] \subseteq U^n[C_i^n] \subseteq U^n[cU^\nu[cQ_i]] \subseteq Q_i \subseteq O_i.$$

Finally, each $x \in X$ has a neighborhood which intersects only finitely many of the sets R_i^n $(n = 1, 2, \ldots$ and $i \in I)$: For if $x \in C_i^n$, then $U^{n+3}[x] \subseteq U^{n+3}[C_i^n] \subseteq C^n$ and so $U^{n+3}[x]$ does not intersect any R_i^m with $m > n$. If $m \leqslant n$, then $U^{n+3}[x] \subseteq O^{n+2} \subseteq O^{m+2}$ for some $O^{n+2} \in \mathcal{O}^{n+2}$ and $O^{m+2} \in \mathcal{O}^{m+2}$. Hence $U^{n+3}[x]$ can meet R_j^m for at most one value of $j \in I$. Therefore $U^{n+3}[x]$ meets at most n sets of the refined cover. This completes the proof of Theorem 1.

Theorem 3. *Every paracompact* (T_3) *space is fully* (T_4).

Proof. By Theorem 6.1 every paracompact (T_3) space is a (T_4) space. Since every open cover of a paracompact space can be replaced by a locally finite refinement the result follows from the following:

Lemma 1. *Every locally finite open cover of a* (T_4) *space admits a star refinement.*

Proof. Let $\{O_i\}$ $(i \in I)$ be a locally finite open cover of X. By Theorem II.4.1 there exists an open cover $\{Q_i\}$ $(i \in I)$ such that $\bar{Q}_i \subseteq O_i$ for every $i \in I$. Every $x \in X$ has an open neighborhood O_x which intersects only a finite number of the O_i's. Let $I(x)$ be the set of those indices $i \in I$ for which $x \in \bar{Q}_i$ and let $J(x)$ be the set of those indices j for which O_x meets O_j but $x \notin \bar{Q}_j$. We let

$$Q_x = O_x \cap \bigcap \{O_i : i \in I(x)\} \cap \bigcap \{c\bar{Q}_j : j \in J(x)\}.$$

We show that the open cover $\{Q_x\}$ $(x \in X)$ is a star refinement of $\{O_i\}$ $(i \in I)$: Given $x \in X$ there is a Q_i which contains x. If $x \in Q_y$, then Q_i intersects Q_y and so $i \notin J(y)$. Since $x \in Q_y \cap Q_i$ we have $O_y \cap O_i \neq \emptyset$ and so $i \in I(y)$ and $Q_y \subseteq O_i$. Therefore $x^* \subseteq O_i$.

As a result of Theorems 1 and 3 we have the following:

Theorem 4. *A space is fully normal if and only if it is paracompact and regular.*

EXERCISES

1. Show that every regular σ-compact space is fully normal.

(This follows immediately from Theorems 6.2 and 3.)

2. Let X be a linearly ordered set with the property that every increasing sequence has a least upper bound in X. Show that in the interval topology for each open cover of X there is a point $x \in X$ such that $\{y : x < y\} \subseteq x^*$.

[Given any open cover and a strictly increasing sequence (x_n) $(n = 1, 2, \ldots)$ let x be its least upper bound. There is a set O_i in the open cover which contains x and so there is a proper interval $(a, x] \subseteq O_i$. Since x is the least upper bound, (a, x) contains an x_n. Hence x_n, $x_{n+1} \in O_i$ and $x_{n+1} \in (x_n)^*$. This shows that no strictly increasing sequence exists with $x_{n+1} \notin (x_n)^*$ for every $n = 1, 2, \ldots$.]

3. Let X be a topological space and let $\mathscr{U}$ be the filter of all neighborhoods of the diagonal of $X \times X$. Show that $\mathscr{U}$ is a uniform structure for some topology on X if and only if X has the following property: Every open cover $\{O_i\}$ $(i \in I)$ of X has a refinement $\{Q_j\}$ $(j \in J)$ such that the union $Q_{j_1} \cup Q_{j_2}$ of two intersecting Q_{j_1}, Q_{j_2} is contained in a suitable O_i.

[Every neighborhood of the diagonal contains some neighborhood of the form $U = \bigcup(O_i \times O_i)$ where $\{O_i\}$ $(i \in I)$ is an open cover of X. For every U there is a $V = \bigcup(Q_j \times Q_j)$ such that $V \circ V \subseteq U$ if and only if $Q_{j_1} \cap Q_{j_2} \neq \emptyset$ implies that $Q_{j_1} \cup Q_{j_2} \subseteq O_i$ for some O_i. Hence if $\mathscr{U}$ is a uniform structure, then the condition must hold. Conversely, if the condition is satisfied, then the sets of the form $\bigcup(O_i \times O_i)$ fulfill the axioms for a base of a uniform structure.]

4. If X is a (T_3) space, the uniform topology associated with the uniform structure $\mathscr{U}$ defined above is the original topology of X. In other words, if X is uniformizable at all, then $\mathscr{U}$ is compatible with the topology of X.

8. Metrizable Uniform Structures and Structure Gages

The uniform structure associated with a pseudometric d is the structure generated by the base $\{U_\epsilon\}$ ($\epsilon > 0$) where $U_\epsilon = \{(x, y) : d(x, y) < \epsilon\}$. A uniform structure $\mathscr{U}$ is called a *pseudometric structure* if there is a pseudometric d such that the uniform structure associated with d is the given $\mathscr{U}$. If $\mathscr{U}$ is associated with d, then the sets

$$U_{1/n} = \{(x, y) : d(x, y) < 1/n\}$$

form a countable structure base for $\mathscr{U}$. We are going to prove the converse statement so that we have the following:

Theorem 1. *A uniform structure $\mathscr{U}$ for a set X is a pseudometric structure if and only if $\mathscr{U}$ has a countable base.*

Proof. The necessity is obvious from the above remark. By Lemma I.8.1 every countable structure base can be replaced by another such that the elements of the new base are symmetric uniformities and they form a decreasing sequence with respect to inclusion. Hence in the proof of the sufficiency we can assume that a given $\mathscr{U}$ has a countable base $\mathscr{U}_B = \{U^{(n)}\}$ ($n = 1, 2, \ldots$) consisting of symmetric uniformities $U^{(n)}$ and $U^{(n)} \supseteq U^{(n+1)}$ for every $n \geqslant 1$. We set $U_1 = U^{(1)}$ and select by induction a sequence of uniformities $U_{2^{-n}}$ from the base $\mathscr{U}_B$ such that $U_{2^{-n}} \circ U_{2^{-n}} \subseteq U_{2^{-(n-1)}}$ and $U_{2^{-n}} \subseteq U^{(n)}$. Next we extend the definition of U_d for every diadic rational $d = 2^{-n_1} + \ldots + 2^{-n_k}$ where $0 < n_1 < \ldots < n_k$ as follows:

$$U_d = U_{2^{-n_1}} \circ \ldots \circ U_{2^{-n_k}}.$$

Then by Lemma II.10.1, for every pair of positive rationals d_1, d_2 satisfying $d_1 + d_2 < 1$ we have $U_{d_1} \circ U_{d_2} \subseteq U_{d_1+d_2}$.

Now we introduce a pseudometric d on the set X:

$$d(x, y) = \begin{cases} 1 & \textit{if } (x, y) \notin U_1 \\ \text{glb}\{d : (x, y) \in U_d\} & \textit{if } (x, y) \in U_1 . \end{cases}$$

The axioms $d(x, x) = 0$ and $d(x, y) = d(y, x)$ are obviously satisfied. If $d(x, y) + d(y, z) \geqslant 1$, then the triangle inequality holds because $d(x, z) \leqslant 1$ for every point pair $x, z \in X$. If $d(x, y) + d(y, z) < 1$, then we can choose arbitrarily small positive real numbers δ_1 and δ_2 such that $d(x, y) + \delta_1$ and $d(y, z) + \delta_2$ are diadic rationals and

$$d(x, y) + d(y, z) + \delta_1 + \delta_2 < 1.$$

This follows from the fact that the diadic rationals are dense in the space of reals under the half-open interval topology $\mathscr{T}_+$. From the definition of the pseudometric d we see that

$$(x, y) \in U_{d(x,y)+\delta_1} \quad \text{and} \quad (y, z) \in U_{d(y,z)+\delta_2}.$$

Hence

$$(x, z) \in U_{d(x,y)+\delta_1} \circ U_{d(y,z)+\delta_2} \subseteq U_{d(x,y)+d(y,z)+\delta_1+\delta_2}.$$

This proves that $d(x, z) \leqslant d(x, y) + d(y, z) + \delta_1 + \delta_2$. Since the positive reals δ_1 and δ_2 can be chosen arbitrarily small we obtain the triangle inequality.

The family $\{U_d\}$ is a structure base because $U_{2^{-n}} \circ U_{2^{-n}} \subseteq U_{2^{-(n-1)}}$ and by $U_{2^{-n}} \subseteq U^{(n)}$ it is equivalent to $\mathscr{U}_B$. We show that the structure generated by $\{U_d\}$ and the structure associated with the pseudometric d are identical. In fact, if d $(0 < d \leqslant 1)$ is a diadic rational and $d(x, y) < d$, then $(x, y) \in U_d$ so that $\{(x, y) : d(x, y) < d\} \subseteq U_d$. On the other hand, let d $(0 < d < 1)$ be a diadic rational and let $(x, y) \in U_{d/2}$. Then $d(x, y) \leqslant d/2$ and so $U_{d/2} \subseteq \{(x, y) : d(x, y) < d\}$. Thus the structure bases $\mathscr{U}_B$, $\{U_d\}$, and $\{\{(x, y) : d(x, y) < d\}\}$ are equivalent. The proof is complete.

The least upper bound $\text{lub}\{\mathscr{U}_\alpha\}$ of a family of uniform structures $\mathscr{U}_\alpha$ $(\alpha \in A)$ for the same set X is the coarest structure which is at least as fine as any of the $\mathscr{U}_\alpha$'s. Its existence is shown in Section I.9. Let $\mathscr{U}_n$ $(n = 1, 2, \ldots)$ be pseudometric structures for a common set X and let $\mathscr{B}_n = \{U_n{}^k\}$ $(k = 1, 2, \ldots)$ be countable bases for the structures $\mathscr{U}_n$ $(n = 1, 2, \ldots)$. The family of all finite intersections $\bigcap U_n^{k_n}$ is a structure base for $\text{lub}\{\mathscr{U}_n\}$. This is a countable family and so by the present theorem $\text{lub}\{\mathscr{U}_n\}$ is a pseudometric structure. Hence we have:

Lemma 1. *The least upper bound of a countable collection of pseudometric structures is pseudometric.*

If the collection is finite, say, it corresponds to the pseudometrics $d_1, \ldots, d_n$, then both $d = d_1 + \ldots + d_n$ and $d = \text{lub}\{d_1, \ldots, d_n\}$ are pseudometrics for the least upper bound of $\mathscr{U}_{d_1}, \ldots, \mathscr{U}_{d_n}$. This follows from the inclusion relations

$$\{(x, y) : d_k(x, y) < \epsilon\} \supseteq \{(x, y) : d(x, y) < \epsilon\}$$

and $\{(x, y) : d_k(x, y) < \epsilon/n \quad \text{for} \quad k = 1, \ldots, n\} \subseteq \{(x, y) : d(x, y) < \epsilon\}$.

The next theorem states that every uniform structure $\mathscr{U}$ can be described by specifying a family of pseudometrics d. The result gives a connection between uniform structures and metric structures which can be very useful in applications.

Theorem 2. *For every uniform structure $\mathscr{U}$ for a set X there is a family of pseudometrics d on X such that $\mathscr{U} = \operatorname{lub}\{\mathscr{U}_d\}$ where $\mathscr{U}_d$ is the pseudometric structure generated by d.*

Note. There is a largest family of pseudometrics satisfying the requirements. It consists of all pseudometrics d for which $\mathscr{U}_d \leqslant \mathscr{U}$. This particular family is called the *gage* of the structure $\mathscr{U}$. Another family which satisfies the requirements consists of those pseudometrics d for which $d(X)$ is finite and $\mathscr{U}_d \leqslant \mathscr{U}$. This is called the *bounded gage* of $\mathscr{U}$. (See Exercise I.8.9.)

Proof. Given $U \in \mathscr{U}$ we construct a sequence $\mathscr{U}_B = \mathscr{U}_B(U)$ of symmetric uniformities $U_{2^{-n}} \in \mathscr{U}$ such that $U_{1/2} \subseteq U$ and

$$U_{2^{-(n+1)}} \circ U_{2^{-(n+1)}} \subseteq U_{2^{-n}}$$

for every $n = 1, 2, \ldots$. Then $\mathscr{U}_B(U)$ is a denumerable base for a uniform structure $\mathscr{U}(U)$ and $U \in \mathscr{U}(U)$ because $U_{1/2} \subseteq U$. By Theorem 1 there is a pseudometric $d = d(U)$ on X such that the uniform structure $\mathscr{U}_d$ generated by d is $\mathscr{U}(U)$. Hence $\mathscr{U}_d = \mathscr{U}(U) \leqslant \mathscr{U}$ and so

$$\operatorname{lub}\{\mathscr{U}_d\} \leqslant \mathscr{U}$$

where the least upper bound is taken for all pseudometrics $d(U)$ $(U \in \mathscr{U})$. On the other hand, $U \in \operatorname{lub}\{\mathscr{U}_d\}$ for every $U \in \mathscr{U}$ because

$$U \in \mathscr{U}_{d(U)} \subseteq \operatorname{lub}\{\mathscr{U}_d\}\,.$$

Therefore $\mathscr{U} = \operatorname{lub}\{\mathscr{U}_d\}$ and the theorem is proved.

EXERCISES

1. Suppose that the compact Hausdorff space X has a countable base. Show that X is metrizable.

[Let $\mathscr{B}$ be a countable base for the open sets of X. Associate with every finite open covering $\{O_i\}$ of X by sets $O_i \in \mathscr{B}$ the following set in $X \times X$:

$$S(\{O_i\}) = \{(x, y) : x, y \in O_i \quad \textit{for some} \quad O_i \in \{O_i\}\}.$$

Then by the statement given in Exercise 3.7 the family of all sets $S(\{O_i\})$ is a structure base for X which is compatible with the topology of X. Since this family is countable it generates a pseudometric structure which by the Hausdorff property is a metric structure.]

2. Show that if X has a countable base and it is a fully (T_4) space satisfying axiom (T_3), then it is pseudometrizable.

[Consider for each pair of base elements $Q \subseteq \bar{Q} \subseteq O$ the open cover $\{O, c\bar{Q}\}$. Let $\mathcal{O}_1$ be a star refinement of this cover and let $\mathcal{O}_{n+1}$ for every $n = 1, 2, \ldots$ be a star refinement of $\mathcal{O}_n$. Let

$$U_n = \{(x, y) : x, y \in O_n \quad \textit{for some} \quad O_n \in \mathcal{O}_n\}.$$

Here U_n depends on the choice of O and Q. The denumerable family of all these U_n is a subbase for a uniform structure which is compatible with the topology of X.]

3. Show that every (T_3) space X having a countable base is pseudometrizable.

[X is a Lindelöf space and so by Theorem 6.2 it is paracompact. Theorem 7.3 implies that X is a fully (T_4) space, hence by the preceding exercise it is pseudometrizable.]

4. Let X be a locally compact Hausdorff space having a countable base. Show that it is metrizable.

(X is regular; hence the result stated in the preceding exercise applies to X.)

5. Let X be compact and suppose that the diagonal I of $X \times X$ is the intersection of countably many closed neighborhoods of I. Show that the filter of all neighborhoods of I has a countable base.

[Let $N_I^1, N_I^2, \ldots$ be closed neighborhoods of I and let $\bigcap N_I^n = I$. Using the compactness of closed subspaces of X we see that given any open N_I and a point $x \in X$ there is an O_x and a suitable N_I^n where n depends on x such that $N_I^n[\xi] \subseteq N_I[\xi]$ for every $\xi \in O_x$. Since X is compact $\{O_x\}$ $(x \in X)$ contains a finite subcover. If N_I^p is the intersection of the corresponding N_I^n sets, then $N_I^p[\xi] \subseteq N_I[\xi]$ for every $\xi \in X$.]

6. Let X be a compact (T_3) space and suppose that the diagonal I of $X \times X$ is the intersection of countably many open sets of $X \times X$. Show that X is metrizable.

[By Exercise 3.7 the neighborhoods of I form a uniform structure for the topology of X. Using the (T_3) property and the compactness of I, one proves that every open neighborhood O_I contains a suitable closed neighborhood N_I. Thus by the preceding exercise this uniform structure has a countable base.]

9. Metrizability Conditions

We shall prove here a theorem on the metrization of arbitrary topological spaces which was discovered independently by Jun-iti Nagata and

Yu. M. Smirnov. A weaker form is also due to R. H. Bing. A number of sufficient conditions on metrizability were introduced earlier. The most significant of these is the metrization theorem of Urysohn: *If* X *satisfies axiom* (T_3) *and has a countable base, then it is pseudometrizable.* The first hypothesis of Urysohn's theorem is a necessary condition. It is the second hypothesis which Nagata and Smirnov replaced by a necessary condition.

Theorem 1. *A* (T_3) *space is pseudometrizable if and only if it has a base which is the union of countably many locally finite systems.*

By Theorem 7.2 every pseudometrizable space X is paracompact, hence the open cover formed by the balls $S_{1/n}[x]$ $(x \in X)$ has a locally finite refinement $\mathcal{O}^n$. The union of the locally finite covers $\mathcal{O}^n$ is obviously a base for X. Hence the conditions stated in the preceding theorem are necessary. Let us now prove that they are also sufficient. By $\mathcal{O}^n$ $(n = 1, 2, ...)$ we denote the locally finite systems whose union is a base for X.

Lemma 1. *Every* (T_3) *space* X *satisfying the Nagata-Smirnov condition is a* (T_4) *space.*

Note. Since every subspace of X satisfies the condition it follows that X is a (T_5) space.

Proof. Let A and B be disjoint closed sets. Since X is a (T_3) space we can determine for every $a \in A$ and $b \in B$ open sets $O_a{}^m \in \mathcal{O}^m$ and $O_b{}^n \in \mathcal{O}^n$ such that $a \in O_a{}^m \subseteq \overline{O_a{}^m} \subseteq cB$ and $b \in O_b{}^n \subseteq \overline{O_b{}^n} \subseteq cA$. Of course the indices m and n depend on the choice of the points a and b, respectively. We set $Q_A{}^k = \bigcup\{O_a{}^m : m = k\}$ and $Q_B{}^k = \bigcup\{O_b{}^n : n = k\}$. The family $\mathcal{O}^k$ is locally finite and so Lemma 6.1 can be applied to obtain

$$\overline{Q_A{}^k} = \bigcup\{\overline{O_a{}^m} : m = k\} \subseteq cB \qquad \text{and} \qquad \overline{Q_B{}^k} = \bigcup\{\overline{O_b{}^n} : n = k\} \subseteq cA.$$

Now we define the open sets $O_A{}^k = Q_A{}^k \cap c\overline{Q_B{}^1} \cap ... \cap c\overline{Q_B{}^k}$ and $O_B{}^k = Q_B{}^k \cap c\overline{Q_A{}^1} \cap ... \cap c\overline{Q_A{}^k}$. The sets $O_A = \bigcup O_A{}^k$ and $O_B = \bigcup O_B{}^k$ are disjoint open neighborhoods of A and B, respectively.

Proof of Theorem 1. We construct a countable family of pseudometrics for X. Let the positive integers m and n be fixed. Given any $O_i{}^n \in \mathcal{O}^n$ we consider

$$O_i = \bigcup\{O^m : O^m \in \mathcal{O}^m \quad \textit{and} \quad \overline{O^m} \subseteq O_i{}^n\}.$$

Since $\mathcal{O}^m$ is a locally finite system, by Lemma 6.1 we have $\bar{O}_i \subseteq O_i{}^n$. Using Theorem II.9.3 we can find a diadic scale $(O_{d.i})$ $(d \in D)$ such that $\bar{O}_i \subseteq O_{d.i} \subseteq O_i{}^n$ for every diadic rational $0 \leqslant d < 1$. For $d = 1$ we set $O_{1,i} = X$. Now we define the real-valued function $f_i = f_{i,m,n}$ by the formula $f_i(x) = \text{glb}\{d: x \in O_{d,i}\}$ so that $0 \leqslant f_i(x) \leqslant 1$ for every $x \in X$. We define the pseudometric $d_{m,n}$ by setting

$$d_{m,n}(x, y) = \text{lub}\{|f_i(x) - f_i(y)| : O_i^n \in \mathcal{O}^n\}.$$

By the local finiteness of $\mathcal{O}^n$ there are open neighborhoods O_x and O_y which intersect only finitely many O_i^n sets. Hence $f_i(x) < 1$ or $f_i(y) < 1$ only for finitely many indices i and so the expression lub $|f_i(x) - f_i(y)|$ indicates a minimum.

By Lemma 8.1 the least upper bound of the topologies generated by the pseudometrics $d_{m,n}$ is itself a pseudometric topology. Hence it is sufficient to prove that this least upper bound $\mathcal{T}$ is identical with the original topology of X. For we first show that if a set O is open relative to the pseudometric $d_{m,n}$, then it is open in the original sense. Essentially this means that $d_{m,n}$ is a continuous function on $X \times X$. If $x \in O$, then there is an $\epsilon > 0$ such that $S_\epsilon[x] \subseteq O$. We show that $S_\epsilon[x]$ contains a neighborhood N_x: Let O_x be an open neighborhood of x which intersects only finitely many of the sets O_i^n. Let i be an index such that O_x intersects O_i^n. If $0 < f_i(x) < 1$, we can determine diadic rationals d_1 and d_2 in D such that

$$f_i(x) - \frac{\epsilon}{2} < d_1 < f_i(x) < d_2 < f_i(x) + \frac{\epsilon}{2}.$$

It follows that $|f_i(x) - f_i(y)| < \epsilon/2$ for every point y in the open set $Q_i = c\overline{O_{d_1,i}} \cap O_{d_2,i}$. If $f_i(x) = 0$ or if $f_i(x) = 1$, the open set Q_i can be determined in a similar manner. Hence if y belongs to the intersection N_x of these finitely many sets Q_i, then $d_{m,n}(x, y) = \text{lub}\{|f_i(x) - f_i(y)|\} \leqslant \epsilon/2$. It follows that $N_x \subseteq S_\epsilon[x]$ and O is open in X.

Now we show that if O is open in the original sense, then it is open relative to the topology $\mathcal{T}$. Let O and $x \in O$ be given. Since X is a (T_3) space and $\bigcup \mathcal{O}^n$ is a base for X, there are open sets $O_j^m \in \mathcal{O}^m$ and $O_i^n \in \mathcal{O}^n$ such that $x \in \overline{O_j^m} \subseteq O_i^n \subseteq O$. Therefore we have

$$\overline{O_j^m} \subseteq \overline{O}_i \subseteq O_{d,i} \subseteq O$$

for every diadic rational $0 < d < 1$. Hence, on the one hand, $f_i(x) = 0$ and, on the other hand, $f_i(y) = 1$ for every $y \in cO$. It follows that $d_{m,n}(x, y) = 1$ for every $y \in cO$ and so $\{y : d_{m,n}(x, y) < 1\} \subseteq O$. Hence O is open relative to $\mathcal{T}$.

EXERCISE

Consider the space described in Exercise II.5.3 and discussed again in Exercise II.12.6. Using this example show the existence of a uniformizable space on a denumerable set X such that no uniform structure compatible with this topology has a countable base.

NOTES

The Bourbaki group and several other people call a space compact only if it is a Hausdorff space. The hereditary Lindelöf property was introduced by Lindelöf [1]. The Lindelöf property itself is due to Kuratowski and Sierpinski [2]. (See also Sierpinski [3].) A generalization of compactness, countable compactness, and Lindelöf property, called (m, n)-compactness, was first published in a famous article dated 1929 [4]. A new edition of this article is available in Russian [5]. These spaces were extensively studied by Smirnov [6]. A modification of the notion is due to Gaal [7]. Cantor's principle of nested intervals appears in a paper [8] where the existence of transcendental numbers is proved by using a countability argument. The importance of the finite intersection property was first realized by F. Riesz [9].

Theorem 1.5 was proved by Haar and König [10]. It was stated several times later. (For instance see Murty [11].) It is a special case of the following theorem due to Frink [11a]: *Every complete lattice is compact in its interval topology*.

Lemma 2.2 is a special case of Ramsey's theorem [12]. (See also Specker [13].) The result stated in Exercise 2.6 is the so-called Cantor-Bendixon theorem. This was first published by Cantor in a paper [14] where perfect sets are introduced. Soon it was extended by Bendixon [15] and by Lindelöf [16].

By Exercise 1.7 a compact Hausdorff topology $\mathcal{T}$ exists on any nonvoid set X. Exercise 3.4 shows that $\mathcal{T}$ is maximal among the compact topologies of X and minimal among its Hausdorff topologies. Ramanathan [17] proved the existence of minimal noncompact Hausdorff topologies. These are all necessarily semiregular. Similarly, there exist maximal non-Hausdorff compact spaces. (See Balachadran [18] and Ramanathan [19].)

Tychonoff's product theorem first occurs in an incomplete form in the paper where the notion of complete regularity is introduced [20]. Later the theorem was stated explicitly when it was applied to the space of bounded real-valued functions on an interval [21]. The equivalence of Tychonoff's product theorem and the axiom of choice was conjectured by S. Kakutani and proved by Kelley [22]. The product of two countably compact spaces need not be countably compact even if they are both regular. An example was published for instance by Novak [23].

Alexander's subbase theorem was first proved by using the well-ordering principle [24]. The present proof is due to Kelley. It is essentially a proof based on the use of filter theory: The complements of the elements of the family $\mathfrak{A}$ form a filter base and we wish to show that its adherence is not void. We first find an ultrafilter containing this filter base and verify the "ultrafilter theorem" for this particular ultrafilter.

The result given in Exercise 4.2 and related theorems were published by several people. (See for instance Michael [25].) They are special cases of a theorem of Smirnov [26]. The theorem given in Exercise 4.3 is due to A. D. Wallace and the result stated in Exercise 4.5 is a special case of a theorem of Fáry [27].

Sometimes one speaks about locally n-compact or $(1, n)$-compact spaces. In general one can define what is meant by a locally (m, n)-compact topological space. If X is such a product space, then every factor is locally (m, n)-compact and all but finitely many of the factors are (m, n)-compact spaces.

Paracompactness was introduced by Dieudonné [28]. The notion of countable paracompactness is due to Dowker [29]. Many people call a space paracompact only if it is a Hausdorff space. The notion of full normality was introduced by Tukey [30]. A fully (T_4) space is often called fully normal. Spaces with the property that every open cover admits a strong star refinement may be called *strongly paracompact*, but this is a reformulation of paracompactness.

It is not known whether every normal space is countably paracompact. This problem was first raised by Dowker. ω-fully normal spaces were introduced by Goldman [31]. Hypocompact spaces were introduced by S. Hu. The result stated in Exercise 7.6 is due to Iséki [32]. Sorgenfrey [33] proved first that the right half-open interval topology of the reals is paracompact but the product of this space by itself is not normal, hence it is not paracompact. Exercise 7.7 gives a generalization of the first half of this result. The proposition stated in Exercise 7.9 is due to Dowker who proved it in the special case $n = \omega$. His proof has the disadvantage that it does not generalize easily to the case $n > \omega$. The notion of a collectionwise (T_4) space was introduced by Bing [34]. The equivalence of the propositions stated in Exercise 9.12 was proved by Dowker in the above-mentioned article [34]. He also proved that every perfectly (T_4) space is countably paracompact. Additional information on paracompact spaces can be found in Kelley's book [35], in Michael's note [25], and in two further papers of Michael [36]. See also Arens and Dugundji [37]. The notion of metacompactness was introduced in this paper. One can find several interesting propositions there on paracompactness and metacompactness: "X is countably compact if and only if every infinite cover of X by open sets admits a proper subcovering." "X is compact if and only if it is countably compact and metacompact." "If X is paracompact, then X is compact if and only if it is countably compact."

All results of Section 7 are due to Stone [38]. Stone also proves in the same paper the following: *The following statements about a product of metric spaces are equivalent*: (i) *the product is normal*; (ii) *the product is fully normal*; (iii) *all but countably many factors are compact*. The result stated in Exercise 7.2 is a lemma of Dieudonné and it was used by Mansfield [39] to prove that the interval topology of every linearly ordered set is ω-fully normal. The result given in Exercise 7.3 was first enounced by Cohen [40].

Theorem 8.1 is due to Weil [41]. The proposition given in Exercise 8.3 is Urysohn's metrization theorem. (See Urysohn [42] and Tychonoff [43].) The special case stated in Exercise 8.4 was proved earlier also by Urysohn [44]. The theorem stated in Exercise 8.6 was published by Sneider [45]. A third proof was given by I. Namioka who actually rediscovered it.

The Nagata-Smirnov metrization theorem was discovered independently by Nagata [46] and by Smirnov [47]. The proof given in the text was published in Kelley's book [35]. Smirnov also wrote another note [48] on the metrization of locally compact Hausdorff spaces. His main results were published in an expository paper which is available also in English [49]. Bing's paper concerning paracompactness and collectionwise normality was mentioned earlier [34]. There are many other results on metrization of which we mention the following: *A compact space X is metrizable if and only if $X \times X \times X$ is completely normal.* (See Katetov [50].)

REFERENCES

1. E. Lindelöf, Sur quelques points de la théorie des ensembles. *C. R. Acad. Sci. Paris* **137**, 697–700 (1903).
2. C. Kuratowski and W. Sierpinski, Le théorème de Borel-Lebesgue dans la théorie des ensembles abstraits. *Fund. Math.* **2**, 172–178 (1921).
3. W. Sierpinski, Sur l'équivalence de trois propriétés des ensembles abstraits. *Fund. Math.* **2**, 179–188 (1921).

4. P. Alexandroff and P. Urysohn, Mémoire sur les espaces topologiques compact dédié à Monsieur D. Egoroff. *Verh. Amsterdam* **14**, No. 1 (1929).
5. P. Alexandroff and P. Urysohn, On compact topological spaces. *Trudy Mat. Inst. Steklov* **31**, 95 (1950).
6. Yu. M. Smirnov, On topological spaces compact in a given interval of powers. *Izv. Akad. Nauk SSSR. Ser. Mat.* [N.S.] **14**, 155–178 (1950); On normally disposed sets of normal spaces. *Mat. Sb.* [N.S.] **29** (71), 173–176 (1951).
7. I. S. Gál, On a generalized notion of compactness, I, II. *Nederl. Akad. Wetensch.* **60** *Indag. Math.* **19**, 421–430, 431–435 (1957); On the theory of (m, n)-compact topological spaces. *Pacific. J. Math.* **8**, 721–734 (1958).
8. G. Cantor, Über eine Eigenschaft des Inbegriffs aller algebraischen Zahlen. *J. Reine Angew. Math.* **77**, 258–262 (1874).
9. F. Riesz, Stetigkeitsbegriff und abstrakte Mengenlehre. *Atti 4 Congr. Internat. Mat. Roma* **2**, 18–24 (1908); footnote b on p. 21.
10. A. Haar and D. König, Über einfach geordnete Mengen. *Crelle's J.* **139**, 16–28 (1910).
11. A. S. N. Murty, Simply ordered spaces. *J. Indian Math. Soc.* [N.S.] **13**, 152–158 (1949).
11a. O. Frink, Topology in lattices. *Trans. Amer. Math. Soc.* **51**, 567–582 (1942).
12. F. P. Ramsey, On a problem of formal logic. *Proc. London Math. Soc.* **30**, 264–286 (1930).
13. E. Specker, Teilmengen von Mengen mit Relationen. *Comm. Math. Helv.* **31**, 302–314 (1956).
14. G. Cantor, Ueber unendliche, lineare Punktmannichfaltigkeiten. *Math. Ann.* **21**, 545–586 (1883).
15. I. Bendixon, Quelques théorèmes de la théorie des ensembles. *Acta Math.* **2**, 415–429 (1883).
16. E. Lindelöf, Remarques sur un théorème fondamental de la théorie des ensembles. *Acta Math.* **29**, 183–190 (1905).
17. A. Ramanathan, Maximal-Hausdorff spaces. *Proc. Indian Acad. Sci. Sect. A* **26**, 31–42 (1947).
18. V. K. Balachadran, Minimal-bicompact space. *J. Indian Math. Soc.* [N.S.] **12**, 47–48 (1948).
19. A. Ramanathan, Minimal-bicompact spaces. *J. Indian Math. Soc.* [N.S.] **12**, 40–46 (1948).
20. A. Tychonoff, Über die topologische Erweiterung von Raumen. *Math. Ann.* **102**, 544–561 (1929).
21. A. Tychonoff, Über einen Funktionenraum. *Math. Ann.* **111**, 762–766 (1935).
22. J. L. Kelley, The Tychonoff product theorem implies the axiom of choice. *Fund. Math.* **37**, 75–76 (1950).
23. J. Novak, On the Cartesian product of two compact spaces. *Fund. Math.* **40**, 106–112 (1953).
24. J. S. Alexander, Ordered sets, complexes, and the problem of compactification. *Proc. Nat. Acad. Sci. U.S.A.* **25**, 296–298 (1939).
25. E. Michael, A note on paracompact spaces. *Proc. Amer. Math. Soc.* **4**, 831–838 (1953).
26. Yu. M. Smirnov, On the theory of finally compact spaces. *Ukrain. Mat. Ž.* **3**, 52–60 (1951).
27. I. Fáry, Un critère de compacité pour les fonctions continues. *C. R. Acad. Sci. Paris* **224**, 992–993 (1947).
28. J. Dieudonné, Une généralisation des espaces compacts. *J. Math. Pures Appl.* [9] **23**, 65–76 (1944).

29. C. H. Dowker, On countably paracompact spaces. *Canad. J. Math.* **3**, 219–224 (1951).
30. J. W. Tukey, Convergence and uniformity in topology. *Ann. of Math. Studies* No. 2 (1940).
31. A. J. Goldman, A Čech fundamental group. Summary of lectures and seminars of the Summer Institute on Set Theoretic Topology, Madison, Wisconsin, Amer. Math. Soc. Summer Inst. pp. 107–109 (1955).
32. K. Iséki, A note on hypocompact spaces. *Math. Japon.* **3**, 46–47 (1953).
33. R. H. Sorgenfrey, On the topological product of paracompact spaces. *Bull. Amer. Math. Soc.* **53**, 631–632 (1947).
34. R. H. Bing, Metrization of topological spaces. *Canad. J. Math.* **3**, 175–186 (1951).
35. J. L. Kelley, "General Topology." Van Nostrand, Princeton, New Jersey, 1955.
36. E. Michael, Another note on paracompact spaces. *Proc. Amer. Math. Soc.* **8**, 822–828 (1957); Yet another note on paracompact spaces. *Ibid.* **10**, 309–314 (1959).
37. M. Arens and J. Dugundji, Remark on the concept of compactness. *Portugal. Math.* **9**, 141–143 (1950).
38. A. H. Stone, Paracompactness and product spaces. *Bull. Amer. Math. Soc.* **54**, 977–982 (1948).
39. M. J. Mansfield, Some generalizations of full normality. *Trans. Amer. Math. Soc.* **86**, 489–505 (1957).
40. H. J. Cohen, Sur un problème de M. Dieudonné. *C. R. Acad. Sci. Paris* **234**, 290–292 (1952).
41. A. Weil, "Sur les espaces à structure uniforme." Hermann, Paris, 1937.
42. P. Urysohn, Zum Metrisation problem. *Math. Ann.* **94**, 309–315 (1925).
43. A. Tychonoff, Über einen Metrisationsatz von P. Urysohn. *Math. Ann.* **95**, 139–142 (1926).
44. P. Urysohn, Über die Metrisation der kompakten topologischen Räume. *Math. Ann.* **92**, 275–293 (1924).
45. V. E. Sneider, Continuous images of Suslin and Borel sets. Metrization theorems. *Dokl. Acad. Nauk SSSR* [N.S.] **50**, 77–79 (1945).
46. J. Nagata, On a necessary and sufficient condition of metrizability. *J. Inst. Polytech. Osaka City Univ. Ser. A* **1**, 93–100 (1950).
47. Yu. M. Smirnov, A. necessary and sufficient condition for metrizability of a topological space. *Dokl. Acad. Nauk SSSR* [N.S.] **77**, 197–200 (1951).
48. Yu. M. Smirnov, On a problem connected with the metrizability of topological spaces. *Ukrain. Math. Ž.* **3**, 161–163 (1951).
49. Yu. M. Smirnov, On metrization of topological spaces. *Uspehi Mat. Nauk* [N.S.] **6**, 100–111 (1951); *Amer. Math. Soc. Transl.* No. 91 (1953).
50. M. Katetov, Complete normality of Cartesian products. *Fund. Math.* **35**, 271–274 (1948).

CHAPTER IV

Continuity

1. Functional Relations and Functions

In order to give a satisfactory definition of functions in general we must start from the theory of sets. First we define what is meant by the *graph* of a function or by a *functional relation.* As usual we denote by $(X \times Y)$ or more simply by $X \times Y$ the product of the sets X and Y, i.e., $X \times Y$ is the set of all ordered pairs (x, y) where $x \in X$ and $y \in Y$. Any subset $R \subseteq X \times Y$ is called a *binary relation* on $X \times Y$. Functional relations are special binary relations:

Definition 1. *A nonvoid subset F of the product $X \times Y$ of the nonvoid sets X and Y is called a functional relation with domain in X and range in Y if for every $x \in X$ there is at most one point $y \in Y$ such that $(x, y) \in F$.*

The sets $D = \{x : (x, y) \in F \ \ \textit{for some} \ \ y \in Y\}$ and $R = \{y : (x, y) \in F$ *for some* $x \in X\}$ are called the *domain of definition* or *domain* and the *range* of F. They are the projections of F into X and Y, respectively. In practice the set F is called the *graph* of the function f and that point y which occurs in the pair $(x, y) \in F$ for a given x is denoted by $f(x)$, fx, or f_x instead of the standard notation $F[x]$. The expressions "$y = f(x)$ corresponds to x" or "f maps x into $y = f(x)$" or "$y = f(x)$ is the image of x" or "f transforms x into $y = f(x)$" all mean that $(x, y) = (x, f(x)) \in F$. If f has domain X and if its range is in Y we write $f : X \to Y$.

The way we defined choice functions and product sets a function $f : X \to Y$ is the same as a point in the product set $Y^X = \prod\{Yx: x \in X\}$ where $Y_x = Y$ for every $x \in X$. But even if the power set Y^X is defined differently there is a natural one-to-one correspondence between functional relations F with domain X and functions f; namely, F corresponds to f if and only if $F[x] = f_x$ for every $x \in X$ where f_x denotes the projection of the point f into the coordinate space Y_x. This one-to-one correspondence allows us to prove theorems on functions and interpret them in terms of functional relations or vice versa.

We shall also use the expressions "f maps the set D into Y" or "f maps D onto R." If A is an arbitrary subset of D, then $f(A)$ will denote the set of all image points $y = f(x)$ $(x \in A)$. The definition is meaningful also when A is an arbitrary subset of X; in general $f(A)$ is the set $F[A \cap D]$ where F denotes the functional relation associated with $f \in Y^D$. If $f(D) = R = Y$, then we say that f maps D onto Y or f is *surjective*; otherwise if only $f(D) = R \subseteq Y$ is known, then f is said to map D into Y. In particular, if the domain D is X, then f maps X onto Y or into Y according as $f(X) = R = Y$ or $f(X) = R \subseteq Y$. The expressions "f is a function from X onto Y" or "into Y" are also used. When it is not important to distinguish between a function and its graph we shall use the word function and the symbol f to denote either one of them. When the graph interpretation is essential we shall speak of the function F or the functional relation F.

If X' and Y' are sets containing X and Y, respectively, and if $F' \subseteq X' \times Y'$ is a function from X' into Y' such that $F \subseteq F'$, then F' is called an *extension* of F. We can also say that F is a *restriction* of F'. If F is an arbitrary function from X into Y with domain D and if A is any subset of X, then $F \cap (A \times Y)$ is the graph of a function with domain $D \cap A$ and having the same values as F. It is called the restriction of F to A. Similarly, if B is a subset of Y, then $F \cap (X \times B)$ furnishes a restriction of F such that its range is in B.

Let F be a function from X into Y such that for every $y \in Y$ there is at most one point x with $y = f(x)$. Then F^{-1}, the set of all ordered pairs (y, x) for which $(x, y) \in F$ is itself a function from Y into X. The domain of definition of F^{-1} is R and its range is D. We say that F is *invertible* on X and its *inverse* is F^{-1}. The point which corresponds to y under F^{-1} is denoted by $f^{-1}(y)$. Using our previous terminology, we denote $f^{-1}(A)$ as the set of all image points $x = f^{-1}(y)$ as y varies over A. It is clear that F^{-1} is also invertible and its inverse is $(F^{-1})^{-1} = F$. If F is invertible, then F defines a one-to-one correspondence between the sets D and R. If f is invertible, then we call f *injective*. Moreover if f: $X \to Y$ is both surjective and injective, i. e., if f is a one-to-one map of X onto Y, then f is called *bijective*.

The symbol $f^{-1}(A)$ is defined independently whether F is invertible or not as the set of those points $x \in X$ for which there is a y in A such that $y = f(x)$. The set $f^{-1}(A)$ is called the *inverse image* of A. For any family of sets $A_i \subseteq Y$ we have

$$f^{-1}(\bigcap A_i) = \bigcap f^{-1}(A_i) \qquad \text{and} \qquad f^{-1}(\bigcup A_i) = \bigcup f^{-1}(A_i).$$

Let X, Y, and Z be arbitrary sets and let f and g be functions from X into Y and from Y into Z, respectively. Then one can define a function from X into Z by forming the composition of the graphs of f and g: If F and G are the graphs of f and g, then $F \circ G$ is the graph of a function from X into Z; namely, (x, z) belongs to $F \circ G$ if there is a $y \in Y$ such that $(x, y) \in F$ and $(y, z) \in G$, or in other words $z = g(y)$ and $y = f(x)$. The function associated with $F \circ G$ is called the *composition* of the functions f and g and is denoted by $f \circ g$. Of course it is possible that the range of f and the domain of g do not intersect. In this case $f \circ g$ is nowhere defined. The composition $f \circ g$ is sometimes called the *function theoretical* or *diagrammatical product* of f and g and is often denoted by gf.

Let f and g be functions from X into Y and from Y into Z, respectively. The inverse of G is in general a relation on $Z \times Y$ and similarly F^{-1} is a relation on $Y \times X$. The composition of these relations determines a relation on $Z \times X$ called $G^{-1} \circ F^{-1}$. Clearly we have $(F \circ G)^{-1} = G^{-1} \circ F^{-1}$ and so if A is any set in Z, then $(F \circ G)^{-1}[A]$, the inverse image of A under $F \circ G$, is the set $(G^{-1} \circ F^{-1})[A]$.

EXERCISES

1. If the function f establishes a one-to-one correspondence between the sets X and Y, verify that $f(cA) = cf(A)$ for every $A \subseteq X$.

2. Show that $f^{-1}(cA) = cf^{-1}(A)$ for any f mapping X onto Y.

3. Verify the relations $f^{-1}(\bigcup A_i) = \bigcup f^{-1}(A_i)$ and $f^{-1}(\bigcap A_i) = \bigcap f^{-1}(A_i)$ where f is a function from a set X into a set Y and $A_i \subseteq Y$ for every $i \in I$.

2. Local Continuity

The general concept of a function is too wide to be of interest in analysis or in topology because in these fields one usually deals with functions which are continuous on certain sets of points. This implies that the sets X, containing the domain of f, and Y, into which f maps the domain, are sets with some special structure such that continuity is meaningful.

The classical concept of continuity is that of *metric continuity*: Let X and Y be metric spaces and let f be a function from X into Y. Then f is said to be continuous at a point x of its domain if for every "allowance" $\epsilon > 0$ there is a "tolerance" $\delta > 0$ such that $d(f(x), f(\xi)) < \epsilon$ for every ξ satisfying $d(x, \xi) < \delta$. The continuity of f at x is independent of the metrics used on X and Y and it depends only on the topologies which are determined by these metrics. This observation leads to the notion of *topological continuity*: Let X and Y be topological spaces and let f be a function from X into Y. Then f is continuous at the point x of its domain if for every open set O_y in Y which contains $y = f(x)$ there is an open set O_x in X which contains x and is such that $f(\xi) \in O_y$ for every $\xi \in O_x$ or in other words $f(O_x) \subseteq O_y$. This definition includes metric continuity as a special case.

The continuity of f at x is a *local property*: f can be continuous at x and discontinuous at every other point of its domain. The definition involves only open sets which contain x and $y = f(x)$. In analysis one uses an even more general continuity concept which can be derived from topological continuity by rephrasing its definition: The function f from the topological space X into the topological space Y is continuous at x if for every neighborhood N_y of $y = f(x)$ there is a neighborhood N_x of x such that $f(N_x) \subseteq N_y$. The family of all neighborhoods of x is a filter on the set X and $x \in N_x$ for every $N_x \in \mathcal{N}(x)$. In the definition of continuity only the filters $\mathcal{N}(x)$ and $\mathcal{N}(y)$ occur and so the definition is meaningful even if no topologies are given on the sets but only these filters are defined. We arrive at the following definition:

Definition 1. *Let f be a function from a set X into a set Y and let x be a point in the domain of f. Suppose that two filters $\mathcal{F}(x)$ and $\mathcal{F}(y)$ are given on X and Y such that $x \in F_x$ for every $F_x \in \mathcal{F}(x)$ and $y = f(x) \in F_y$ for every $F_y \in \mathcal{F}(y)$. If for every $F_y \in \mathcal{F}(y)$ there is an $F_x \in \mathcal{F}(x)$ such that $f(F_x) \subseteq F_y$, then f is said to be continuous at x with respect to the filters $\mathcal{F}(x)$ and $\mathcal{F}(y)$.*

Note. If X and Y are topological spaces and if f is continuous at x with respect to the neighborhood filters of x and y, then f is called continuous at x with respect to the topologies given on X and Y.

The continuity of a function is usually verified by considering filter bases instead of the filters $\mathcal{F}(x)$ and $\mathcal{F}(y)$. This is the situation for instance when we are dealing with metric continuity and in particular with real-valued functions of a real variable: According to the so-called "ϵ - δ *definition of continuity*" f is continuous at x if for every $\epsilon > 0$ there is a $\delta > 0$ such that $|f(\xi) - f(x)| < \epsilon$ whenever $|\xi - x| < \delta$, i.e., $\eta = f(\xi)$ belongs to the interval $(y - \epsilon, y + \epsilon)$ whenever ξ belongs to $(x - \delta, x + \delta)$. As ϵ and δ vary over the positive real numbers these open intervals give bases for the neighborhood filters of x and y. As is well known, ϵ and δ can be restricted to the set of positive rationals or to the set consisting of all rationals of the form 2^{-n} ($n = 1, 2, \ldots$). The reason for the possibility of such restrictions is:

Lemma 1. *If $\mathcal{B}(x)$ and $\mathcal{B}(y)$ are bases for the filters $\mathcal{F}(x)$ and $\mathcal{F}(y)$, and if for every $B_y \in \mathcal{B}(y)$ there is a $B_x \in \mathcal{B}(x)$ such that $f(B_x) \subseteq B_y$, then f is continuous at x with respect to the filters $\mathcal{F}(x)$ and $\mathcal{F}(y)$.*

Proof. Given any $F_y \in \mathcal{F}(y)$ there is a $B_y \in \mathcal{F}(y)$ such that $B_y \subseteq F_y$.

By hypothesis there is a $B_x \in \mathscr{B}(x)$ such that $f(B_x) \subseteq B_y$. Since $F_x = B_x \in \mathscr{F}(x)$ we found a set $F_x \in \mathscr{F}(x)$ such that $f(F_x) \subseteq F_y$.

Upper and lower semicontinuity of real-valued functions are the best-known instances where metric continuity is not sufficient but filters must be used. Let X be a linearly ordered set and for a given $x \in X$ let $\mathscr{N}_<(x)$ be the neighborhood filter of x relative to the order topology generated by the intervals $(a, +\infty)$ $(a \in X)$. Similarly, let $\mathscr{N}_>(x)$ be the neighborhood filter of x in the order topology generated by the intervals $(-\infty, b)$ $(b \in X)$. By $\mathscr{N}_-(x)$ and $\mathscr{N}_+(x)$ we denote the neighborhood filters with respect to the half-open interval topologies $\mathscr{T}_-$ and $\mathscr{T}_+$.

Now we consider functions f into a linearly ordered set Y from some set X with some filter $\mathscr{F}(x)$ defined on it or with some topology given on X. We say that f is *upper semicontinuous* at x with respect to $\mathscr{F}(x)$ or with respect to the topology given on X if f is continuous with respect to the filters $\mathscr{F}(x)$ and $\mathscr{N}_>(y)$ where $y = f(x)$. Similarly, we can speak about the *lower semicontinuity* of f at x. If X itself is also a linearly ordered set, then we can introduce four different types of continuities at x according as we consider continuity at x with respect to the filters $\mathscr{N}_-(x)$ and $\mathscr{N}_<(y)$, or $\mathscr{N}_-(x)$ and $\mathscr{N}_>(y)$, or $\mathscr{N}_+(x)$ and $\mathscr{N}_<(y)$, or $\mathscr{N}_+(x)$ and $\mathscr{N}_>(y)$. These are called in order *left lower, left upper, right lower*, and *right upper semicontinuities*. It is easy to see the following: If f is left upper semicontinuous and also right upper semicontinuous at x, then f is upper semicontinuous at x with respect to the interval topology of X. In particular, if both X and Y are the reals and if f is left upper semicontinuous and right upper semicontinuous at x, then f is upper semicontinuous at x with respect to the usual topology of the reals. If X is linearly ordered and Y is an arbitrary set with a filter $\mathscr{F}(y)$ or if Y is an arbitrary topological space, then we may speak about the *left and right semicontinuity* of f at x: The function f from X into Y is left semicontinuous at x if it is continuous with respect to the filters $\mathscr{N}_-(x)$ and $\mathscr{F}(y)$ or if it is continuous with respect to $\mathscr{N}_-(x)$ and the filter $\mathscr{N}(y)$ of the point $y = f(x)$. The right semicontinuity of f can be defined similarly.

Lower and upper semicontinuous functions on a metric space are pointwise limits of monotonic sequences of continuous functions. This characterization of semicontinuity was discovered by Baire who proved the following:

Theorem 1. *A real-valued function f is lower (upper) semicontinuous on a metric space X if and only if there exists an increasing (decreasing) sequence of continuous functions f_n $(n = 1, 2, \ldots)$ on X such that $f_n(x) \to f(x)$ for every $x \in X$.*

Note. The convergence can be interpreted as convergence in the topological sense. For the pointwise convergence of the sequence (f_n) is equivalent to the convergence of the elementary filter associated with (f_n) in the product topology of Y^X where Y denotes the space of the reals.

Proof. The sufficiency of the condition follows easily: Let $\epsilon > 0$ be given. Since the functions f_n are continuous we can determine open sets $O_x{}^n$ such that $|f_n(\xi) - f_n(x)| < \epsilon/2$ for every $\xi \in O_x{}^n$. By hypothesis, $f_n(x) \to f(x)$ as $n \to \infty$ and so we can choose n so large that $f_n(x) > f(x) - \epsilon/2$. Hence $f_n(\xi) > f_n(x) - (\epsilon/2) > f(x) - \epsilon$ for every $\xi \in O_x{}^n$. The sequence of functions f_n being increasing we have $f(\xi) \geqslant f_n(\xi)$ for every $\xi \in X$ and so $f(\xi) > f(x) - \epsilon$ for every $\xi \in O_x{}^n$. This shows that f is lower semicontinuous at x.

The necessity can be proved as follows: Let d be a metric on X. Using the lower semicontinuity of f we can fix a $\delta_x > 0$ such that $f(\xi) \geqslant f(x) - 1$ whenever $d(\xi, x) \leqslant \delta_x$. We define the functions f_n $(n = 1, 2, ...)$ at every $x \in X$ as

$$f_n(x) = \text{glb}\{f(\xi) + nd(\xi, x) : d(\xi, x) \leqslant \delta_x\}.$$

We show that these f_n's satisfy all the requirements. For simplicity let $\varphi_n(\xi, x) = f(\xi) + nd(\xi, x)$. Then $\varphi_n(\xi, x) \leqslant \varphi_{n+1}(\xi, x)$ and so $f_n(x) \leqslant f_{n+1}(x)$. Moreover, $f_n(x) \leqslant f(x)$ and so $\lim f_n(x)$ exists for every $x \in X$ and $\lim f_n(x) \leqslant f(x)$. We have

$$\varphi_n(\xi, x_1) - \varphi_n(\xi, x_2) = n(d(\xi, x_1) - d(\xi, x_2)) \leqslant nd(x_1, x_2)$$

and so $f_n(x_1) \leqslant \varphi_n(\xi, x_2) + nd(x_1, x_2)$ and $f_n(x_1) \leqslant f_n(x_2) + nd(x_1, x_2)$. Hence by symmetry $|f_n(x_1) - f_n(x_2)| \leqslant nd(x_1, x_2)$. This shows the continuity of f_n. Now we show that $f_n(x) \to f(x)$ for every $x \in X$. By hypothesis, f is lower semicontinuous at x and so there is a $\delta > 0$ for every $\epsilon > 0$ such that $f(\xi) > f(x) - \epsilon$ for every $\xi \in X$ satisfying $d(\xi, x) < \delta$. Hence $f(\xi) + nd(\xi, x) > f(x) - \epsilon$ for every ξ satisfying $d(\xi, x) < \delta$. If we choose n so large that $n\delta > 1 - \epsilon$ then $f(\xi) + nd(\xi, x) > f(x) - \epsilon$ whenever $\delta_x \geqslant d(\xi, x) \geqslant \delta$, and so for such large values of n we have

$$\varphi_n(\xi, x) = f(\xi) + nd(\xi, x) > f(x) - \epsilon$$

for every $\xi \in X$ satisfying $d(\xi, x) \leqslant \delta_x$. Therefore $f_n(x) \geqslant f(x) - \epsilon$ which shows that $\lim f_n(x) = f(x)$.

The following is a basic result on the continuity of the composition of two functions:

Theorem 2. *Let f be a function from the set X into the set Y and let g be a function from Y into the set Z such that the composition $f \circ g$ is defined at a point $x \in X$. If f is continuous at x with respect to the filters $\mathscr{F}(x)$ and $\mathscr{F}(y)$ and if g is continuous at $y = f(x)$ with respect to the filters $\mathscr{F}(y)$ and $\mathscr{F}(z)$, then the composition $f \circ g$ is continuous at x with respect to $\mathscr{F}(x)$ and $\mathscr{F}(z)$.*

Proof. The result follows immediately from the definition of continuity: Given any $F_z \in \mathscr{F}(z)$ there is an $F_y \in \mathscr{F}(y)$ such that $g(F_y) \subseteq F_z$ because g is continuous at y. Similarly, f being continuous at x we can find a set $F_x \in \mathscr{F}(x)$ such that $f(F_x) \subseteq F_y$. Therefore

$$(f \circ g)(F_x) = g(f(F_x)) \subseteq g(F_y) \subseteq F_z$$

and so $f \circ g$ is continuous at x with respect to $\mathscr{F}(x)$ and $\mathscr{F}(z)$.

EXERCISES

1. Determine all continuous functions $f: X \to Y$ when
   (a) the topology of X is discrete;
   (b) X is nondiscrete and Y is a (T_1) space;
   (c) Y has the discrete topology and X is connected;
   (d) the topology of Y is nondiscrete.

[(a) and (d): Every f is continuous everywhere. (b): If f is continuous somewhere, then f is a constant. (c) For every $\xi \in X$ there is an O_ξ such that f is constant on O_ξ. Given any point $x \in X$ consider the set $S_x = \{\xi : f(\xi) = f(x)\}$. It follows that S_x is open and has no boundary points. Hence $S_x = X$ and f is constant.]

2. Determine all real-valued functions which are continuous on the noncountable set X when the topology of X is the topology of finite complements.

[If f is continuous at x, then $f(\xi) = f(x)$ for all but countably many values of $\xi \in X$: For $|f(\xi) - f(x)| < 1/n$ for all but finitely many values of ξ for each fixed value of $n = 1, 2, \ldots$. Hence f is a constant.]

3. Let $X = \{\infty\} \cup S$ where S is a noncountable set and ∞ is not an element of S. We say that $O \subseteq X$ is open if $\infty \notin O$ or if card $cO \leqslant \omega$. Show that this topology is completely normal.

(If $x \in S$, then $\{x\}$ is both open and closed. X is a Hausdorff space because if $x, y \in S$, then $O_x = \{x\}$ and $O_y = \{y\}$ are disjoint open neighborhoods of the distinct points x and y, and if $x \in S$, then $O_x = \{x\}$ and $O_\infty = c\{x\}$ are disjoint open sets containing x and ∞, respectively. Disjoint separated sets can be separated by open sets: If $\infty \notin A \cup B$, then we may choose $O_A = A$ and $O_B = B$. Suppose that $\infty \in A$ and A and B are separated. We show that B is closed and so we can choose $O_A = cB$ and $O_B = B$. Indeed $\bar{B} \subseteq \{\infty\} \cup B$ and $A \cap \bar{B} = \emptyset$ so that $\infty \in A$ implies $\bar{B} = B$.)

4. Let X be the space constructed in the preceding exercise. Show that a real-valued f is lower semicontinuous at ∞ if and only if there is an open set O_∞ such that $f(\xi) \geqslant f(\infty)$ for every $\xi \in O_\infty$.

[If f is lower semicontinuous at ∞, then for every $n = 1, 2, \ldots$ there is an open set $O_\infty{}^n$ such that $f(\xi) > f(\infty) - (1/n)$ for every $\xi \in O_\infty{}^n$. Hence $f(\xi) > f(\infty) - (1/n)$ for every $\xi \in \bigcap O_\infty{}^n$ and for every $n = 1, 2, \ldots$. It follows that $f(\xi) \geqslant f(\infty)$ for every $\xi \in O_\infty = \bigcap O_\infty{}^n$. Since cO_∞ is countable, O_∞ is open.]

5. Using the results stated in the preceding two exercises show that Baire's theorem cannot be extended to arbitrary uniform spaces.

[A function f is continuous everywhere if and only if f is continuous at ∞. Let $f(\infty) = 0$ and $f(x) = 1$ for every $x \in S$. Then f is lower semicontinuous everywhere. Let (f_n) be an increasing sequence of continuous functions on X such that $f_n(\infty) \to 0$. Since f_n is continuous there is an open set $O_\infty{}^n$ such that $f_n(\xi) = f_n(\infty)$ for every $\xi \in O_\infty{}^n$. Hence $f_n(\xi) = f_n(\infty)$ for every $n = 1, 2, \ldots$ and every $\xi \in O_\infty = \bigcap O_\infty{}^n$. Thus $f_n(\xi) \to f(\infty) = 0$ for every $\xi \in O_\infty$ and f is not the pointwise limit of the squence (f_n) $(n = 1, 2, \ldots)$.]

6. Let f map X into Y and let N_x be a neighborhood of $x \in X$. Show that if the restriction of f to N_x is continuous at x, then f is continuous at x.

[Given O_y, there is an O_x such that $f(O_x \cap N_x) \subseteq O_y$. Since $O_x \cap N_x$ is a neighborhood of x the function f is continuous at x.]

7. Prove the following: If f is a function of a complex variable and if $f(\zeta) \to f(z)$ as $\zeta \to z$ along every convex path, then f is continuous at z.

[Suppose there is an N_y such that there are points ζ arbitrarily close to z for which $f(\zeta) \notin N_y$. Then determine a polar angle φ such that for any small positive angle ϵ there are points ζ arbitrarily close to z for which $\varphi \leqslant \arg \zeta < \varphi + \epsilon$ and $f(\zeta) \notin N_y$. Either on the ray $\{\zeta : \arg \zeta = \varphi\}$ or on a convex path tangent to it, $f(\zeta) \notin N_y$ for points ζ arbitrarily close to z.]

3. Continuous Functions

The local continuity or continuity at a point is of interest primarily in analysis. In subjects dealing with more primitive mathematical structures one is interested only in functions which are continuous everywhere in their domain of definition. Then we may as well assume that the domain is the entire set X because points not in the domain can be discarded without affecting the definitions and the continuity of the function. These special types of functions are called continuous functions so that continuity in this context means continuity at every $x \in X$:

Definition 1. *A function f mapping a topological space X into another Y is called continuous if it is continuous at every $x \in X$.*

More generally we can speak about the continuity of a function $f: X \to Y$ with respect to two systems of filters $\{\mathscr{F}(x)\}$ $(x \in X)$ and $\{\mathscr{F}(y)\}$ $(y \in Y)$ given in the domain set X and the range set Y. We say that such a function $f: X \to Y$ is continuous with respect to these filter systems if f is continuous at every $x \in X$ with respect to the filters $\mathscr{F}(x)$ and $\mathscr{F}(y)$. If no confusion can arise, we say simply that f is continuous on X. In topology X and Y are usually topological spaces and the filter systems are the neighborhood filters of the points $x \in X$ and $y \in Y$. Then f is called continuous with respect to the topologies given on X and Y. Many authors reserve the words *map* and *mapping* for everywhere continuous functions.

For continuous functions defined on topological spaces there is a very useful continuity criterion:

Theorem 1. *A function mapping one topological space into another is continuous if and only if the inverse image of open sets is open.*

Proof. The sufficiency of the condition is clear: Given any $x \in X$ and any open set O_y containing the point $y = f(x)$ we can find an open set O_x such that $f(O_x) \subseteq O_y$; namely, $O_x = f^{-1}(O_y)$ satisfies the requirement. As a matter of fact, we have exactly $f(O_x) = O_y$. The condition is also necessary: Let f be continuous on X and let Q be an open set in Y. If $y \in Q$ and $y = f(x)$ for some point $x \in X$, then by the continuity of f at x there is an open set O_x in X such that $x \in O_x$ and $f(O_x) \subseteq Q$. This implies $O_x \subseteq f^{-1}(Q)$ and so $f^{-1}(Q)$ contains with each point x an open set O_x. Hence $f^{-1}(Q)$ is open.

There is a corollary of Theorem 1 which is an immediate consequence of the identity $f^{-1}(cA) = cf^{-1}(A)$; namely, choosing A to be open in Y we obtain:

Theorem 2. *A function mapping one topological space into another is continuous if and only if the inverse image of closed sets is closed.*

A simple application of these theorems leads to a useful lemma:

Lemma 1. *If f is a continuous map of X into Y, then for every set S in Y we have*

$$\overline{f^{-1}(S)} \subseteq f^{-1}(\bar{S}) \quad \text{and} \quad f^{-1}(S)^i \supseteq f^{-1}(S^i).$$

Conversely, if either one of these inclusions holds, then f is a continuous function.

Proof. Clearly, $f^{-1}(S) \subseteq f^{-1}(\bar{S})$ where $f^{-1}(\bar{S})$ is a closed set in the space X. Similarly, $f^{-1}(S) \supseteq f^{-1}(S^i)$ where $f^{-1}(S^i)$ is an open set in X. Hence the first half of the result follows from the definition of the closure and of the interior. The converse statement is a consequence of Theorems 1 and 2: In the first case we let S be an arbitrary closed set C of Y and obtain $\overline{f^{-1}(C)} \subseteq f^{-1}(\bar{C}) = f^{-1}(C)$ which show that the inverse image of each closed set is closed. The second case is similar but S has to be chosen open.

Lemma 2. *A function f from a topological space X into another Y is continuous if and only if for every set S in X we have $f(\bar{S}) \subseteq \overline{f(S)}$.*

Proof. Suppose that f is continuous. By Theorem 2, the inverse image of the closed set $\overline{f(S)}$ is a closed set in X. Since we have

$$S \subseteq f^{-1}(f(S)) \subseteq f^{-1}(\overline{f(S)})$$

where $f^{-1}(\overline{f(S)})$ is closed, it follows that $\bar{S} \subseteq f^{-1}(\overline{f(S)})$ and so

$$f(\bar{S}) \subseteq \overline{f(S)}.$$

Now suppose that the condition is satisfied. Let C be a closed set in Y and let $S = f^{-1}(C)$. Then

$$\bar{S} \subseteq f^{-1}(f(\bar{S})) \subseteq f^{-1}(\overline{f(S)}) \subseteq f^{-1}(\bar{C}) = f^{-1}(C) = S.$$

So $\bar{S} \subseteq S$ and the inverse image of closed sets is closed.

Lemma 3. *A function $f : X \to Y$ is continuous if and only if there is a base $\mathscr{B}$ for the open sets of Y such that $f^{-1}(B)$ is open in X for every $B \in \mathscr{B}$.*

Proof. Every open set $Q \subseteq Y$ can be expressed in the form $Q = \bigcup B_i$. Hence if f satisfies the condition then $f^{-1}(Q) = \bigcup f^{-1}(B_i)$ is open for every open set $Q \subseteq Y$. The necessity is obvious from Theorem 1.

Suppose that f maps X onto Y such that f is invertible, i.e., f defines a one-to-one correspondence between the elements of X and Y. It is possible that f is continuous on X, nevertheless, f^{-1} is not a continuous function. As a matter of fact, in some cases f can be chosen such that f^{-1} is discontinuous at every point $y \in Y$; for example, let Y be any set containing at least two elements and let the open sets of Y be $\varnothing$ and Y itself. We let $X = Y$ but the topology of X we choose to be discrete. Then the identity map $f : x \to x$ is continuous on X but f^{-1} is discontinuous at every $y \in Y$. The reader will notice that the topology on Y is not a (T_0) topology. It is possible to find similar examples where Y satisfies some axioms of separation; for instance, one can choose Y to be a Hausdorff space.

EXERCISES

1. Let $\mathscr{S}$ be a subbase for the open sets of Y. Show that if $f^{-1}(S)$ is open for every $S \in \mathscr{S}$, then f is continuous.

2. Let f be a real-valued function and let $\{x : f(x) > a\}$ be open for every real a. Show that f is lower semicontinuous.

3. Show that a real-valued f is continuous if and only if the sets $\{x : f(x) < a\}$ and $\{x : f(x) > a\}$ are open for every point $a \in A$ of a dense subset A of the space of real numbers.

4. Use the definition of local continuity to prove that if $f : X \to Y$ is continuous, then $f(\bar{S}) \subseteq \overline{f(S)}$ for every $S \subseteq X$.

[Let $x \in \bar{S}$ and let $y = f(x)$. For every O_y there is an O_x such that $f(O_x) \subseteq O_y$. Since O_x intersects S it follows that O_y intersects $f(S)$ and so $y \in \overline{f(S)}$.]

5. Let f and g be lower semicontinuous real-valued functions on X. Show that $f \cap g : x \to \min \{f(x), g(x)\}$ is lower semicontinuous.

6. Let f and g be continuous real-valued functions. Show that $f \cup g : x \to \max \{f(x), g(x)\}$ and $f \cap g : x \to \min \{f(x), g(x)\}$ are continuous functions.

7. Show that $f : X \to Y$ is continuous with respect to the topologies $\mathscr{T}_x$ and $\mathscr{T}_y$ if and only if $f^{-1}(\mathscr{T}_y) \leqslant \mathscr{T}_x$.

8. Prove that if $f : X \to Y$ is continuous with respect to $\mathscr{T}_x^i$ and $\mathscr{T}_y$ for each $i \in I$, then f is continuous also with respect to $\text{glb}\{\mathscr{T}_x^i\}$ and $\mathscr{T}_y$.

4. Homeomorphisms, Open and Closed Maps

Of particular interest are those invertible functions f for which f and f^{-1} are both continuous.

Definition 1. *A homeomorphism of one topological space X onto another Y is an invertible function f mapping X onto Y such that f and f^{-1} are both continuous.*

This definition is in complete accordance with Definition I.1.2 because in view of Theorem 3.1 a one-to-one map of X onto Y is a homeomorphism if and only if it gives rise to a one-to-one correspondence between the open sets of X and Y.

Evidently if f is a homeomorphism, then so is f^{-1}, and so if X is homeomorphic to Y, then also Y is homeomorphic to X. In symbols, if $X \sim Y$, then $Y \sim X$. Clearly, we have $X \sim X$ for every topological space X because the identity map $f : x \to x$ is a homeomorphism of X onto itself. Furthermore, in view of Theorem 2.2, if $X \sim Y$ and $Y \sim Z$, then $X \sim Z$. Therefore the homeomorphism is a symmetric, reflexive, and transitive relation between topological spaces. For this reason homeomorphic spaces are often called *topologically equivalent*.

One of the fundamental problems of topology is to supply a great number of useful criteria which can be used to determine whether or not two topological spaces are homeomorphic. The importance of this question is due to the following: If $X \sim Y$, then there is a one-to-one correspondence between the elements of X and Y which induces a one-to-one correspondence between the open sets of X and Y. In fact, if f is a homeomorphism of X onto Y, then by Theorem 3.1 the set $f^{-1}(Q)$ is open for every open set Q in Y and also $f(O)$ is open for every open set O in X. Consequently, from a purely topological point of view, X and Y are indistinguishable; if a topological theorem holds for X, then a corresponding theorem holds also for Y and conversely.

We shall use the expression "*topological isomorphism*" for another relation between topological spaces which in general is weaker than homeomorphism:

Definition 2. *The topological spaces X with open sets $O \in \mathcal{O}$ and Y with open sets $Q \in \mathcal{Q}$ are called topologically isomorphic if there is a one-to-one correspondence $f : \mathcal{O} \to \mathcal{Q}$ such that*

(i) $$f(\emptyset) = \emptyset \quad \text{and} \quad f(X) = Y;$$

(ii) $$f(O_1 \cap O_2) = f(O_1) \cap f(O_2);$$

(iii) $$f(\bigcup O_i) = \bigcup f(O_i)$$

for every choice of the open sets O_1, O_2, and O_i $(i \in I)$ in the family $\mathcal{O}$.

A one-to-one correspondence $f : \mathcal{O} \to \mathcal{Q}$ satisfying the requirements (i), (ii), and (iii) is called a topological isomorphism of the space X onto the space Y. It is obvious that if f is a topological isomorphism of X

onto Y, then f^{-1} is a topological isomorphism of Y onto X. Hence the topological isomorphism is a symmetric relation between X and Y. The transitivity can be easily verified. If X and Y are topologically isomorphic and a statement involving only open and closed sets holds in X, then a corresponding statement holds in Y.

Lemma 1. *Let X and Y be topological spaces, let $\mathscr{B}$ be a base for X and let f be a one-to-one map of X onto Y. Then f is a homeomorphism if and only if the family $f(\mathscr{B}) = \{f(B)\}$ $(B \in \mathscr{B})$ is a base for the topology on Y.*

Proof. First let f be a homeomorphism. If Q is an open set in Y, then $O = f^{-1}(Q)$ is an open set in X. Therefore O is the union of some sets $B \in \mathscr{B}$ and so $Q = f(O)$ is a union of some of the sets $f(B) \in f(\mathscr{B})$. Hence $f(\mathscr{B})$ is a base for the topology on Y. Now let us suppose that $f(\mathscr{B})$ is a base for Y. Then the image of every open set $O \subseteq X$ is open in Y because $O = \bigcup B_i$ for some sets $B_i \in \mathscr{B}$ and so $f(O) = \bigcup f(B_i)$. Similarly, if $Q \subseteq Y$ is an open set, then $Q = \bigcup f(B_i)$ and so $f^{-1}(Q) = f^{-1}(\bigcup f(B_i)) = \bigcup B_i$ is open in X. Consequently f is a homeomorphism.

There are a few more standard expressions which we shall use in the future: If f is a homeomorphism, then so is f^{-1}, the *inverse homeomorphism*. A homeomorphism is sometimes called a *topological transformation*. If f is invertible and both f and f^{-1} are continuous functions, then f is called a *bicontinuous transformation* from X into Y.

Definition 3. *An open (closed) map of a topological space X into another Y is a function f on X into Y such that the image of open (closed) sets under f is open (closed).*

For example, the projection π_s of the product space $X = \prod X_s$ onto the coordinate space X_s is an open map. If f is an open map and C is a closed set in X, then $f(C)$ need not be closed. A simple example for this situation can be found in product spaces: Let $X = X_1 \times X_2$ where X_1 and X_2 are copies of the space of real numbers, let f be the projection $\pi_1 : X \to X_1$ and let $C = \{(x_1, x_2) : x_1 x_2 \geqslant 1\}$. Hence not every continuous open map is closed. Similarly, a continuous map can be closed without being open. We can also find open and closed maps which are not continuous.

Continuous open and closed maps can be characterized by using the following notion:

Definition 4. *A map f of X onto Y is called quasi compact if S is open in Y whenever $f^{-1}(S)$ is open in X.*

Since $cf^{-1}(S) = f^{-1}(cS)$ for any set S we have:

Lemma 2. *A surjective map $f : X \twoheadrightarrow Y$ is quasi compact if and only if S is closed in Y whenever $f^{-1}(S)$ is closed in X.*

It is clear that every open or closed map is quasi compact but quasi compactness alone does not imply that f is open or closed. The additional property needed for a complete characterization of surjective open and closed maps concerns the decomposition of X into the disjoint sets $f^{-1}(y)$ $(y \in Y)$.

Definition 5. *Let $\mathcal{S} = \{S_i\}$ $(i \in I)$ be a partition of the topological space X into disjoint sets S_i $(i \in I)$. $\mathcal{S}$ is called an upper semicontinuous decomposition if $\bigcup \{S_i : S_i \subseteq O\}$ is open for every open set O. $\mathcal{S}$ is called a lower semicontinuous decomposition if $\bigcup \{S_i : S_i \cap O \neq \varnothing\}$ is open for every open set O.*

Since $\mathcal{S}$ is a cover of X for every set $A \subseteq X$ we have

$$X = \bigcup S_i = \bigcup \{S_i : S_i \cap A \neq \varnothing\} \cup \bigcup \{S_i : S_i \subseteq cA\}$$

where the two unions on the right-hand side are disjoint sets. Choosing A to be closed or open we obtain:

Lemma 3. *A partition $\mathcal{S} = \{S_i\}$ $(i \in I)$ is an upper semicontinuous decomposition of X if and only if $\bigcup \{S_i : S_i \cap C \neq \varnothing\}$ is closed whenever C is a closed set in X and it is lower semicontinuous if and only if $\bigcup \{S_i : S_i \subseteq C\}$ is closed when C is a closed set in X.*

We have the following theorem on continuous open and closed maps:

Theorem 1. *A continuous mapping f of X onto Y is open (closed) if and only if it is quasi compact and $\{f^{-1}(y)\}$ $(y \in Y)$ is a lower (upper) semi-continuous decomposition of X.*

Proof. Suppose first that the decomposition $\{f^{-1}(y)\}$ $(y \in Y)$ generated by f is lower semicontinuous. Then for every open O in X the set $f^{-1}(\{y : f^{-1}(y) \cap O \neq \varnothing\})$ is open. If f is quasi compact, then

$$\{y : f^{-1}(y) \cap O \neq \varnothing\} = f(O)$$

is also open. Hence f is an open map and the conditions are sufficient. Now suppose that f is a continuous open map of X onto Y. Then f is clearly quasi compact and also $f^{-1}(f(O))$ is open for every open set $O \subseteq X$. However,

$$f^{-1}(f(O)) = \{x : f^{-1}(f(x)) \cap O \neq \varnothing\} = \bigcup \{f^{-1}(y) : f^{-1}(y) \cap O \neq \varnothing\}.$$

Hence f generates a lower semicontinuous decomposition of X. In the case of closed maps the proof is similar.

Every map $f: X \to Y$ generates an equivalence relation R on X: We say that $x_1 \sim x_2$ if $f(x_1) = f(x_2)$, so the equivalence classes are the sets $f^{-1}(y)$ $(y \in Y)$. If f is an open or closed continuous map, then up to a homeomorphism Y is determined by X and the equivalence relation R:

Theorem 2. *If f is an open (closed) continuous map of X onto Y, then Y is homeomorphic to the quotient space X/R where R denotes the equivalence relation generated by f.*

Proof. We associate with each $y \in Y$ the equivalence class $f^{-1}(y)$. Under this one-to-one map a set $Q \subseteq Y$ corresponds to the set

$$\{f^{-1}(y): y \in Q\} \subseteq X/R.$$

By the definition of the quotient topology, $\{f^{-1}(y) : y \in Q\}$ is open in X/R if and only if $\bigcup \{f^{-1}(y) : y \in Q\} = f^{-1}(Q)$ is open in X. Since f is continuous and open, $f^{-1}(Q)$ is open if and only if Q is an open set in Y. Therefore the correspondence $y \to f^{-1}(y)$ between the elements of Y and X/R is a homeomorphism.

We finish with a simple lemma concerning closed maps:

Lemma 4. *If f is a closed map of X onto Y, then for every open set $O \subseteq X$ and for every $y \in Y$ satisfying $f^{-1}(y) \subseteq O$ we have $y \in f(O)^i$.*

Proof. Since $f^{-1}(y) \subseteq O$ we see that $y \in cf(cO)$ where $f(cO)$ is a closed set. Moreover, $cf(cO) \subseteq f(O)$ and so $cf(cO)$ being open we have $y \in f(O)^i$.

Corollary. *If f is a closed map of X onto Y, then for every set $S \subseteq X$ and for every point $x \in X$ satisfying $f^{-1}(f(x)) \subseteq S^i$ we have $f(x) \in f(S^i)^i$.*

EXERCISES

1. Show that the cross formed by the union of the coordinate axes is a subspace of the plane which is not homeomorphic to the real line.

[If we omit any point from the real line the subspace obtained has two components. If we omit the point $(0, 0)$ of the given space, then the remaining subspace has four components.]

2. Show that the real line is homeomorphic to the space $(-1, 1)$.

[A homeomorphism of the space of reals onto $(-1, 1)$ is given by the correspondence $x \to x/(1 + |x|)$.

3. Show that the intervals (0,1) and [0,1] are not homeomorphic subspaces of the reals.

(The closed interval is compact while the open interval is not.)

4. Show that the intervals (0,1] and (0,1) are not homeomorphic.

[If we omit 1 from (0,1], then the rest is connected. If we omit any point x of (0,1), the rest is not connected.]

5. Show that $X = [0,1]$ and $Y = \{(x_1, x_2) : 0 \leqslant x_1, x_2 \leqslant 1\}$ are not homeomorphic spaces.

(If we omit three points of X, the remaining subspace will not be connected. Omitting any three points of Y we always obtain a connected space.)

6. Show the existence of a one-to-one map of the unit interval $X = [0,1]$ onto the unit square $Y = \{(x_1, x_2) : 0 \leqslant x_1, x_2 \leqslant 1\}$ by using the correspondence $x \to y = (x_1, x_2)$ where $x = \sum \epsilon_k 2^{-k}$; $x_1 = \sum \epsilon_{2k-1} 4^{-k}$; $x_2 = \sum \epsilon_{2k} 4^{-k}$, and $\epsilon_k = 0, 1$ for every $k = 1, 2, \ldots$.

[Every $x \in (0, 1]$.can be expressed uniquely in the form $x = \sum \epsilon_k 2^{-k}$ where $\epsilon_k \neq 0$ for infinitely many indices $k = 1, 2, \ldots$. The given correspondence leads to a one-to-one map between X and a subset of Y. Construct a similar one-to-one correspondence between Y and a subset of X. Then apply the Cantor-Bernstein theorem.]

7. Show by an example that there is no Cantor-Bernstein theorem for homeomorphisms.

[The problem is to find topological spaces X and Y such that X is homeomorphic to a subspace of Y and Y is homeomorphic to a subspace of X but X and Y are not homeomorphic. Let $X = \{0\} \cup [1, 2]$ and let $Y = [1, 2]$, or let $X = (0, 1)$ and let $Y = (0, 1) \cup (2, 3)$. In the second case X is homeomorphic to an open subspace of Y and Y is homeomorphic to an open subspace of X.]

8. Show that a one-to-one map f of X onto Y is a homeomorphism if and only if $f(\bar{A}) = \overline{f(A)}$ for every $A \subseteq X$.

[Use Lemma 3.2: By the continuity of f we have $f(\bar{A}) \subseteq \overline{f(A)}$ and by the continuity of f^{-1} we have $f^{-1}(\overline{f(A)}) \subseteq \overline{f^{-1}(f(A))} = \bar{A}$ so that $\overline{f(A)} \subseteq f(\bar{A})$.]

9. Let f be a closed continuous map of X onto Y. Show that if X is a (T_4) space, then so is Y.

[Let A and B be disjoint closed sets in Y. The closed sets $f^{-1}(A)$ and $f^{-1}(B)$ can be separated by open sets O_A and O_B. Let

$$Q_A = \{y : f^{-1}(y) \subseteq O_A\} \quad \text{and} \quad Q_B = \{y : f^{-1}(y) \subseteq O_B\}.$$

These are disjoint sets containing A and B, respectively. By Theorem 1 the sets Q_A and Q_B are open.]

10. Show by an example that the closed continuous image of a regular topological space need not satisfy axiom (T_3).

[Let $X = Y$ be the set of reals. Let the topology of X be the discrete topology and let Y be topologized in the way as is given in Example II.2.1. Then X is completely normal, Y is a Hausdorff space but it is not a (T_3) space.]

11. Show that if $f : X \to Y$ is a one-to-one transformation, then the following properties are equivalent: (i) f is open; (ii) f is closed; (iii) f is quasi compact; (iv) f^{-1} is continuous.

[(i) → (ii) because $f(C) = cf(cC)$; (ii) → (iii) because every surjective open or closed transformation is quasi compact; (iii) → (iv) since $(f^{-1})^{-1}(O) = f(O)$ where $f^{-1}(f(O)) = O$ is open; (iv) → (i) because $f(O) = (f^{-1})^{-1}(O)$ and the inverse image of the open set O under the continuous f^{-1} is open.]

12. Suppose that f maps X onto Y and g maps Y onto Z. Show that if f is continuous and if $f \circ g$ is open (or closed or quasi compact), then g is also open (or closed or quasi compact).

[We have $g(S) = (f \circ g)(f^{-1}(S))$ and $(f \circ g)^{-1}(S) = f^{-1}(g^{-1}(S))$. Set $S = O$ or $S = C$.]

13. Let $h : Y \to Z$ be a homeomorphism and let $f : X \to Y$ and $g : X \to Z$ be functions such that $(f \circ h)(x) = h(f(x)) = g(x)$ for every $x \in X$. Show that f is continuous at x if and only if g is continuous there.

14. For any real-valued function f let $[f]$ be defined by the formula

$$[f](x) = \frac{f(x)}{1 + |f(x)|}$$

Show that f is continuous at a point x if an only if $[f]$ is continuous at x.

[By Exercise 2 the map $h : y \to y/(1 + |y|)$ is a homeomorphism. Hence the result stated in the preceding exercise can be applied.]

15. Extend the result of the preceding exercise to upper and lower semicontinuous functions.

5. Real-Valued Functions

Real-valued functions defined on topological spaces are particularly important and deserve special attention. If the topology of the range space is the usual topology of the reals, these functions are often called *functionals.* This name is especially common for real-valued functions defined on linear spaces. Real-valued functions of a real variable will be referred to as *functions of a real variable.* This section contains some of the elementary results concerning functionals and functions of a real variable.

Theorem 1. *Let f be a real-valued function on a topological space X. Then*

(i) *f is lower semicontinuous if and only if $f^{-1}((a, \infty)) = \{x : a < f(x)\}$ is an open set in X for every real a.*

(ii) *f is upper semicontinuous if and only if $f^{-1}((-\infty, a)) = \{x : f(x) < a\}$ is open for every real a.*

(iii) *If f is continuous, then the closures of the open sets $\{x : a < f(x)\}$ and $\{x : f(x) < a\}$ are contained in the closed sets $\{x : a \leqslant f(x)\}$ and $\{x : f(x) \leqslant a\}$, respectively.*

Proof. The first two statements follow from Lemma 3.3. The third is an immediate consequence of the first and the second.

Theorem 2. *A real-valued function $f : X \to \mathfrak{R}$ is lower semicontinuous if and only if the set $L(f) = \{(x, r) : f(x) > r\}$ is open in $X \times \mathfrak{R}$ and it is upper semicontinuous if and only if $U(f) = \{(x, r) : f(x) < r\}$ is an open set of $X \times \mathfrak{R}$.*

Proof. Let f be lower semicontinuous and let $f(x) > r$. Then there is an open neighborhood O_x such that $f(\xi) > [f(x) + r]/2$ for every $\xi \in O_x$. Hence $L(f)$ contains with each point (x, r) a whole open set

$$\left\{(\xi, \rho) : \xi \in O_x \quad \textit{and} \quad \frac{f(x) + r}{2} > \rho\right\}.$$

Conversely, if $L(f)$ is open in $X \times \mathfrak{R}$, then given $x \in X$ and $\epsilon > 0$, there is an O_x and a $\delta > 0$ such that $\{(\xi, \rho) : \xi \in O_x \ \ \textit{and} \ \ f(x) - \epsilon + \delta > \rho\}$ is a subset of $L(f)$. Choosing $\rho = f(x) - \epsilon$ we see that $f(\xi) > f(x) - \epsilon$ for every $\xi \in O_x$. Hence f is lower semicontinuous at x. Considering $-f$ in place of f we obtain the corresponding result on upper semicontinuous functions.

Lemma 1. *If a real-valued function $f : X \to \mathfrak{R}$ is continuous, then its graph $F = \{(x, r) : f(x) = r\}$ is a closed set in the product space $X \times \mathfrak{R}$.*

Proof. Since $L(f) \cup U(f)$ is the complement of F in $X \times \mathfrak{R}$, the lemma follows from the preceding theorem. It can also be proved directly from the continuity of f.

It is simple to show that a linear combination of continuous real-valued functions is again a continuous function. For it is evident from the definition of continuity that together with f and g the sum $f+g$ and the constant multiple λf of f are continuous functions. The deeper meaning of this simple result will be discussed later in Section 9. (See Lemma 9.1.)

Now we turn to real-valued functions of a real variable. The simplest ones are the monotonic functions and these have a number of interesting continuity properties. Some properties hold also when the domain X and the range Y of the monotonic function f are arbitrary linearly ordered sets and their topology is the interval topology. There are of course some obvious properties. For instance, an increasing function is left upper and right lower semicontinuous and a decreasing function is right upper and left lower semicontinuous. A somewhat less obvious result concerns the continuity of the inverse function:

Theorem 3. *Let f be a strictly monotonic function mapping a linearly ordered set X onto another linearly ordered set Y. If f is continuous at a point $x \in X$ with respect to the interval topologies of X and Y, then f^{-1} is continuous at $y = f(x)$ with respect to the same topologies.*

Note. The point of the theorem is that it is sufficient to suppose continuity at x. If f is continuous everywhere, the result holds also for other types topological spaces X and Y. (See Lemma 7.3.)

Corollary. *If X and Y are linearly ordered and if $f : X \to Y$ is a continuous one-to-one map, then f is a homeomorphism.*

Proof. We suppose that f is a strictly increasing function. Then f^{-1} is also strictly increasing. Let $y = f(x)$ and let (a, b) be an open interval which contains x. If we set $c = f(a)$ and $d = f(b)$, then f being strictly increasing we have $c < y < d$. For the same reason if $\eta \in (c, d)$, then $\xi = f^{-1}(\eta) \in (a, b)$. Hence (c, d) is an open set which contains y and its image $f^{-1}((c, d))$ is a subset of (a, b). Therefore f^{-1} is continuous at y.

It is well known that a monotonic function of a real variable can be discontinuous at not more than countably many points. (See the exercises following this section.) Functions of bounded variation and the so-called regular functions of a real variable also have this property. We now define a class of functions called nearly regular functions which includes all the preceding ones and its elements still have this continuity property. For the sake of simplicity we shall use the expression "*nearly everywhere*" to abbreviate the phrase "with the possible exception of at most countably many points."

We assume that the reader is familiar with the concept of one-sided limits of real-valued functions of a real variable. These notions can also be defined for real-valued functions on any linearly ordered set X as follows:

Definition 1. *The real number l is called the left limit of f at x and is denoted by* $\lim_{\xi\to x-0} f(\xi)$ *if for every $\epsilon > 0$ there is a point $a < x$ such that $|f(\xi) - l| < \epsilon$ for every $\xi \in [a, x)$.*

The right limit of f can be defined similarly.

Definition 2. *The real number l is the left limit superior of f at x and is denoted by* $\limsup_{\xi\to x-0} f(\xi)$ *if*

(i) *for every $\epsilon > 0$ there is a point $a < x$ such that $f(\xi) < l + \epsilon$ for each $\xi \in [a, x)$ and*

(ii) *for every $\epsilon > 0$ and $a < x$ there is a point $\xi \in [a, x)$ such that $f(\xi) > l - \epsilon$.*

Similarly we can define the right limit superior of f at x and also the left and right limit inferior of f at x. These notions are specially interesting when X is densely ordered at x; that is, when no interval (a, x) $(a < x)$ or (x, b) $(x < b)$ is void. The deeper topological meaning of these notions will be discussed in the next chapter.

Definition 3. *A real-valued function f defined on a linearly ordered set X is called (nearly) regular if the one-sided limits* $\lim_{\xi\to x-0} f(\xi)$ *and* $\lim_{\xi\to x+0} f(\xi)$ *exist (nearly) everywhere in X.*

Now we are going to prove two theorems which will imply that nearly regular functions of a real variable are continuous nearly everywhere. For functions of a real variable these theorems were first proved by W. H. Young in 1908.

Theorem 4. *Let f be a real-valued function defined on a linearly ordered set X. If X is a hereditary Lindelöf space with respect to its half-open interval topologies, then the points x at which* $\limsup_{\xi\to x-0} f(\xi)$ *and* $\limsup_{\xi\to x+0} f(\xi)$ *are not equal form a countable set.*

Note. Notice that there are no conditions given on f. The hereditary Lindelöf property can be replaced by the following weaker hypothesis: Every noncountable set $S \subseteq X$ contains an accumulation point of S.

Several corollaries can be obtained by putting various conditions on f and on X and also by replacing f by $-f$. We mention only one of these which immediately implies that monotonic functions of a real variable are continuous nearly everywhere:

Corollary. *If f is a function of a real variable which is left or right upper semicontinuous and also left or right lower semicontinuous nearly everywhere in an interval I, then f is continuous nearly everywhere in I.*

Note. The direction of approach of the one-sided semicontinuity can vary from point to point. The herditary Lindelöf property of the half-open interval topologies of the reals is discussed in Exercise II.12.9.

Proof. For simplicity we use the symbols $\Lambda^- = \Lambda^-(x)$ and $\Lambda^+ = \Lambda^+(x)$ to denote the one-sided limit superiors of f as $\xi \to x$; thus $\Lambda^\pm(x) = \limsup_{\xi\to x\pm 0} f(\xi)$. We prove the theorem by contradiction. We suppose that $\Lambda^-(x) \neq \Lambda^+(x)$ for every point of a noncountable set $X_1 \subseteq X$. We assume that $\Lambda^-(x) < \Lambda^+(x) < +\infty$ for the points of a noncountable set $X_2 \subseteq X_1$.

If $\Lambda^+(x) < \Lambda^-(x) < +\infty$ for every point of a noncountable set $X_2 \subseteq X_1$, then the contradiction follows by a similar reasoning. If neither situation takes place, then there is a noncountable set X_2 such that $\Lambda^-(x) < \Lambda^+(x) = +\infty$ for every $x \in X_2$ or $\Lambda^+(x) < \Lambda^-(x) = +\infty$ for every $x \in X_2$. Then the reasoning is similar but considerably simpler.

Since X_2 is not countable there is an $\epsilon > 0$ and a noncountable subset X_3 of X_2 such that $\Lambda^-(x) < \Lambda^+(x) - \epsilon$ for every $x \in X_3$: For if this were not true, then we would have $\Lambda^-(x) < \Lambda^+(x) - (1/k)$ for at most countably many points and so considering $k = 1, 2, \ldots$ the inequality $\Lambda^-(x) < \Lambda^+(x)$ could hold for at most countably many points $x \in X_2$. Now we consider the set of values $\Lambda^+(x)$ $(x \in X_3)$. If a particular value Λ^+ is attained for the points of a noncountable subset of X_3, then we choose such a Λ^+ and such a set $X_4 \subseteq X_3$. If this is not the case, then the set of all distinct $\Lambda^+(x)$ values for $x \in X_3$ is not countable and so it has an ω-accumulation point Λ^+, i.e., an accumulation point Λ^+ each neighborhood of which meets the set of $\Lambda^+(x)$ values in uncountably many points. Hence in both cases we can find a noncountable subset $X_4 \subseteq X_3$ such that $|\Lambda^+ - \Lambda^+(x)| < \epsilon/3$ for every $x \in X_4$.

The half-open interval topology $\mathscr{T}_-$ has the hereditary Lindelöf property and so there is an accumulation point x of X_4 with respect to $\mathscr{T}_-$ which belongs to X_4. Let $a < x$ and let $\xi \in [a, x)$. By the definition of $\Lambda^+(\xi)$ there is an $\eta \in (\xi, x]$ such that $\Lambda^+(\xi) - (\epsilon/3) < f(\eta)$. Since x is an accumulation point of X_4 with respect to $\mathscr{T}_-$ we can choose ξ in X_4 and η in (ξ, x). Hence we have $|\Lambda^+ - \Lambda^+(\xi)| < \epsilon/3$. These inequalities imply $\Lambda^+ - (2\epsilon/3) < f(\eta)$. Since $a < x$ is arbitrary this shows that $\Lambda^+ - (2\epsilon/3) \leqslant \Lambda^-(x)$. As $x \in X_4$ we have

$$| \Lambda^+(x) - \Lambda^+ | < \epsilon/3,$$

therefore $\Lambda^+(x) - \epsilon < \Lambda^-(x)$. However, $x \in X_4 \subseteq X_1$ and so $\Lambda^-(x) < \Lambda^+(x) - \epsilon$. This is a contradiction and the theorem follows.

The next theorem was also first proved by W. H. Young for the special case of functions of a real variable:

Theorem 5. *Let f be a real-valued function on a hereditary Lindelöf and metrizable space X. Then the points x for which the closed interval*

$$[\liminf_{\xi \to x} f(\xi), \quad \limsup_{\xi \to x} f(\xi)]$$

does not contain $y = f(x)$ form a countable set.

Note. In view of Exercise II.12.10 X is a metrizable space satisfying the second countability axiom.

Proof. We give an indirect proof. We suppose that there is a noncountable set $X_1 \subseteq X$ such that for every $x \in X_1$ the interval in question does not contain $y = f(x)$. Then there exist $\epsilon(x) > 0$ and $\delta(x) > 0$ for every $x \in X_1$ with the property that $| y - \eta | > \epsilon(x)$ for every $\xi \in X$ ($\xi \neq x$ and $\eta = f(\xi)$) satisfying $d(x, \xi) < \delta(x)$. We may assume that $\delta(x)$ and $\epsilon(x)$ are of the form $\epsilon(x) = 1/m(x)$ and $\delta(x) = 1/n(x)$ where $m(x)$ and $n(x)$ are positive integers. Then on the one hand there are only denumerably many possibilities for the pair $m(x), n(x)$ and on the other hand there are noncountably many choices for $x \in X_1$. Therefore there exist $\epsilon > 0$, $\delta > 0$, and a noncountable subset X_2 of X_1 such that for every $x \in X_2$ we have $| y - \eta | > \epsilon$ whenever $\xi \in X$ ($\xi \neq x$) satisfies $d(x, \xi) < \delta$.

Now we consider the set S of ordered pairs (x, y) where $x \in X_2$ and $y = f(x)$. This is a noncountable set. The product of X and of the space of reals is a hereditary Lindelöf space. (See Lemma II.12.3 and Exercise II.12.11.) Therefore S has an accumulation point (ξ, η). Hence there is a point $x \in X_2$ ($x \neq \xi$) such that $d(x, \xi) < \delta$ and $| y - \eta | < \epsilon$ where $\eta = f(\xi)$. Since $x \in X_2$ and $d(x, \xi) < \delta$ we have $| y - \eta | > \epsilon$ for every $\xi \in X$ ($\xi \neq x$) satisfying $d(x, \xi) < \delta$ where $y = f(x)$ and $\eta = f(\xi)$. This is a contradiction and so the theorem is proved.

The last two theorems imply:

Theorem 6. *A nearly regular function defined on an interval of the real line is continuous nearly everywhere in its domain of definition.*

Proof. By the definition of a regular function the limits $\lim_{\xi\to x-0} f(\xi)$ and $\lim_{\xi\to x+0} f(\xi)$ exist nearly everywhere and by Theorem 4 they are equal nearly everywhere. Therefore $\lim_{\xi\to x} f(\xi)$ exists nearly everywhere. By Theorem 5 we have $f(x) = \lim_{\xi\to x} f(\xi)$ for nearly all of these points and so f is continuous nearly everywhere.

It is possible to extend Theorems 4 and 5 to functions whose domain and range are metric spaces or uniform spaces satisfying some simple requirements. Then the limit inferior and limit superior can also be replaced by more general limiting processes such as, for example, convergence with respect to systems of filters given in the domain space.

EXERCISES

1. Give an example of a real-valued continuous function f such that the closures of the sets $\{x : f(x) > a\}$ and $\{x : f(x) < a\}$ are proper subsets of the closed sets $\{x : f(x) \geqslant a\}$ and $\{x : f(x) \leqslant a\}$, respectively.

2. Let f be a real-valued function on X, for simplicity let $\text{glb}\, f(A) = \text{glb}\{f(a) : a \in A\}$, and let D be that subset of X for which the definition

$$f_1(x) = \text{lub}\{\text{glb}\, f(N_x) : N_x \in \mathscr{N}(x)\}$$

is meaningful. Show that f is lower semicontinuous at x if and only if $x \in D$ and $f_1(x) = f(x)$.

[It is clear that $f_1(x) \leqslant f(x)$ for every $x \in D$. If f is lower semicontinuous at x, then for every $\epsilon > 0$ there is a neighborhood N_x such that $\text{glb}\, f(N_x) \geqslant f(x) - \epsilon$. Hence $x \in D$ and $f_1(x) = f(x)$. Conversely, if $x \in D$ and $f_1(x) = f(x)$, then for every $\epsilon > 0$ there is an N_x such that $\text{glb}\, f(N_x) \geqslant f(x) - \epsilon$.]

3. Show that D is an open set in X.

[We have $x \in D$ if and only if there is an open set O_x such that $\text{glb}\, f(O_x)$ is finite. Hence with x every $\xi \in O_x$ belongs to D.]

4. Show that f_1 is a lower semicontinuous function on D.

[For every $\epsilon > 0$ there is an open set O_x such that $f_1(x) - \epsilon \leqslant \text{glb}\, f(O_x)$. If $\xi \in D \cap O_x$, then $f_1(\xi) \geqslant \text{glb}\, f(O_x)$. Hence $f_1(x) - \epsilon \leqslant f_1(\xi)$ for every $\xi \in D \cap O_x$.]

5. Suppose that the real-valued function f of the real variable x is bounded and its graph $F = \{(x, y) : f(x) = y\}$ is a closed set in $\mathfrak{R} \times \mathfrak{R}$. Show that f is continuous.

[Let $x \in \mathfrak{R}$ be given. If for every $\epsilon > 0$ the integer m can be chosen so large that $f(\xi) > f(x) - \epsilon + (1/m)$ for every ξ satisfying $| x - \xi | < 1/m$, then f is lower semicontinuous at x. If for some $\epsilon > 0$ such an integer

m does not exist, then we can find a sequence of numbers $x_1, \ldots, x_m, \ldots$ such that $f(x_m) \leqslant f(x) - \epsilon + (1/m)$ for every m and $x_m \to x$ as $m \to \infty$. Since the range is a bounded set the function values $f(x_m)$ have an accumulation point y which satisfies the inequality $y \leqslant f(x) - \epsilon$. However, F being closed we must have $y = f(x)$. This would lead to a contradiction. Similarly, we can prove that f is upper semicontinuous.]

6. Prove directly from the definition of continuity: *If the real-valued function of a real variable is monotonic in* $(-\infty, +\infty)$, *then it is continuous nearly everywhere.*

[Let f be increasing and let $v(x) = \lim_{\xi \to x+0} f(\xi) - \lim_{\xi \to x-0} f(\xi)$. If $a < x_1 < \ldots < x_n < b$, then $v(x_1) + \ldots + v(x_n) \leqslant f(b) - f(a)$. Hence for every $\epsilon > 0$ there are at most finitely many points in (a, b) for which $v(x) \geqslant \epsilon$. Choose $\epsilon = 1/m$ for $m = 1, 2, \ldots$ and $a = -n$, $b = n$ for $n = 1, 2, \ldots$.]

7. Prove the theorem on the continuity of regular functions by a direct argument.

[As usual let $f(x \pm 0) = \lim_{\xi \to x \pm 0} f(\xi)$. Suppose that the set of those points x at which $f(x + 0) \neq f(x)$ or $f(x - 0) \neq f(x)$ is not countable. Then there is an $\epsilon > 0$ and a noncountable set S such that

$$| f(x + 0) - f(x) | > \epsilon$$

for every $x \in S$ or $| f(x - 0) - f(x)| > \epsilon$ for every $x \in S$. The set S has a one-sided accumulation point x in the interior of the domain of definition of f. In every neighborhood of x there are points x_1 and x_2 on the same side of x such that $| f(x_1) - f(x_2)| > \epsilon$. Hence f is not regular at x.]

8. Prove: *If f is a real-valued function on a separable metric space X, then the existence of* $\lim_{\xi \to x} f(\xi)$ *at nearly every $x \in X$ implies the continuity of f nearly everywhere in X.*

(Apply Theorem 5 or prove it directly using reasoning similar to that in the p-eceding exercise.)

6. Continuity and Axioms of Separation

Constant functions defined on a topological space X are trivially continuous and so when we are speaking about continuity properties of functions we assume always that there exist nonconstant continuous functions on X. Real-valued functions with continuity relative to the usual topology of the reals form a particularly important instance. Here having fixed the range space the existence of nonconstant continuous functions depends only on the topology of the domain X. If the topology of X is discrete, then of course every real-valued function is continuous on X and if the only open sets of the space are $\emptyset$ and X, then the only continuous real valued functions are the constants.

The object of this section is to classify the relationship between separation axioms and the existence of nonconstant real-valued continuous functions. First Urysohn gave an example of an infinite Hausdorff space X such that every continuous real-valued function is constant on X. Later Hewitt constructed a regular space in which the same situation holds. Urysohn and Tychonoff proved also that on (T)-spaces and on (T_4) spaces there exist nonconstant continuous real-valued functions, and as a matter of fact these functions can be chosen such that some additional requirements are also satisfied. It is remarkable that the existence of such special nonconstant functions implies the corresponding separation axioms (T) and (T_4). The proofs are based on two lemmas:

Lemma 1. *Let X be a topological space and let (O_d) $(d \in D)$ be a scale of open sets in X such that $\bigcup O_d = X$. Then the real-valued function f defined by $f(x) = \text{glb}\{d : x \in O_d\}$ is continuous.*

Note. The condition $\bigcup O_d = X$ is necessary only in order to assure that the domain of f is the entire space X and the condition $\bar{O}_{d_1} \subseteq O_{d_2}$ is needed only to show the lower semicontinuity of f. The function is upper semicontinuous even if (O_d) $(d \in D)$ is any family of open sets whose union is X.

Proof. We show that f is continuous by using the definitions of upper and lower semicontinuity at a point. Let $x \in X$ and $\epsilon > 0$ be given. By the definition of f there is a $d \in D$ such that $d < f(x) + \epsilon$ and $x \in O_d$. If $\xi \in O_d$, then we have $f(\xi) \leqslant d$, hence $f(\xi) < f(x) + \epsilon$ for every $\xi \in O_d$ where O_d contains x. Therefore f is upper semicontinuous at x.

If $f(x) = 0$, then f is lower semicontinuous at x because f is nonnegative. If $f(x) > 0$, then D being dense in $[0, 1]$ we can choose d' and d'' such that $f(x) - \epsilon < d' < d'' < f(x)$, and this implies that $x \notin O_{d''}$. Clearly $x \notin \bar{O}_{d'}$ because $\bar{O}_{d'} \subseteq O_{d''}$ and thus the open set $c\bar{O}_{d'}$ contains x. Now if $\xi \in c\bar{O}_{d'}$, then $\xi \notin O_{d'}$ and so $f(\xi) \geqslant d' > f(x) - \epsilon$ for every ξ in $c\bar{O}_{d'}$. As $c\bar{O}_{d'}$ contains x it follows that f is lower semicontinuous at x. Therefore f is a continuous function.

Lemma 2. *Let A and B be disjoint sets in a topological space X. Then any one of the following propositions implies the others:*

(i) *There exists a scale (O_d) $(d \in D)$ of open sets such that $B \subseteq O_d \subseteq cA$ for every $d \in D$.*

(ii) *There exists a real-valued continuous function f such that $0 \leqslant f(x) \leqslant 1$ for every $x \in X$, $f(a) = 1$ for every $a \in A$, and $f(b) = 0$ for every $b \in B$.*

(iii) *There exists a continuous real-valued function f on X such that $\text{glb}\{f(a) : a \in A\} > \text{lub}\{f(b) : b \in B\}$.*

Proof. It is sufficient to show that (i) implies (ii) and (iii) implies (i) because (iii) is an obvious consequence of (ii). First we suppose (i) and show (ii): If $1 \in D$ we replace O_1 by X so that using the new meaning of O_1 we have $\bigcup O_d = X$. If $1 \notin D$, then we join 1 to the index set D and define O_1 to be X. In both cases we can apply Lemma 1 to show that the real-valued function f defined by $f(x) = \text{glb}\{d : x \in O_d\}$ is continuous. It is clear that $0 \leqslant f(x) \leqslant 1$ for every $x \in X$ and $f(a) = 1$ for every $a \in A$ because $O_d \subseteq cA$ for every $d < 1$. Since $B \subseteq O_d$ for $d \in D$ we have $f(b) = 0$ for every $b \in B$. Therefore f satisfies the requirements of proposition (ii).

Now we suppose (iii) and show that (i) also holds. We prove the existence of a scale with index set $D = I = [0, 1]$; namely, we define for every real number d $(0 \leqslant d \leqslant 1)$ the open set

$$O_d = \left\{x : x \in X \quad \textit{and} \quad f(x) < \frac{\alpha - \beta}{2} d + \frac{\alpha + \beta}{2}\right\}$$

where $\alpha = \text{glb}\{f(a) : a \in A\}$ and $\beta = \text{lub}\{f(b) : b \in B\}$. It is obvious that $B \subseteq O_d$ for every $d \in D$ and also $O_d \subseteq cA$ for every $d \in D$. The remaining requirement "$\bar{O}_{d_1} \subseteq O_{d_2}$ for every pair $d_1 < d_2$" follows immediately from Theorem 5.1.

Now we can combine the results which we obtained in this section with those of Section II.9 to show the existence of nonconstant continuous functions on topological spaces satisfying various separation axioms. First we get the following:

Theorem 1. *There exist nonconstant continuous real-valued functions on a topological space X if and only if there is a scale of open sets (O_d) $(d \in D)$ such that $\varnothing \subset O_d \subset X$ for at least two values of d.*

Note. If there is a continuous function f such that $f(a) \neq f(b)$, then there is another continuous function φ such that

$$\varphi(a) < \varphi(b) \qquad \text{and} \qquad \varphi(a) \leqslant \varphi(x) \leqslant \varphi(b) \qquad \text{for every} \quad x \in X.$$

Proof. The necessity of the condition follows from Theorem 5.1: In fact, if there is a continuous f such that $f(a) \neq f(b)$, then there is a scale of open sets (O_d) $(d \in D)$ such that $b \in O_d$ and $a \notin O_d$ for every $d \leqslant 1$; for instance, if $f(a) > f(b)$, then we may define

$$O_d = \left\{x : f(x) < \frac{f(a) - f(b)}{2} d + \frac{f(a) + f(b)}{2}\right\}$$

for each $d \in [0, 1]$. Conversely, if there is a scale (O_d) $(d \in D)$ and indices $d_1 < d_2$ with the property that $O_{d_1} \neq \varnothing$ and $O_{d_2} \neq X$, then using the transformation $\delta = (d - d_1)/(d_2 - d_1)$ we can replace the indices d $(d_1 \leqslant d \leqslant d_2)$ by indices δ $(0 \leqslant \delta \leqslant 1)$ and hence we obtain a scale (O_δ) $(\delta \in [0,1])$ such that $O_0 \neq \varnothing$ and $O_1 \neq X$. If we choose $a \in O_0$ and $b \in cO_1$, then Lemma 2 gives a continuous function φ which satisfies $\varphi(a) = 0 \leqslant \varphi(x) \leqslant \varphi(b) = 1$.

If X is a (T) space or a (T_4) space, then by Definition II.9.2 and Theorem II.9.3 the conditions of the preceding tneorem are satisfied and so on uniform spaces and on (T_4) spaces there exist nonconstant continuous functions. Actually we can prove much stronger results which hold if and only if X is a (T) space or a (T_4) space, respectively:

Theorem 2. *The space X is a* (T) *space if and only if for every closed set A and every point $b \notin A$ there exists a real-valued continuous function f such that $0 \leqslant f(x) \leqslant 1$ for every $x \in X$, $f(a) = 1$ for every $a \in A$, and $f(b) = 0$.*

Proof. We combine Definition II.9.2 and Lemma 2.

We notice the following:

Corollary. *If $\mathscr{T}_1$ and $\mathscr{T}_2$ are* (T) *topologies on the set X and if the same real-valued functions are continuous relative to $\mathscr{T}_1$ and $\mathscr{T}_2$, then $\mathscr{T}_1$ and $\mathscr{T}_2$ are identical.*

Proof. Let $O \neq \varnothing$, X be open relative to a (T) topology $\mathscr{T}$. For every $x \in O$ define the continuous function $f = f_x$ such that $0 \leqslant f(\xi) \leqslant 1$ everywhere, $f(x) = 0$, and $f(\xi) = 1$ for each $\xi \in cO$. Then $\{\xi : f(\xi) < 1\}$ lies in O and contains x. Conversely, if for each point x of a set O there is a continuous function $f = f_x$ such that $x \in \{\xi : f(\xi) < 1\} \subseteq O$, then O is open. Hence O is open relative to $\mathscr{T}$ if and only if for every $x \in O$ there is a continuous function $f = f_x$ such that

$$x \in \{\xi : f(\xi) < 1\} \subseteq O.$$

This characterization of an open set depends only on the set of real-valued functions which are continuous relative to $\mathscr{T}$.

We can easily derive the following useful result:

Lemma 3. *If X is a* (T) *space, A is a closed set, and $b \notin A$, then there is a closed neighborhood C_b of b and a real-valued continuous function φ such that $0 \leqslant \varphi(x) \leqslant 1$ for every $x \in X$, $\varphi(a) = 1$ for every $a \in A$, and $\varphi(x) = 0$ for every $x \in C_b$.*

Proof. Let f denote the function in Theorem 2. We let $\varphi = 2(f \cup \frac{1}{2}) - 1$ so that $\varphi(x) = 2 \max\{f(x), \frac{1}{2}\} - 1$ for every $x \in X$. The requirements are satisfied with $C_b = \{x : f(x) \leqslant \frac{1}{2}\}$.

Theorem 3. *The space X is a* (T_4) *space if and only if for every pair of disjoint closed sets A and B there exists a real-valued continuous function f such that $0 \leqslant f(x) \leqslant 1$ everywhere on X, $f(a) = 1$ for every $a \in A$, and $f(b) = 0$ for every $b \in B$.*

Proof. We combine Theorem II.9.3 and Lemma 2.

It is natural to ask whether it is possible to choose the function f in Theorem 3 such that it takes the value 1 exactly on A and the value 0 exactly on B. Similar questions can be raised concerning the functions which occur in some of the earlier theorems, e.g., in Theorem 2. We shall prove that this can be done whenever every closed set can be represented as the intersection of at most denumerably many open sets. *If a set A is the intersection of countably many open sets, then A is called a G_δ-set.* The complement of a G_δ-set is called an *F_σ-set* so that *F_σ-sets are the unions of countably many closed sets.* The proof is based on the following lemma which extends a well-known real variable theorem to arbitrary topological spaces:

Lemma 3. *If the sequence of the real-valued continuous functions f_n $(n = 1, 2, \ldots)$ is uniformly convergent on the topological space X then the limit function f is continuous.*

Proof. The object is to find a neighborhood N_x of the arbitrary point $x \in X$ such that $|f(\xi) - f(x)| < \epsilon$ for every ξ in N_x. By the uniform convergence there is an index n such that $|f_n(\xi) - f(\xi)| < \epsilon/3$ for every ξ in X. By the continuity of f_n at x there is a neighborhood N_x such that $|f_n(\xi) - f_n(x)| < \epsilon/3$ for every $\xi \in N_x$. Hence if $\xi \in N_x$, then

$$|f(\xi) - f(x)| \leqslant |f(\xi) - f_n(\xi)| + |f_n(\xi) - f_n(x)| + |f_n(x) - f(x)| < \epsilon .$$

Thus f is continuous at x.

Let A and B be disjoint closed sets of the (T_4) space X. We prove that if A is a G_δ-set, then there is a real-valued continuous function φ on X

such that $0 \leqslant \varphi(x) \leqslant 1$ everywhere, $\varphi(b) = 0$ for every $b \in B$, and φ takes the value 1 exactly on A. For let O_n $(n = 1, 2, \ldots)$ be open neighborhoods of A such that $A = \bigcap O_n$. Since X is a (T_4) space by Theorem 3 there are continuous functions φ_n on X having the following property: $0 \leqslant \varphi_n(x) \leqslant 2^{-n}$ everywhere, $\varphi_n(x) = 2^{-n}$ on A and $\varphi_n(x) = 0$ on cO_n. The infinite series $\Sigma\, \varphi_n(x)$ is uniformly convergent and so it defines a continuous function φ. By the construction $\varphi(x) = 1$ only if $\varphi_n(x) = 2^{-n}$ for every $n = 1, 2, \ldots$ and so only if $x \in c \bigcup cO_n = A$. The remaining requirements are obviously satisfied. If B is also a G_δ-set, then there is a function ψ such that $0 \leqslant \psi(x) \leqslant 1$ everywhere on X, $\psi(a) = 1$ for every $a \in A$ and ψ takes the value 0 exactly on B. The continuous function $f = \frac{1}{2}(\varphi - \psi + 1)$ takes the value 1 exactly on A, the value 0 exactly on B and satisfies $0 \leqslant f(x) \leqslant 1$ everywhere on X.

The next result shows that the preceding remarks do apply in many cases:

Lemma 4. *If X is metrizable then every closed set is a G_δ-set and every open subset is an F_σ-set.*

Proof. Let d be a metric for X and let C be a nonvoid closed set. For $n = 1, 2, \ldots$ we define

$$O_n = \left\{x : x \in X \quad and \quad d(x, c) < \frac{1}{n} \quad for\ some \quad c \in C\right\}.$$

Then O_n is an open set containing C and so $C \subseteq \bigcap O_n$. If $x \notin C$, then C being closed there is an $\epsilon > 0$ such that $d(x, c) \geqslant \epsilon$ for every $c \in C$ and so $x \notin O_n$ for sufficiently high values of $n = 1, 2, \ldots$. Therefore $C = \bigcap O_n$ and C is a G_δ-set. Taking complements we see that open sets are F_σ-sets.

We give one more characterization of (T_4) spaces in terms of continuous real-valued functions. This result is known as *Tietze's extension theorem:*

Theorem 4. *A topological space X is a (T_4) space if and only if every real-valued function which is defined and continuous on a closed subset of X has a continuous extension to the entire space X.*

Note. As far as the sufficiency is concerned it is sufficient to suppose that bounded continuous functions can be extended. If f is bounded, then there is an extension ψ which does not increase the least upper bound of the absolute value of f on the given closed set C. Actually

$$|\psi(\xi)| < \mathrm{lub}\{|f(x)| : x \in C\}$$

for every $\xi \notin C$.

Proof. The sufficiency is immediate: Let A and B be disjoint closed sets and let f be 0 on A and 1 on B. By hypothesis f can be extended to a real-valued continuous function on X. The open sets $O_A = \{x : f(x) < \frac{1}{2}\}$ and $O_B = \{x : f(x) > \frac{1}{2}\}$ are disjoint neighborhoods of A and B, respectively. Hence X is a (T_4) space. To prove the necessity of the condition let us suppose that X satisfies axiom (T_4) and f is a real-valued continuous function on the closed set C. First we suppose that f is bounded on C, for the sake of simplicity let $|f(x)| \leqslant 1$ on C. Since $A = \{x : f(x) \leqslant -\frac{1}{3}]$ and $B = \{x : f(x) \geqslant \frac{1}{3}\}$ are disjoint closed subsets of the closed subspace C they are closed in X. Hence by Theorem 3 there is a real-valued continuous function φ on X such that $\varphi(x) = -\frac{1}{3}$ on A, $\varphi(x) = \frac{1}{3}$ on B, and $|\varphi(x)| \leqslant \frac{1}{3}$ for every $x \in X$. We let $f = f_0$, $\varphi = \varphi_0$, and $f_1 = f_0 - \varphi_0$. By the construction we have $|f_1(x)| \leqslant \frac{2}{3}$ on C. Using the same process we construct a real-valued continuous function φ_1 such that $|\varphi_1(x)| \leqslant \frac{1}{3} \cdot \frac{2}{3}$ on X and $|f_2(x)| \leqslant (\frac{2}{3})^2$ where $f_2 = f_1 - \varphi_1$. In general we can determine φ_n ($n = 0, 1, \ldots$) such that $|\varphi_n(x)| \leqslant \frac{1}{3}(\frac{2}{3})^n$ on X and $|f_{n+1}(x)| \leqslant (\frac{2}{3})^{n+1}$ on C where f_{n+1} denotes $f_n - \varphi_n$. We can easily see that the infinite series

$$\varphi_0(x) + \varphi_1(x) + \ldots + \varphi_n(x) + \ldots$$

converges to a finite limit $\psi(x)$ for every $x \in X$ and the limit function ψ is continuous on X: In fact,

$$|\varphi_m(x) + \ldots + \varphi_n(x)| \leqslant \tfrac{1}{3}((\tfrac{2}{3})^m + \ldots + (\tfrac{2}{3})^n) < (\tfrac{2}{3})^m$$

and so the series is uniformly convergent. By the preceding Lemma 3 the limit function is continuous and satisfies $|\psi(x)| \leqslant 1$ on X. Moreover, on C we have

$$|f(x) - \varphi_0(x) - \ldots - \varphi_n(x)| = |f_{n+1}(x)| \leqslant (\tfrac{2}{3})^{n+1}$$

and so $f(x) = \psi(x)$ on C. This shows that bounded continuous functions can be extended to the entire space without increasing the least upper bound of their absolute value.

If f is not bounded on C, it can be replaced by $g = f/(1 + |f|)$ without destroying its continuity. Let φ be a continuous extension of g to X such that $|\varphi(x)| \leqslant 1$ for every $x \in X$. Since φ is continuous $A = \{x : \varphi(x) = \pm 1\}$ is closed and by $|g(x)| < 1$ it is disjoint from C. By the (T_4) axiom there is a real-valued continuous function ψ on X such that $0 \leqslant \psi(x) \leqslant 1$ everywhere, $\psi(a) = 0$ for every $a \in A$, and $\psi(c) = 1$ for every $c \in C$. Thus $|\varphi(x)\psi(x)| < 1$ everywhere on X and $\varphi(x)\psi(x) = g(x)$ on C. Hence $\varphi\psi/(1 - |\varphi\psi|)$ is defined and is continuous on X and on C it coincides with f.

As an application of Theorem 3 we prove the following result which is due to C.H. Dowker:

Theorem 5. *Let X be a countably paracompact* (T_4) *space. Then we have: If h is upper semicontinuous and g is lower semicontinuous on X and if $h(x) < g(x)$ for every $x \in X$, then there is a continuous function f on X such that $h(x) < f(x) < g(x)$ for every $x \in X$.*

By Theorems II.5.2 and III.7.2 the condition holds for every pseudometric space. Hence a continuous function can be inserted between any upper semicontinuous h and lower semicontinuous g satisfying the inequality $h(x) < g(x)$ for every point $x \in X$ of the pseudometrizable space X.

Note. Dowker proves also that the conclusion implies that X is a countably paracompact (T_4) space.

Proof. We may suppose that the ranges of h and g are subsets of $[0, 1]$. For every rational number $r \in (0, 1)$ let $O_r = \{x : h(x) < r < g(x)\}$. The family $\{O_r\}$ is a countable open cover of X. Since X is a countably paracompact (T_4) space, by Exercise III.6.12 there is a locally finite open cover $\{Q_r\}$ of X such that $\bar{Q}_r \subseteq O_r$ for every r. Similarly there is an open cover $\{R_r\}$ such that $\bar{R}_r \subseteq Q_r$ for every rational $r \in (0, 1)$. By Theorem 3 there is a continuous function f_r such that $f_r(x) = -1$ if $x \in cQ_r$ and $f_r(x) = r$ if $x \in R_r$. We define f by the formula $f(x) = \text{lub}\{f_r(x)\}$. The cover $\{Q_r\}$ being locally finite there is an open set O_x which meets only finitely many Q_r sets. Hence the restriction of f to O_x is the maximum of finitely many continuous functions f_r and so it is continuous on O_x. By Exercise 2.6 f is continuous at x relative to the entire space X. Hence f is a continuous function on X. If $x \in \bar{R}_r$, then $x \in O_r$ and so $f_r(x) = r < g(x)$. Therefore $\{Q_r\}$ being locally finite, $f(x) < g(x)$. Moreover, $\{R_r\}$ being a cover of X, there is a rational r such that $x \in R_r$ and so $h(x) < r = f_r(x) \leqslant f(x)$. Hence $h(x) < f(x) < g(x)$ for every $x \in X$.

It would take more effort to prove that the conclusion implies countable paracompactness. The fact that X is then a (T_4) space can be proved very easily: Indeed, let A and B be disjoint closed sets in X. We define h to be the characteristic function of $A : h(x) = 1$ if $x \in A$ and $h(x) = 0$ if $x \in cA$. Moreover, we define g by the rule: $g(x) = 1$ if $x \in B$ and $g(x) = 2$ if $x \in cB$. Then h is upper semicontinuous and g is lower semicontinuous on X. Since $h(x) < g(x)$ for every $x \in X$, by the theorem a continuous f can be inserted between h and g. The open sets $O_A = \{x : f(x) > 1\}$ and $O_B = \{x : f(x) < 1\}$ are disjoint neighborhoods of A and B, respectively.

EXERCISES

1. Show that the half-open interval topologies of the real line are completely regular.

[Let A be closed and let $x \notin A$. Choose y such that $[x, y) \subseteq cA$. Define f to be 1 on the complement of $[x, y)$ and define $f(\xi) = (\xi - x)/(y - x)$ for every $\xi \in [x, y)$. Then f is continuous everywhere relative to the half-open interval topology $\mathscr{T}_+$. We have $0 \leqslant f(\xi) \leqslant 1$ everywhere, $f(x) = 0$ and $f(\xi) = 1$ for every $\xi \in A$.]

2. Show that the continuous image $Y = f(X)$ can be a (T_i) space where $i = 0, 1, 2, 3, 4$, or 5 while X itself is not a (T_i) space.

[Let Y consist of a single point y.]

3. Let X be completely regular, let A be compact, and let B be closed and disjoint from A. Show the existence of a real-valued continuous f on X such that $0 \leqslant f(x) \leqslant 1$ everywhere, f is 1 on A and it is 0 on B.

[Determine for every $a \in A$ a continuous f_a such that $0 \leqslant f_a(x) \leqslant 2$ everywhere, $f_a(a) = 2$ and $f_a(b) = 0$ for every $b \in B$. We define $O_a = \{x : f_a(x) > 1\}$ and find a finite cover of A consisting of sets O_a. If these correspond to the points $a_1, ..., a_n$ we let $f = (f_{a_1} \cup ... \cup f_{a_n}) \cap 1$. Then f is continuous on X and has the required properties.]

7. Continuity and Compactness

Many important results are known on continuous functions whose domain or range is in a compact space. We collected here some of the simplest and most fundamental of these.

Theorem 1. *The continuous image of a compact space is compact; that is, if f is a continuous map of X onto Y and if X is compact then so is Y.*

Note. Compactness can be replaced by countable compactness, Lindelöf property, or hereditary Lindelöf property.

Proof. Let $\{Q_i\}$ $(i \in I)$ be an open cover of Y and let $O_i = f^{-1}(Q_i)$. Since the domain of f is X the family $\{O_i\}$ $(i \in I)$ is a cover of X and so by the compactness of X there is a finite subcover $\{O_{i_j}\}$ $(j \in J)$. The corresponding finite family $\{Q_{i_j}\}$ $(j \in J)$ is a cover of Y because f maps X

onto Y and so for every $y \in Y$ there is an $x \in X$ such that $y = f(x)$. Hence every open cover of Y contains a finite subcover.

Lemma 1. *The image of a compact space X under a continuous mapping of X into a Hausdorff space Y is a closed subset of Y.*

Proof. This result is an easy consequence of the preceding theorem and of Theorem III.3.4. In fact, if f maps X into Y, then $f(X)$ is a compact subspace of Y and so Y being a Hausdorff space $f(X)$ is a closed set.

Lemma 2. *Let f be a continuous map of the compact space X into the Hausdorff space Y. Then f is a closed map.*

Proof. If $C \subseteq X$ is closed, then by Theorem III.3.1 C is compact and so $f(C)$ is compact. Since Y is a Hausdorff space by Theorem III.3.4 $f(C)$ is closed. Hence f maps closed sets of X onto closed sets of Y.

Lemma 3. *Let X be compact and let Y be a Hausdorff space. If f is a continuous one-to-one map of X onto Y, then f is a homeomorphism.*

Proof. We must show that f is an open map. Since f is one-to-one $f(O) = cf(cO)$ for every open set $O \subseteq X$. By the preceding lemma $f(O)$ is open in Y.

In the next section we shall use the following lemma:

Lemma 4. *Let f be a continuous map of a compact space X onto a Hausdorff space Y. Then for every set $S \subseteq X$ and for every point $y \in Y$ such that $f^{-1}(y) \subseteq S^i$ we have $y \in f(S)^i$.*

Proof. The result is an immediate consequence of Lemma 4.4 and the foregoing Lemma 2.

The next result is not a topological theorem because it involves not only the topology of the spaces X and Y but also the specific structure which is used to define the topology on Y. If Y is the set of reals under its usual topology the result is well known: If f is a real-valued function on the compact space X, then f is bounded on X, i.e., there is a constant $M > 0$ such that $|f(x)| \leqslant M$.

Lemma 5. *Suppose that the domain of the continuous function f is a compact space X and its range lies in a metric space Y. Then given any metric d for the topology of Y there is a positive constant M such that $d(f(x_1), f(x_2)) \leqslant M$ for any pair of points $x_1, x_2 \in X$.*

Note. If the domain of f is a proper subset of the compact space

X, the conclusion is not necessarily true. Of course, if the metric function d is bounded on $Y \times Y$, then the conclusion is trivial.

Proof. Since f is continuous at every $\xi \in X$ there is an open set O_ξ such that $d(y, \eta) \leqslant 1$ for every $x \in O_\xi$ where $y = f(x)$ and $\eta = f(\xi)$. The family $\{O_\xi\}$ $(\xi \in X)$ is an open cover of the compact space X and so there are finitely many points $\xi_1, ..., \xi_n$ such that $\{O_{\xi_i}\}$ $(i = 1, ..., n)$ is a cover of X. If we choose M so large that $M \geqslant 2 + \max\{d(\eta_i, \eta_j)\}$ where $\eta = f(\xi)$, then the requirements are satisfied: Indeed, given $x_1, x_2 \in X$ we have $x_1 \in O_{\xi_i}$ and $x_2 \in O_{\xi_j}$ for some indices i, j and so

$$d(y_1, y_2) \leqslant d(y_1, \eta_i) + d(\eta_i, \eta_j) + d(\eta_j, y_2) \leqslant 1 + \max\{d(\eta_i, \eta_j)\} + 1 \leqslant M.$$

The above lemma can be derived in several other ways; for instance, it is a simple consequence of the following:

Lemma 6. *Let $f: X \to Y$ continuously where X is compact and Y is a uniform space with uniform structure $\mathscr{V}$. Then for every $V \in \mathscr{V}$ there is a decomposition of the range of f into finitely many sets $B_1, ..., B_n$ such that $(y', y'') \in V$ for every pair $y', y'' \in B_k$ $(k = 1, ..., n)$.*

Proof. By Theorem 1 the image space $f(X)$ is compact. Thus it is sufficient to prove the following proposition:

Lemma 7. *If X is compact and uniformizable and U is a uniformity of its unique uniform structure, then X can be represented as the union of finitely many open sets $O_1, ..., O_n$ such that $O_k \times O_k \subseteq U$ for $k = 1, ..., n$.*

Proof. Determine a symmetric uniformity V such that $V \circ V \subseteq U$. By the compactness of X there are finitely many points $x_1, ..., x_n$ such that the union of the open sets $O_k = V[x_k]^i$ is X. Since V is symmetric $V[x_k] \times V[x_k] \subseteq V \circ V \subseteq U$ and so $O_k \times O_k \subseteq U$ for every $k = 1, ..., n$.

Theorem 2. *If f is a real-valued function on a compact space X, then it has a maximum and a minimum value on X.*

Proof. It is sufficient to show that a maximum exists. The existence of the minimum follows by considering the maximum of the function $-f$. By Lemma 5, f has an upper bound on X. Hence there is a least upper bound M of the function values $f(x)$ $(x \in X)$. We must show that $f(\xi) = M$ for some point $\xi \in X$. For every $n = 1, 2, ...$ we can find a point $x_n \in X$ such that $f(x_n) \geqslant M - (1/n)$ and by Theorem III.1.3 the sequence (x_n) $(n = 1, 2, ...)$ has an accumulation point ξ. Since f is continuous at ξ for every $\epsilon > 0$ we can find an open set O_ξ such that $|f(\xi) - f(x)| < \epsilon$ for every $x \in O_\xi$. Consequently $f(\xi) > f(x_n) - \epsilon$ for infinitely many indices n. Hence $f(\xi) \geqslant M - \epsilon$ for every $\epsilon > 0$ which shows that $f(\xi) \geqslant M$.

Theorem 3. *If an upper (lower) semicontinuous real-valued function is constant on the complement of a compact subspace C of the domain space X, then it has a maximum (minimum) value on X.*

Note. If X is compact, we can choose X itself as C. Hence if f is upper (lower) semicontinuous on a compact space, then f has a maximum (minimum) value on X.

Proof. It is sufficient to prove the theorem in the special case when f is upper semicontinuous, X is compact, and $C = X$: In the general case the restriction of f (or $-f$) to C has a maximum value M on C and so the maximum of f (or $-f$) on X is either M or it is the constant value of f (or $-f$) on cC. To prove the theorem in the special case when f is upper semicontinuous and X is compact we first show that f is bounded from above. For every $\xi \in X$ let the open set O_ξ be determined such that $f(x) < 1 + f(\xi)$ for every $x \in O_\xi$. Then $\{O_\xi\}$ $(\xi \in X)$ is an open cover of X and hence there is a finite subcover, say, $\{O_{\xi_1}, ..., O_{\xi_n}\}$. Therefore we have $f(x) < 1 + \max\{f(\xi_i)\}$ for every $x \in X$ and so f is bounded from above. Let M be the least upper bound of f on X. To show that M is a function value at some point $\xi \in X$ we merely repeat the reasoning given in the proof of Theorem 2: There is a sequence of points x_n $(n = 1, 2, ...)$ with the property that $f(x_n) \geqslant M - (1/n)$ and since X is compact this sequence has an accumulation point $\xi \in X$. By the upper semicontinuity of f at ξ we have $f(x) < f(\xi) + \epsilon$ for every point x of some open set O_ξ which may depend on $\epsilon > 0$. Therefore we have

$$M - (1/n) \leqslant f(x_n) < f(\xi) + \epsilon$$

for all but finitely many indices n. This shows that $M \leqslant f(\xi) + \epsilon$ for every $\epsilon > 0$ and so $f(\xi) = M$.

Any continuous one-to-one image of the interval $[0, 1]$ is called a *simple arc*. By Lemma 3 all simple arcs in a Hausdorff space are homeomorphic. We give a few examples which indicate how essential is the one-to-one property of the map. The construction of these examples is based on the following observation:

Let

$$
\begin{array}{cccccccc}
 & Y_0 & & & & Y_1 & & \\
Y_{00} & & Y_{01} & & Y_{10} & & Y_{11} & \\
Y_{000} & Y_{001} & Y_{010} & Y_{011} & Y_{100} & Y_{101} & Y_{110} & Y_{111} \\
Y_{0000} & \cdots & \cdots & \cdots & \cdots & \cdots & \cdots & Y_{1111} \\
\cdots & \cdots & \cdots & \cdots & \cdots & \cdots & \cdots & \cdots
\end{array}
$$

be closed sets in a compact metric space having the following properties:

(i) $$Y_{\epsilon_1 \cdots \epsilon_n} \supseteq Y_{\epsilon_1 \cdots \epsilon_n 0} \cup Y_{\epsilon_1 \cdots \epsilon_n 1}$$

for every $n = 1, 2, \ldots$ and for every choice of the indices $\epsilon_1, \ldots, \epsilon_n = 0,1$.

(ii) The diameter of the sets approaches zero as the number of indices increases to infinity.

(iii) Any two adjacent sets in the same row have a nonvoid intersection.

If we define $Y^1 = Y_0 \cup Y_1$, $Y^2 = Y_{00} \cup Y_{01} \cup Y_{10} \cup Y_{11}$, ... then $Y = \bigcap Y^n$ is the continuous image of the interval $[0, 1]$. Indeed, every $x \in [0, 1]$ has a development $x = \sum \epsilon_k 2^{-k}$ where for $x > 0$ we have $\epsilon_k = 1$ for infinitely many indices k. Since the space is compact by (ii) and by Theorem III.1.1 the intersection $Y_{\epsilon_1} \cap Y_{\epsilon_1 \epsilon_2} \cap \ldots$ consists of a single point y. We define $f(x) = y$. The continuity of the map $f : [0, 1] \to Y$ follows from hypotheses (i) and (ii): For if $| x_1 - x_2 | < 2^{-n}$, then the value of $\epsilon_1, \ldots, \epsilon_{n-1}$ is the same for both x_1 and x_2. Hence $f(x_1), f(x_2) \in Y_{\epsilon_1 \cdots \epsilon_{n-1}}$.

It is now easy to show for instance that a closed right triangle, its interior included, is a continuous image of the interval [0, 1]. For let the right triangle Y be divided into two congruent triangles Y_0 and Y_1 by dropping the perpendicular from the right angle. Similarly, divide Y_0 and Y_1 into congruent triangles Y_{00}, Y_{01} and Y_{10}, Y_{11}, respectively, and label them such that Y_{01} and Y_{10} are adjacent. Continue this process by constructing $Y_{000}, \ldots, Y_{111}$, etc. It follows that Y is the continuous image of the interval [0,1]. It is evident that the same holds for any figure which is homeomorphic to a triangle. Hence for instance a square is the continuous image of the line segment [0, 1].

Any continuous image of the closed interval [0, 1] is called a *continuous curve* or a *Peano curve*. For Peano was the first to show that the unit square is the continuous image of the segment [0, 1]. As can be seen from Exercise 4.6 the square is also a one-to-one image of the segment.

However, no one-to-one continuous map can exist between these figures because by Lemma 3 such a map would be a homeomorphism and this would contradict the result stated in Exercise 4.5. In the next sections we shall see that continuous curves preserve a number of important properties of the line segment [0, 1] such as compactness, connectedness, local connectedness, and rationality.

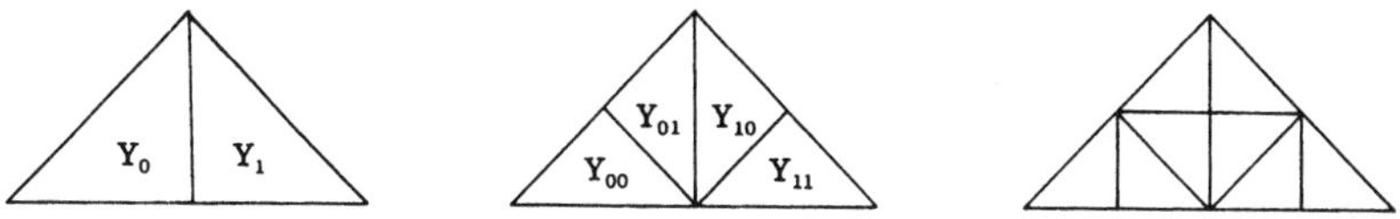

EXERCISES

1. A Hausdorff space X with topology $\mathscr{T}$ is called a *minimal Hausdorff space* if there exists no weaker Hausdorff topology on X than $\mathscr{T}$. Prove that X is a minimal Hausdorff space if and only if continuous Hausdorff one-to-one images of X are necessarily homeomorphic images.

[Suppose that X is a minimal Hausdorff space with open sets $O \in \mathscr{O}$. Let $f : X \to Y$ be a continuous one-to-one map. The sets $f^{-1}(Q)$ $(Q \in \mathscr{Q})$ where Q is open in Y form a Hausdorff topology on X which is at most as strong as $\mathscr{T}$. Hence it is $\mathscr{T}$ and so every O is of the form $O = f^{-1}(Q)$ where $Q \in \mathscr{Q}$. Consequently, $f(O)$ is open for every $O \in \mathscr{O}$ and f is a homeomorphism. Next suppose that every continuous Hausdorff one-to-one map is a homeomorphism. Given a Hausdorff topology which is weaker than $\mathscr{T}$ the identity map must be continuous and so the weaker topology and $\mathscr{T}$ must coincide.]

2. Prove Lemma 3 by using Exercise III.3.4.
(Compare with the preceding exercise.)

3. Let $\mathscr{O}_1$ and $\mathscr{O}_2$ be the families of open sets of the topologies $\mathscr{T}_1$ and $\mathscr{T}_2$ on the set X. If $\mathscr{O}_1 \subseteq \mathscr{O}_2$ we say that $\mathscr{T}_1$ is a *contraction* of $\mathscr{T}_2$ and $\mathscr{T}_2$ is an *expansion* of $\mathscr{T}_1$. Show that no proper contraction of a compact Hausdorff space is a Hausdorff space and no proper expansion of a compact Hausdorff space is compact.
(See Lemma 3 and Exercise III.3.4.)

4. Let X be a metric space with metric d having the property that for any two distinct members of a sequence (x_n) $(n = 1, 2, \ldots)$ we have

$d(x_i, x_j) > \epsilon > 0$. Show the existence of a nonbounded real-valued continuous function on X.

[Let f be defined by $f(x) = (k/\epsilon) - (2k/\epsilon^2)d(x, x_k)$ for every $x \in X$ satisfying $d(x, x_k) \leqslant \epsilon/2$ and by $f(x) = 0$ for the remaining points of X. Then f is continuous on X and $f(x_k) = k/\epsilon$ for each $k = 1, 2, \ldots$.]

5. Let f map X into Y where X is a compact metric space with metric d_X and Y is a metric space with metric d_Y. Show that if f is continuous, then for every $\epsilon > 0$ there is a $\delta > 0$ such that $d_Y(f(A)) < \epsilon$ for every $A \subseteq X$ satisfying $d_X(A) < \delta$.

[For every $x \in X$ there is a $\delta_x > 0$ such that $d(x, \xi) < \delta_x$ implies $d(y, \eta) < \epsilon/2$. Consider a finite subcover of the cover $\{S_{\delta_x/2}[x]\}$ $(x \in X)$ which corresponds, say, to the points $x_1, \ldots, x_n$. Define

$$\delta = \tfrac{1}{2}\min\{\delta_{x_1}, \ldots, \delta_{x_n}\}.$$

If $a \in A$, then there is an x_i such that $d(x_i, a) < \frac{1}{2}\delta_{x_i}$. If $d_X(A) < \delta$, then also $d_X(x_i, \xi) < \frac{1}{2}\delta_{x_i} + \delta \leqslant \delta_{x_i}$ for every $\xi \in A$. Hence $d_Y(y_i, \eta) < \epsilon/2$ for every $\xi \in A$ and so $d_Y(f(A)) \leqslant \epsilon$.]

6. Prove Lemma 5 by using Exercise III.2.1 or the Corollary of Theorem III.2.1.

[Let φ map $X \times X$ into the space of the reals as follows: $\varphi(x_1, x_2) = d(f(x_1), f(x_2))$. Since φ is continuous and $X \times X$ is compact, $\varphi(X \times X)$ is a compact set and so it is bounded.]

7. Let X and Y be metric spaces and let f map X onto Y such that compact sets of X are mapped onto compact sets of Y. Show that if f is discontinuous at a point x, then there is a sequence of distinct points x_n $(n = 1, 2, \ldots)$ and a point $\eta \neq y = f(x)$ such that $x_n \to x$ and $f(x_n) = \eta$ for every $n = 1, 2, \ldots$.

[There is an open set O_y and a sequence (x_n) $(n = 1, 2, \ldots)$ such that $x_n \to x$ and $y_n = f(x_n) \notin O_y$ for every $n = 1, 2, \ldots$. Since $x_n \to x$ the set $S = \{x, x_1, x_2, \ldots\}$ is compact so $f(S) = \{y, y_1, y_2, \ldots\}$ is also compact. We have $y_n \notin O_y$ $(n = 1, 2, \ldots)$, hence $f(S) - \{y\} = \{y_1, y_2, \ldots\}$ is also compact. If $f(S) - \{y\}$ were an infinite set it would contain an accumulation point, say, y_m. Since $S - \{x_m\}$ is compact, $f(S) - \{y, y_m\}$ would be a compact set. However $f(S) - \{y\}$ contains a convergent subsequence whose limit is y_m. Hence $f(S) - \{y, y_m\}$ cannot be compact. Therefore $f(S)$ is finite and $f(x_n)$ is the same point η for infinitely many indices $n = 1, 2, \ldots$.]

8. Let X and Y be metric spaces and let f map X onto Y such that $f(C)$ is compact for every compact $C \subseteq X$ and $f^{-1}(y)$ is closed for every $y \in Y$. Show that f is continuous.

[This is an immediate consequence of the proposition stated in the preceding exercise: If f is not continuous at x, then $f^{-1}(\eta)$ is not closed in X. An interesting special case is obtained by assuming that f is a one-to-one map: *If the one-to-one maps f and f^{-1} preserve compactness, then the metric spaces X and Y are homeomorphic.*]

9. Exhibit a map to show that the $\perp$ shaped plane figure is a continuous curve.

10. Show that no continuous map of [0, 1] onto a triangle can be a one-to-one map.

(If it were, Lemma 3 would apply.)

11. Let $f_1, f_2, \ldots$ be real-valued nondecreasing functions on a compact interval $[a, b]$. Prove that if the family is uniformly bounded, then there is a subsequence which is convergent on $[a, b]$.

(Let C be the set of real-valued nondecreasing functions f: $[a, b] \to [-M, M]$. Then C is a closed subspace of $[-M, M]^{[a,b]}$ in the product topology. Thus C is a compact Hausdorff space. Its sequential compactness can be proved by showing that it satisfies the first axiom of countability.)

8. Continuity and Connectedness

A theorem in classical analysis known as the intermediate value theorem states that if f is a continuous function of a real variable on an interval $[a, b]$ and if $f(a) \leqslant f(b)$, then in that interval f assumes every value η satisfying $f(a) \leqslant \eta \leqslant f(b)$ at least once. This theorem can be generalized in many directions but here we are interested only in the topological nature of the theorem and in purely topological generalizations of it. We describe two proofs of the intermediate value theorem.

First there is the classic proof which leads to the desired result in a completely elementary way with the least use of topology. Clearly it is sufficient to show the theorem in the special case when $f(a) < 0 < f(b)$ and $\eta = 0$. Otherwise f can be replaced by $f - \eta$. Using the same halving process which we applied in Section III.2 to prove the Bolzano-Weierstrass theorem we can construct a sequence $((a_k, b_k))$ $(k = 0, 1, \ldots)$ of nested intervals such that $a_0 = a$, $b_0 = b$, $b_k - a_k = (b - a)2^{-k}$ $(k = 0, 1, \ldots)$, and $f(a_k) \leqslant 0 \leqslant f(b_k)$ for every $k \geqslant 0$. The monotonic sequences (a_k) and (b_k) have a common limit ξ in $[a, b]$ and hence by the continuity of f at this point ξ we have $\lim f(a_k) = \lim f(b_k) = f(\xi)$. The inequalities $f(a_k) \leqslant 0 \leqslant f(b_k)$ imply that $f(\xi) = 0$.

The topological proof of the intermediate value theorem is shorter but no simpler because a number of topological results are used. It does, however, show the topological nature of the theorem. We need Theorem II.7.4 and the following:

Lemma 1. *The continuous image of a connected space is connected.*

Proof. Let $f(X) = A \cup B$ where A and B are disjoint open sets in Y. Since f is continuous $f^{-1}(A)$ and $f^{-1}(B)$ are open sets. They are disjoint and their union is X. By hypothesis X is connected and so $f^{-1}(A) = \varnothing$ or $f^{-1}(B) = \varnothing$. Consequently, $A = \varnothing$ or $B = \varnothing$.

Now let f be continuous on the interval $[a, b]$ and let $f(a) < f(b)$. Then by the preceding lemma, $f([a, b])$ is a connected set lying in the space of the real numbers. Hence by Theorem II.7.4, $f([a, b])$ is an interval and so it contains the closed interval $[f(a), f(b)]$. Therefore given any η in this interval there is a $\xi \in [a, b]$ such that $f(\xi) = \eta$. This topological proof can be adapted to show:

Theorem 1. *A topological space X is connected if and only if every real-valued continuous function f defined on X assumes for each pair of values $f(a) < f(b)$ also every η satisfying $f(a) \leqslant \eta \leqslant f(b)$.*

Proof. First suppose that there is a continuous f which leaves out a value η lying between $f(a)$ and $f(b)$. Then $f(X)$ is not an interval and so by the corollary of Theorem II.7.4 it is not connected. Since the continuous image of a connected space is connected this shows that X is not a connected space. Therefore the condition is necessary. To show its sufficiency let us suppose that X is not connected. Then there exist nonvoid open sets O_1 and O_2 such that $X = O_1 \cup O_2$ and $O_1 \cap O_2 = \varnothing$. We define $f(x) = 1$ for every $x \in O_1$ and $f(x) = 2$ for every $x \in O_2$. Then f is a continuous function because for every $x \in X$ there is an open set O_x, namely, O_1 or O_2 according as $x \in O_1$ or $x \in O_2$, such that $f(\xi) = f(x)$ for every $\xi \in O_x$. Since O_1 and O_2 are not void f assumes the values 1 and 2 but it leaves out every $\eta \in (1, 2)$. Therefore if X is not connected then the condition fails.

The connectedness of intervals on the real line can be used in several other ways to show that a topological space X is connected. For instance we have:

Lemma 2. *If X is a topological space and if for every pair of distinct points $a, b \in X$ there is a continuous function on a closed interval $[\alpha, \beta]$ of the real line with values in X such that $f(\alpha) = a$ and $f(\beta) = b$, then X is connected.*

Note. The condition of this lemma is not necessary. If X satisfies the condition, then it is called *pathwise connected.* If f can be chosen a one-to-one map, then X is *arcwise connected.*

Proof. Let us suppose that X is not connected, say $X = A \cup B$ where A and B are nonvoid separated sets. Let $a \in A$ and $b \in B$ and let f be a function on an interval $[\alpha, \beta]$ with values in X and such that $f(\alpha) = a$, $f(\beta) = b$. Then the sets $f([\alpha, \beta]) \cap A$ and $f([\alpha, \beta]) \cap B$ are nonvoid subsets of the separated sets A and B and hence they are separated. Their union $f([\alpha, \beta])$ is therefore not a connected set. Since the interval $[\alpha, \beta]$ is connected and the continuous image of a connected set is connected, we see that f cannot be a continuous function. This shows that the condition is sufficient.

Extensions of these theorems can be obtained by replacing the space of reals by some other topological spaces.

Lemma 3. *If any two points of X can be enclosed in some connected set, then X is connected.*

Proof. Let $x \in X$ be fixed and for every $y \in X$ let S_{xy} be a connected subspace of X which contains the points x and y. By Lemma II.7.2 the set $\bigcup \{S_{xy} : y \in X\}$ is connected and by hypothesis it is the entire space X. Hence X is connected.

Theorem 2. *A normed vector space over a connected normed field F is connected.*

Note. The topologies of the vector space and of the normed field F are determined by metrics $\| x - y \|$ and $| \lambda - \mu |$. The connectedness of F is not a necessary condition.

Proof. We associate with every field element $\lambda \in F$ the vector λx where x is an arbitrary fixed vector in X. The mapping $f : \lambda \to \lambda x$ is continuous because if $| \mu - \lambda | \leqslant \epsilon$, then $\| \mu x - \lambda x \| \leqslant \epsilon \| x \|$. Therefore the set $\{\lambda x : \lambda \in F\}$ is connected and it contains $0 \in X$. Now given any two vectors x and y the sets $\{\lambda x : \lambda \in F\}$ and $\{\lambda y : \lambda \in F\}$ are connected and both contain 0. Hence by Lemma II. 7.2 the union of these sets is also connected. The result follows from Lemma 3.

The continuous image of a locally connected space need not be locally connected. However, if the space is compact the situation is different and we have the following:

Theorem 3. *If a Hausdorff space Y is the continuous image of a compact locally connected topological space X, then Y is locally connected.*

Proof. Let f be a continuous function which maps X onto Y. Let a point $y \in Y$ be given and let N_y be a neighborhood of y. The object is to find a connected neighborhood of y which is contained in N_y. We choose a point x which belongs to the inverse image of $y : x \in f^{-1}(y) \subseteq f^{-1}(N_y)$. Since f is continuous and $y \in N_y{}^i$ the set $f^{-1}(N_y)$ is a neighborhood of x. By the local connectedness of X there is a connected neighborhood N_x which is contained in $f^{-1}(N_y)$. By Lemma 1, $f(N_x)$ is a connected set in Y and $f(N_x) \subseteq N_y$. We have $x \in N_x{}^i$, so $y \in f(N_x{}^i) \subseteq N_y$ for every $x \in f^{-1}(y)$. Hence Lemma II.7.2 implies that

$$N = \bigcup \{f(N_x) : x \in f^{-1}(y)\}$$

is a connected set in Y. We must show that y belongs to the interior of the connected set N. For this purpose we can use Lemma 7.4. The set $S = f^{-1}(N)$ contains the complete inverse image of y in its interior because if $x \in f^{-1}(y)$, then by our construction $x \in N_x{}^i \subseteq f^{-1}(N)^i$. Therefore we have $y \in f(S)^i = N^i$. Thus N is a connected neighborhood of y and $N \subseteq N_y$. This shows that Y is locally connected.

The results of this section and the principles which we used in their proofs are special instances of important techniques and results of topology. The process of halving an interval has its counterpart for higher dimensional Euclidean spaces known as the elementary subdivision of a simplex. In the plane for instance a simplex is a triangle and by its ***elementary subdivision*** we understand a subdivision into four triangles such that the three new vertices are the midpoints of the sides of the original triangle. The process of halving an interval is a special case of another operation on simplices which is called the ***baricentric subdivision*** of a simplex. In the plane for example the baricentric subdivision of a triangle consists of the six triangles obtained by connecting its vertices and the midpoints of its sides to the center of the triangle.

The Corollary of Theorem II.7.4 implies: If a point is omitted from the real line, then the remainder is not connected but consists of two components and both of these are homeomorphic to the real line and their common boundary is the omitted point. This has important generalizations for the Euclidean plane and also for the n-dimensional Euclidean space known as the *Jordan curve theorem* and the *Jordan-Brouwer theorem*.

Let f be a continuous function of a real variable which maps the interval $[-1, +1]$ onto itself. Since $f(x_1) = -1$ and $f(x_2) = +1$ for some points $-1 \leqslant x_1, x_2 \leqslant 1$ we see that either $\varphi(x) = f(x) - x$ vanishes for $x = x_1$ or $x = x_2$, or φ takes on both positive and negative values in $[-1, +1]$. If we apply the intermediate value theorem to φ we obtain a point x such that $\varphi(x) = f(x) - x = 0$. Hence if f is a continuous map of the closed interval $[-1, +1]$ onto itself, then there is a fixed point, that is, a point $x \in [-1, +1]$, such that $f(x) = x$. The same proposition holds even if f is not surjective. Instead of the interval $[-1, +1]$ we can consider a disk in the Euclidean plane or more generally the ball $\{x : \| x \| \leqslant 1\}$ in the n-dimensional Euclidean space and ask whether or not every continuous map of this set into itself has a fixed point. The affirmative answer to this question is given by the *Brouwer fixed point theorem.*

EXERCISES

1. Show that every polynomial function with real coefficients and of odd degree has a real root.

2. Prove the following statement: Let X be connected and let Y_1 and Y_2 be separated sets in Y. If f maps X continuously into Y and if $f(x_1) \in Y_1$ and $f(x_2) \in Y_2$ for suitable $x_1, x_2 \in X$, then there is a point $x \in X$ such that $f(x) \notin Y_1 \cup Y_2$.

[By Lemma 1 the set $f(X)$ is connected.]

3. Show that a metric space X can be connected only if it contains exactly one element or if it contains more then denumerably many elements.

[Let $a \in X$. The function $d_a : d_a(x) = d(a, x)$ $(x \in X)$ is continuous and so its image $d_a(X)$ is a connected set in the space of the reals. Hence either $d_a(X)$ consists of the single point 0 or it is an interval and so X is not countable.]

4. Give a simple proof of the following proposition: If f is a continuous open map of the locally connected space X onto the space Y, then Y is locally connected.

[Let $y = f(x)$ and let N_y be a neighborhood of y. Then $f^{-1}(N_y)$ is a neighborhood of x and so there is a connected neighborhood N_x contained in $f^{-1}(N_y)$. Then $y \in f(N_x) \subseteq N_y$ and f being open $f(N_x)$ is a connected neighborhood of y.]

5. Give an example of a connected, locally connected but not arcwise connected Hausdorff space.

6. Show by an example that the condition given in Theorem 2 is not necessary.

(Let X be the space of reals normed as usual and let F be the field of the rational numbers.)

7. Let C denote the cross $C = \{(\xi, 0) : | \xi | \leqslant 1\} \cup \{(0, \eta) : | \eta | \leqslant 1\}$ and let $f(C)$ be a homeomorphic image of C in the same plane. Define

the distance $d(i, f) = \text{lub}[d(p, f(p)) : p \in C\}$ and show that if $d(i, f) < 1$, then C and $f(C)$ intersect.

[The image of $\{(\xi, 0) : |\xi| \leqslant 1\}$ under f belongs entirely to the rectangle $\{(\xi, \eta) : |\xi| \leqslant 2$ *and* $|\eta| \leqslant 1\}$ and it intersects both

$$R_1 = \{(\xi, \eta) : -2 \leqslant \xi < 0 \quad \textit{and} \quad |\eta| \leqslant 1\}$$

and

$$R_2 = \{(\xi, \eta) : 0 < \xi \leqslant 2 \quad \textit{and} \quad |\eta| \leqslant 1\}.$$

Since R_1 and R_2 are separated while the image of $\{(\xi, 0) : |\xi| \leqslant 1\}$ is connected this image must intersect the segment $\{(0, \eta) : |\eta| \leqslant 1\}$ which is a part of C.]

8. Let X be a locally connected metric space and let f map X onto the metric space Y such that compact sets are mapped onto compact sets and connected sets onto connected sets. Use the result stated in Exercise 7.7 to show that f is continuous.

[Suppose that f is discontinuous at x. Denote by x_n $(n = 1, 2, ...)$ and η the points described in Exercise 7.7. Since X is locally connected at x and $x_n \to x$ there is a subsequence (x_{n_k}) and connected neighborhoods $N_x{}^k$ $(k = 1, 2, ...)$ such that $x_{n_k} \in N_x{}^k \subseteq S_{1/k}[x]$. We have $y, \eta \in f(N_x{}^k)$ and so by the connectedness of $f(N_x{}^k)$ and Exercise 3 we can determine a point $\eta_k = f(\xi_k)$ where $\xi_k \in N_x{}^k$ such that $0 < d(\eta_k, \eta) < 1/k$. The set $\{x, \xi_1, \xi_2, ...\}$ is compact but its image $\{y, \eta_1, \eta_2, ...\}$ is not.]

9. Continuity in Product Spaces

Several theorems on the continuity of functions from a product of topological spaces into another topological space are of general interest. The simplest concerns the restriction of such functions to various subproducts:

Theorem 1. *Let $X = \Pi X_i$ be the product of an arbitrary family $\{X_i\}$ $(i \in I)$ of topological spaces and let f be a function from X into a topological space Y. Denote by φ the function from a subproduct $\Pi_j X_{i_j}$ into Y defined by the rule*

$$\varphi((x_{i_j})) = f((x_i))$$

where x_i for $i \neq i_j$ is a fixed point of X_i and x_{i_j} for each j varies over X_{i_j}. Then the continuity of f at a point (x_i) implies that φ is continuous at the corresponding (x_{i_j}).

Proof. Consider the subspace S formed by points ξ whose i_jth coordinates are arbitrary for every $j \in J$ but every other coordinate ξ_i coincides with the ith coordinate x_i of the point $x = (x_i)$. The restriction of f to this subspace S is continuous at x. However, by Theorem I.11.2, S is homeomorphic to the subproduct $\prod X_{i_j}$ and at the corresponding points f and φ have the same function values. Therefore φ is continuous at the point (x_{i_j}).

Simple applications of this theorem frequently occur in the theory of functions of several real variables: For instance, f with $f((x_1, x_2)) = x_1x_2 + x_2^2$ is continuous and so φ with $\varphi(x_2) = 3x_2 + x_2^2$ is a continuous function of the real variable x_2. One can raise the question whether the continuity of all restrictions of f to proper subproducts would imply the continuity of the function f itself. This need not be the case even if the product space X has only a finite number of factors. For example, let f be the real-valued function of the real variables x_1 and x_2 such that $f((0, 0)) = 0$ and $f((x_1, x_2)) = x_1x_2/(x_1^2 + x_2^2)$ for every $(x_1, x_2) \neq (0, 0)$. Then the restrictions of f to points with $x_1 = 0$ or to points with $x_2 = 0$ vanish identically and so these are continuous functions. However, f is discontinuous at $(0, 0)$ because $f((x, x)) = \frac{1}{2}$ for every $x \neq 0$.

The next theorem is a convenient tool in proving theorems such as the continuity of the sum of continuous functions. It also gives us a deeper insight to the older proofs of these theorems and leads to important new mathematical concepts.

Theorem 2. *Let X, Y_i $(i \in I)$, and Z be topological spaces. Let f_i $(i \in I)$ be functions from X into Y_i which are continuous at a fixed $x \in X$ and let g be a function from $Y = \prod Y_i$ into Z which is continuous at the point $y = (y_i)$ where $y_i = f_i(x)$. Then the composite function h defined by $h(\xi) = g((\eta_i)) = g((f_i(\xi))$ is continuous at x.*

Proof. Let an open set O_z be given in Z such that it contains $z = h(x)$. The object is to find an open set O_x in X with the property that $h(O_x) \subseteq O_z$. Since g is continuous at $y = (y_i)$ there is an O_y in $Y = \prod Y_i$ such that $g(O_y) \subseteq O_z$. By the definition of the product topology there are open sets $O_{y_i} \subseteq Y_i$ such that $\prod O_{y_i} \subseteq O_y$ and all but finitely many factors satisfy $Y_i = O_{y_i}$. As f_i is continuous at x we can find an open set $O_{xi} \subseteq X$ such that $f_i(O_{xi}) \subseteq O_{y_i}$. For all but finitely many factors we may choose $O_{xi} = X$. Then $O_x = \bigcap O_{xi}$ contains only a finite number of distinct factors and so O_x is an open set in X. Now if $\xi \in O_x$, then $\xi \in O_{xi}$ for every $i \in I$, hence $f_i(\xi) \in O_{y_i}$ and so $(f_i(\xi)) \in O_y$ which shows that $h(\xi) \in O_z$. Therefore O_x is an open set containing x with the property that $h(O_x) \subseteq O_z$. This shows that h is continuous at x.

We choose X, Y_i ($i = 1, ..., n$), and Z to be the space of reals under the usual topology. For g we choose the linear function $g((y_i)) = y_1 + ... + y_n$ which is clearly continuous on $Y_1 \times ... \times Y_n$. An application of the theorem gives: If the functions $f_1, ..., f_n$ of a real variable are continuous at x, then their sum $f_1 + ... + f_n$ is also continuous at x. If we choose the continuous g by $g((y_i)) = y_1 \cdots y_n$, then we obtain the corresponding theorem on the continuity of the product of continuous functions. Similarly, if we let g be λy, $|y|$, $y \cup 0 = \max\{y, 0\}$, and $y \cap 0 = \min\{y, 0\}$, we see that λf, $|f|$, $f^+ = f \cup 0$, and $f^- = f \cap 0$ are continuous at every point where f is continuous.

It is easy to find natural conditions such that the preceding theorems on the continuity of sums and products can be generalized. First it is necessary that sums and products be meaningful, so for instance we can assume that Y is a *semigroup*, i.e., an associative binary operation is defined on Y with values in Y. In order to assure the continuity of the product $f_1 \cdots f_n$ of continuous functions $f_1, ..., f_n : X \to Y$ at a point $x \in X$, it is sufficient to suppose that $y_1 \cdots y_n$ maps $Y \times ... \times Y$ continuously into Y. Since the operation is associative we may restrict ourselves to a product of two factors:

Lemma 1. *If a topological space Y is a semigroup such that the product $y_1 y_2$ maps $Y \times Y$ continuously into Y, then $y_1 \cdots y_n$ is a continuous mapping from $Y \times ... \times Y$ into Y for every $n \geqslant 1$.*

Proof. For $n = 1$ the result is trivial and for $n = 2$ it reduces to the hypothesis. For $n > 2$ we apply induction: Let

$$y = y_1 \cdots y_n = (y_1 \cdots y_{n-1}) y_n = y' y_n$$

where $y' = y_1 \cdots y_{n-1}$. Given O_y by hypothesis there are open sets $O_{y'}$ and O_{y_n} such that $\eta' \eta_n \in O_y$ for every $\eta' \in O_{y'}$ and $\eta_n \in O_{y_n}$. By the induction hypothesis we can find open sets $O_{y_1}, ..., O_{y_{n-1}}$ such that $\eta_1 \cdots \eta_{n-1} \in O_{y'}$ whenever $\eta_1 \in O_{y_1}, ..., \eta_{n-1} \in O_{y_{n-1}}$. Hence if $\eta_1 \in O_{y_1}$, ..., $\eta_{n-1} \in O_{y_{n-1}}$ and $\eta_n \in O_{y_n}$, then $\eta = (\eta_1 \cdots \eta_{n-1})\eta_n \in O_y$. Therefore $\eta_1 \cdots \eta_n$ is continuous at $y_1 \cdots y_n$.

The continuity of $y_1 y_2$ depends on both the algebraic and the topological structure of the set Y and it must be verified in each specific instance. A mathematical system which is both a semigroup and a topological space and in which $y_1 y_2$ is continuous is called a *topological semigroup*. These are among the simplest of the so-called *topological algebraic structures*.

We conclude this section with the proof and applications of the following simple theorem:

Theorem 3. *Let X_s $(s \in S)$ and Y_s $(s \in S)$ be topological spaces and let f_s map X_s continuously onto Y_s for every $s \in S$. Then the function f: $\prod X_s \to \prod Y_s$ defined by $f((x_s)) = (f_s(x_s))$ maps the product $\prod X_s$ continuously onto $\prod Y_s$.*

Proof. Given $x = (x_s)$ and a neighborhood N_y of $y = (y_s) = f(x) = f((x_s)) = (f_s(x_s))$ we wish to determine N_x such that $f(N_x) \subseteq N_y$. We can find open sets O_{y_s} $(s \in S)$ with the property that $O_{y_s} = X_s$ for all but finitely many indices s and $\prod O_{y_s} \subseteq N_y$. Since $f_s : X_s \to Y_s$ is continuous at x_s an open set O_{x_s} can be determined such that $f(O_{x_s}) \subseteq O_{y_s}$. If $O_{y_s} = X_s$, we may choose $O_{x_s} = X_s$. Then $N_x = \prod O_{x_s}$ is an open neighborhood of x and clearly $f(N_x) \subseteq \prod O_{y_s} \subseteq N_y$.

As usual, we denote by X^S the product $\prod X_s$ where $X_s = X$ for every index $s \in S$. Then we have the following:

Corollary. *If Y is the continuous image of X, then Y^S is the continuous image of X^S.*

The map f constructed in the last theorem is a surjective and injective provided f_s is a one-to-one transformation of X_s onto Y_s for every $s \in S$. Moreover, if f_s and f_s^{-1} are continuous functions, then by the theorem f and f^{-1} are both continuous. This shows that the product topology has the following invariant property:

Theorem 4. *If X_s and Y_s are homeomorphic for every $s \in S$, then $\prod X_s$ and $\prod Y_s$ are homeomorphic spaces.*

Moreover, we have:

Theorem 5. *If all factor spaces X_s $(s \in S)$ and Y_σ $(\sigma \in \Sigma)$ are homeomorphic and if card $S =$ card Σ, then $\prod X_s$ is homeomorphic to $\prod Y_\sigma$.*

Proof. By the preceding theorem it is sufficient to investigate the special case when $X_s = Y_\sigma = X$ for every $s \in S$ and $\sigma \in \Sigma$ where X is a fixed topological space. If $\varphi : S \to \Sigma$ is a one-to-one correspondence between the index sets S and Σ, then $f : (x_s) \to (x_{\varphi(s)})$ is a one-to-one map of $\prod X_s$ onto $\prod Y_\sigma$ which leads to a one-to-one correspondence between their open sets. For f gives a one-to-one map between the generalized cubes $\prod O_s$ and $\prod O_\sigma$ where $O_s = X$ and $O_\sigma = X$ for all but finitely many indices.

As an application of Theorem 3 we shall prove that the n-dimensional unit cube I^n, where $I = [0, 1]$, is the continuous image of the interval I for every $n = 1, 2, \ldots$. This gives a generalization of the result which we discussed at the end of Section 7 and it shows that not only the unit square but any one of the sets I^n $(n \geqslant 1)$ can be considered as a continuous curve. Moreover, it turns out that $I \times I \times \dots$ is also a continuous curve which is homeomorphic to the *Hilbert cube*

$$H = \left\{ x : x = (x_n) \quad \textit{where} \quad |x_n| \leqslant \frac{1}{n} \quad \text{for every} \quad n = 1, 2, \dots \right\}.$$

The topology of H is of course the induced topology of the separable Hilbert space so that $d(x, y) = \| x - y \|$ where $\| x \|^2 = x_1{}^2 + x_2{}^2 + \dots$. Using earlier results it follows that H is compact, connected, and locally connected.

First we define what is meant by the *Cantor set* or by the *triadic set of Cantor:* This is a noncountable subset P of the interval $[0, 1]$, namely, it consists of all real numbers x which can be represented in the form $x = \Sigma \epsilon_n 3^{-n}$ where $\epsilon_n = 0$ or 2 for every $n = 1, 2, \dots$. Although we speak about the Cantor set P it is understood that P is topologized by the induced topology of the reals. It is easy to see that P is a closed subset of the space of real numbers: For if $x \in [0, 1]$ but $x \notin P$, then in all possible triadic developments of x there is an $\epsilon_n = 1$. But $\epsilon_n = 1$ for every ξ satisfying $x - \delta < \xi < x + \delta$ where $\delta > 0$ is sufficiently small and so $(x - \delta, x + \delta) \subseteq cP$. It is clear that every neighborhood of a point $x \in P$ contains uncountably many other points of P. Hence the Cantor set is a perfect set.

Lemma 2. *The triadic set of Cantor is homeomorphic to the product space $X^\omega = X \times X \times \dots$ where X is a set of two elements topologized by the discrete topology.*

Proof. We choose $X = \{0, 2\}$. If $p \in X^\omega$, then $p = (\epsilon_1, \epsilon_2, \dots)$ where $\epsilon_n = 0, 2$ $(n = 1, 2, \dots)$. We define $f(p) = \Sigma \epsilon_n 3^{-n}$ so that $f : X^\omega \to P$ gives a one-to-one correspondence between X^ω and P. The map f is continuous because given $\epsilon > 0$ and $f(p) \in P$ there is an index n such that we have $|f(q) - f(p)| < \epsilon$ for every $q = (\eta_1, \eta_2, \dots)$ satisfying $\eta_1 = \epsilon_1, \dots, \eta_n = \epsilon_n$. Moreover, if ϵ is sufficiently small, then $|f(p) - f(q)| < \epsilon$ implies that the first n terms of the triadic development of $f(q)$ are the same as the corresponding terms in the development of $f(p)$. Hence f^{-1} is also continuous and X^ω and P are homeomorphic. (Since P is a Hausdorff space and X^ω is compact the continuity of f^{-1} can be seen also by applying Lemma 7.3.)

Lemma 3. *The closed interval $I = [0, 1]$ is a continuous image of the Cantor set P.*

Proof. We map $x = \Sigma \epsilon_n 3^{-n} \in P$ into $f(x) = \Sigma (\frac{1}{2}\epsilon_n) 2^{-n} \in I$. Then f is continuous and every point of I is an image point.

We need one more lemma:

Lemma 4. *For every $n = 1, 2, \ldots$ and for every topological space X the spaces X^ω, $(X^\omega)^n$, and $(X^\omega)^\omega$ are homeomorphic.*

Proof. We show that X^ω and $(X^\omega)^\omega$ are homeomorphic by exhibiting a bicontinuous one-to-one map as follows: If

$$x = ((x_m{}^n)) = ((x_m{}^1), (x_m{}^2), \ldots) \in (X^\omega)^\omega,$$

then we define $\varphi(x) = (x_{mn})$ $(m, n = 1, 2, \ldots)$ by setting $x_{mn} = x_m{}^n$. By Theorem 5 the nature of the denumerable index set is irrelevant, hence $(X^\omega)^\omega$ is homeomorphic to X^ω. The homeomorphism of $(X^\omega)^n$ and X^ω can be proved similarly.

Now we can draw some interesting conclusions: Since $I = [0, 1]$ is the continuous image of the Cantor set P, by the corollary of Theorem 3, I^n and P^n (I^ω and P^ω) are related in a similar way. However, P is of the form X^ω and so by the last lemma P^n and P^ω are homeomorphic to P. Consequently the n-dimensional unit cube I^n and also the infinite-dimensional product I^ω are continuous images of the Cantor set P. Moreover, we see that every I^n $(n = 1, 2, \ldots)$ and also I^ω are continuous curves. For P is a continuous image of the interval I under the map $\sum \epsilon_n 2^{-n} \to \sum 2\epsilon_n 3^{-n}$ where $\epsilon_n = 0$ or 1.

EXERCISES

1. Show that the graph of a continuous function $f : X \to Y$ considered as a subspace of the product space $X \times Y$ is homeomorphic to the domain space X.

[Let F denote the graph of f. The one-to-one map $\varphi : F \to X$ given by the formula $\varphi((x, y)) = \varphi((x, f(x)) = x$ is a homeomorphism: For given $(x, y) \in F$ and a neighborhood N_x we have $\varphi(F \cap (N_x \times Y)) = N_x$ where $F \cap (N_x \times Y)$ is a neighborhood of (x, y) in F. The continuity of f is used in the proof of the continuity of φ^{-1}: Given a neighborhood of the form $F \cap (N_x \times N_y)$ of $(x, y) \in F$ there is a neighborhood $O_x \subseteq N_x$ of x in X such that $f(O_x) \subseteq N_y$. Hence $\varphi^{-1}(O_x) \subseteq F \cap (N_x \times N_y)$.]

2. Let X_s and Y_s be homeomorphic for every $s \in S$. Show that the Boolean products $\prod X_s$ and $\prod Y_s$ are homeomorphic spaces.

(The map constructed in Theorem 3 is one-to-one provided f_s is one-to-one for every $s \in S$. By a similar proof f and f^{-1} are continuous if every f_s and f_s^{-1} is continuous.)

3. Show that if $Y = X^\omega$, then $X \times Y$ is homeomorphic to Y.

4. Let $X_s = X$ for every $s \in S$ where X is a fixed Hausdorff space and let I be the diagonal set $I = \{(x_s) : x_s = x \text{ for every } s \in S \text{ for some } x \in X\}$. Show that I is a closed set in $\prod X_s$ which is homeomorphic to X.

[We show that cI is open: If $(x_s) \in cI$, there are $s_1, s_2 \in S$ such that $x_{s_1} \neq x_{s_2}$. Let the open sets O_{s_1} and O_{s_2} be disjoint neighborhoods of x_{s_1} and x_{s_2}, respectively. The cylinder $Z(s_1, O_{s_1}) \cap Z(s_2, O_{s_2})$ does not intersect I. A homeomorphism is given by the map $(x_s) \to x$ where x denotes the common value of the projection of (x_s) into the factor spaces X_s.]

5. Let X be the space of irrational numbers. Show that X^ω is homeomorphic to X.

(By Exercise I.12.3 X is homeomorphic to a product space of the form Y^ω.)

6. Let the topology on $I = [0, 1]$ be the usual topology of the reals. Show that I^ω is homeomorphic to the Hilbert cube.

[Let $I_n = [0, 1/n]$ be topologized by the usual topology of the reals. The object is to show that the topology induced on $H = \prod I_n$ by the metric $\| x - y \|$ is the product topology. The induced topology is clearly at least as fine as the product topology. The converse follows from the inequality

$$\| x - y \|^2 \leqslant \sum_1^N (x_n - y_n)^2 + \sum_{n>N} n^{-2}.]$$

7. Let the real-valued function f be defined on the interval $(0, 1)$ by the formula

$$f(x) = \limsup(\epsilon_1 + \dots + \epsilon_n)/2n$$

where $x = \sum \epsilon_n 3^{-n}$, $\epsilon_n = 0, 1,$ or 2, and $\epsilon_n \neq 0$ for infinitely many indices $n = 1, 2, \dots$. Show that $\bar{F} = (0, 1) \times (0, 1)$.

[Given $\xi \in (0, 1)$ we can bring x into any neighborhood of ξ by fixing finitely many indices ϵ_n in a suitable way. Given $\eta \in (0, 1)$ we can make $f(x) = \eta$ by choosing the remaining ϵ_n's suitably.]

10. Uniform Continuity and Equicontinuity

Uniform continuity is one of the most important concepts of classical analysis. A function f of a real variable is called uniformly continuous on a set S if for every $\epsilon > 0$ there is a $\delta > 0$ with property that $|f(x_1) - f(x_2)| < \epsilon$ for every pair of points x_1, x_2 in S which satisfy

$|x_1 - x_2| < \delta$. For functions of a complex variable and for functions whose domain and range is a metric space the definition is analogous. It is obvious from the definition that uniform continuity is a property which is stronger than the continuity of the restriction of f to S in the topological sense: If f is uniformly continuous, then given any $x \in S$ there is a $\delta > 0$ such that $|f(\xi) - f(x)| < \epsilon$ for every $\xi \in S$ satisfying $|\xi - x| < \delta$. The point in the definition of uniform continuity is that such a value of δ can be determined without first specifying x.

The difference between topological continuity and uniform continuity is deeper than appears at first sight. Topological continuity is a purely topological concept and as such it is determined by the topologies of the domain and the range. Although some special concepts which are used to describe these topologies might appear in the definition of topological continuity, these concepts are used only for convenience. For instance, as long as it generates the same topology, a change in the metric does not affect the continuity of a function. Uniform continuity on the other hand is not a topological concept but depends intrinsically on the metrics or other structures which are used to determine the topologies of the domain space X and of the range space Y. There are only a few special cases when the two notions of continuity coincide.

The true importance of uniform structures is that they can be used to define uniform continuity for a wider class of topological spaces than the class of metric spaces:

Definition 1. *Let X and Y be sets and let $\mathscr{U}$ and $\mathscr{V}$ be uniform structures for X and Y, respectively. A function f from X into Y is called uniformly continuous on a set $S \subseteq X$ with respect to the uniform structures $\mathscr{U}$ and $\mathscr{V}$ if for every $V \in \mathscr{V}$ there is a $U \in \mathscr{U}$ such that $(y_1, y_2) = (f(x_1), f(x_2)) \in V$ whenever $x_1, x_2 \in S$ and $(x_1, x_2) \in U$.*

There are a number of equivalent definitions and there are also several useful criteria for uniform continuity. Of these we mention the following one which concerns the whole space X:

Lemma 1. *A function f is uniformly continuous on the space X with respect to the uniform structures $\mathscr{U}$ and $\mathscr{V}$ if and only if the sets*

$$f^{-1}(V) = \{(x_1, x_2) : x_1, x_2 \in X \quad \text{and} \quad (f(x_1), f(x_2)) \in V\}$$

belong to $\mathscr{U}$ for every $V \in \mathscr{V}_S$ of a subbase $\mathscr{V}_S$ for $\mathscr{V}$.

Proof. If f is uniformly continuous on X with respect to $\mathscr{U}$ and $\mathscr{V}$, then for every $V \in \mathscr{V}$ there is a $U \in \mathscr{U}$ such that $(y_1, y_2) \in V$ for every $x_1, x_2 \in X$ satisfying $(x_1, x_2) \in U$. Thus U is contained in $f^{-1}(V)$ and so by axiom (U.2) the set $f^{-1}(V)$ belongs to $\mathscr{U}$. Conversely, if $f^{-1}(V)$ belongs to $\mathscr{U}$ for every $V \in \mathscr{V}_S$, then by (U.2) and (U.3) the same holds for every $V \in \mathscr{V}$ and so f is uniformly continuous on X.

It is easy to verify that the sets $f^{-1}(V)$ $(V \in \mathscr{V})$ form a base for a uniform structure independently whether f is continuous or not. Indeed, it is simple to check that

$$f^{-1}(V) \cap f^{-1}(W) = f^{-1}(V \cap W) \qquad \text{for any} \quad V, W \in \mathscr{V},$$

$$f^{-1}(V)^{-1} = f^{-1}(V^{-1}) \qquad \text{for every} \quad V \in \mathscr{V},$$

and if $W \circ W \subseteq V \in \mathscr{V}$, then also

$$f^{-1}(W) \circ f^{-1}(W) \subseteq f^{-1}(V).$$

The uniform structure generated by the base $\{f^{-1}(V)\}$ $(V \in \mathscr{V})$ is called the *inverse image* of $\mathscr{V}$ under f and is denoted by $f^{-1}(\mathscr{V})$. An immediate consequence of Lemma 1 is the following:

Lemma 2. *A function $f : X \to Y$ is uniformly continuous with respect to the uniform structures $\mathscr{U}$ and $\mathscr{V}$ if and only if $f^{-1}(\mathscr{V}) \leqslant \mathscr{U}$.*

The relationship between uniform continuity and topological continuity is the same as in the special case of metric spaces:

Theorem 1. *If f is uniformly continuous on the set S with respect to the uniform structures $\mathscr{U}$ and $\mathscr{V}$, then the restriction of f to S is continuous with respect to the uniform topologies associated with the structures $\mathscr{U}$ and $\mathscr{V}$.*

Proof. We first notice that $\{U \cap (S \times S)\}$ $(U \in \mathscr{U})$ is a uniform structure for the uniform topology of the subspace S. Given any $V \in \mathscr{V}$ there is a $U \in \mathscr{U}$ such that $(y_1, y_2) \in V$ for every x_1, x_2 satisfying $(x_1, x_2) \in U \cap (S \times S)$ and so $f(U \cap (S \times S)[x]) \subseteq V[y]$ for every $x \in S$. This means that f is continuous at x with respect to the filters $\{(U \cap (S \times S))[x]\}$ $(U \in \mathscr{U})$ and $\{V[y]\}$ $(V \in \mathscr{V})$. Since these are the neighborhood filters $\mathscr{N}(x)$ and $\mathscr{N}(y)$ relative to the uniform topologies associated with $\mathscr{U}$ and $\mathscr{V}$ the function f is continuous at x.

There are many simple examples which show that continuity with respect to the uniform topologies associated with the uniform structures $\mathscr{U}$ and $\mathscr{V}$ does not imply uniform continuity. As a matter of fact, we can find uniformizable spaces X and Y and functions f from X into Y such that f is uniformly continuous with respect to some uniform structures $\mathscr{U}$ and $\mathscr{V}$, but these structures can be replaced by others yielding the same topologies and such that the uniform continuity no longer holds. (See the exercises at the end of this section.)

Some of the most useful results on uniform continuity are those theorems which assert that certain continuous functions are necessarily uniformly continuous. The best known among these concerns continuity on compact spaces. By Theorem III.3.8 at most one uniform structure is compatible with the topology of such a space. This uniqueness property does not *a priori* imply that continuous functions should be uniformly continuous, nevertheless, we have the following:

Theorem 2. *Let X be a compact uniform space with uniform structure $\mathscr{U}$ and let Y be a uniform space with a uniform structure $\mathscr{V}$. If $f: X \to Y$ is continuous relative to the uniform topologies associated with $\mathscr{U}$ and $\mathscr{V}$, then f is uniformly continuous relative to these uniform structures.*

Theorem III.3.8 is included as a simple corollary. For we have the following:

Lemma 3. *If $\mathscr{U}$ and $\mathscr{V}$ are uniform structures for X such that the identity map $f: x \to f(x) = x$ is uniformly continuous with respect to $\mathscr{U}$ and $\mathscr{V}$, then $\mathscr{V} \leqslant \mathscr{U}$.*

Proof. This is a special case of the last lemma but it is simpler to give a direct proof. Given any $V \in \mathscr{V}$, by the definition of the uniform continuity there is a $U \in \mathscr{U}$ such that $U \subseteq V$. Hence $V \in \mathscr{U}$ and $\mathscr{V} \leqslant \mathscr{U}$.

Now if $\mathscr{U}$ and $\mathscr{V}$ are both compatible with a compact topology given on a set X, then by Theorem 2 the identity map is uniformly continuous relative to $\mathscr{U}$ and $\mathscr{V}$ and also relative $\mathscr{V}$ and $\mathscr{U}$. Hence by the lemma we have $\mathscr{V} \leqslant \mathscr{U}$ and also $\mathscr{U} \leqslant \mathscr{V}$.

The proof of Theorem 2 is a straightforward generalization of the proof of Theorem III.3.8 and the same method can be used to obtain the following extension:

Theorem 3. *Let f be a continuous function from one uniform space X into another uniform space Y. If f is constant on the complement of a closed compact set, then f is uniformly continuous with respect to any pair of uniform structures $\mathscr{U}$ and $\mathscr{V}$ compatible with the topologies of X and Y, respectively.*

Proof. Given $V \in \mathscr{V}$ we choose a symmetric uniformity W such that $W \circ W \subseteq V$. By the continuity of f, for every point x of the given closed compact set S there is a $U_x \in \mathscr{U}$ so that $f(U_x[x]) \subseteq W[y]$. We choose a symmetric T_x satisfying $T_x \circ T_x \subseteq U_x$. The family of open neighborhoods $T_x[x]^i$ $(x \in S)$ is a cover of the compact space S and so there is a finite subcover, say $T_{x_1}[x_1]^i, \ldots, T_{x_m}[x_m]^i$. We let

$$U = T_{x_1} \cap \ldots \cap T_{x_m}.$$

We are going to prove that $(\eta_1, \eta_2) \in V$ whenever $(\xi_1, \xi_2) \in U$: If $\xi_1, \xi_2 \notin S$, then $\eta_1 = \eta_2$ and so $(\eta_1, \eta_2) \in V$. For any point $\xi_1 \in S$ we can find an index i such that $(\xi_1, x_i) \in T_{x_i}$. Now if ξ_2 is a second point in X such that $(\xi_1, \xi_2) \in U$, then by $U \subseteq T_{x_i}$ we have

$$(x_i, \xi_2) = (x_i, \xi_1) \circ (\xi_1, \xi_2) \in T_{x_i} \circ U \subseteq T_{x_i} \circ T_{x_i} \subseteq U_{x_i}.$$

Therefore by the definition of U_{x_i} we see that $(y_i, \eta_2) \in W$. Similarly $(\xi_1, x_i) \in T_{x_i} \subseteq U_{x_i}$ implies that $(\eta_1, y_i) \in W$ and so

$$(\eta_1, \eta_2) = (\eta_1, y_i) \circ (y_i, \eta_2) \subseteq W \circ W \subseteq V.$$

Hence f is uniformly continuous on X.

This is perhaps the proper time to introduce the notion of *equicontinuity* of a family of functions. First consider a family E of real-valued functions on a topological space X. If every $f \in E$ is continuous at $x \in X$, then given $\epsilon > 0$ there is a neighborhood N_x^f of x such that $| \eta - y | < \epsilon$ for every $\xi \in N_x^f$. In general the neighborhood N_x^f depends not only on the choice of ϵ but also on the function $f \in E$. The family E is called equicontinuous at x if there is a neighborhood N_x such that $| \eta - y | < \epsilon$ for every $\xi \in N_x$ and for every $f \in E$.

The equicontinuity of a family E at a point x is not a topological property. For when we fix $\epsilon > 0$ we are really specifying a uniformity U_ϵ of the usual uniform structure of the reals and this uniformity plays a vital part in the definition of equicontinuity. This will be clear from the following more general and precise:

Definition 2. *Let E be a family of functions f which map a topological space X into a uniform space Y with uniform structure $\mathscr{V}$. The family E is called equicontinuous at the point $x \in X$ relative to the topology of X and the uniform structure $\mathscr{V}$ if for every $V \in \mathscr{V}$ there is a neighborhood N_x such that $(\eta, y) = (f(\xi), f(x)) \in V$ for every $f \in E$ and $\xi \in N_x$.*

If E is equicontinuous at every point x of a set S, then the family E is called equicontinuous on S.

It is clear that if E is an equicontinuous family, then every $f \in E$ is continuous at x relative to the topology of X and the uniform topology associated with $\mathscr{V}$. On the other hand, we can easily find uniform structures $\mathscr{V}_1$ and $\mathscr{V}_2$ both compatible with the usual topology of the reals and a family E of real-valued continuous functions of a real variable such that E is equicontinuous at a point x with respect to $\mathscr{V}_1$ but not equicontinuous relative to $\mathscr{V}_2$. Every finite family of continuous functions is equicontinuous relative to every uniform structure compatible with the topology of the range space Y. More generally the union of finitely many equicontinuous families is equicontinuous.

Equicontinuous families play the main role in *Arzelà's theorem*:

Theorem 4. *Let* (f_n) $(n = 1, 2, \ldots)$ *be a sequence of real-valued continuous functions on a closed interval* $[a, b]$ *subject to the conditions:*

(i) *the family* $\{f_n\}$ $(n = 1, 2, \ldots)$ *is equicontinuous*;

(ii) *there is a bound* $M > 0$ *such that* $|f_n(x)| \leqslant M$ *for every* $n = 1, 2, \ldots$ *and for every* $x \in [a, b]$.

Then there is a subsequence (f_{n_k}) $(k = 1, 2, \ldots)$ *of* (f_n) *which is uniformly convergent on the interval* $[a, b]$.

This is a very useful theorem and it has many interesting applications. A careful analysis shows that both the hypotheses and the conclusion are essentially statements of topological or uniform nature: Let S and Y be topological spaces and let $C(S, Y)$ denote the set of all those continuous functions which map S into Y. We assume that Y is uniformizable and a uniform structure $\mathscr{V}$ is actually given. In the next section we shall introduce a topology on the set $C(S, Y)$ called the *topology of uniform convergence*. This topology depends not only on the spaces S and Y but also on the choice of the uniform structure $\mathscr{V}$. Arzelà's theorem can be interpreted in terms of the topology of uniform convergence and the new theorem, known as *Ascoli's theorem* can be generalized in several directions. Briefly, Ascoli's theorem gives a necessary and sufficient condition that a family of functions $f : S \to Y$ be a compact set of $C(S, Y)$ when S is compact and $C(S, Y)$ is topologized by the topology of uniform convergence relative to $\mathscr{V}$. We shall see that the new conditions are modifications of Arzelà's (i) and (ii).

EXERCISES

1. Show that if the real-valued functions f and g are uniformly continuous on the set S with respect to some uniform structure, then $f + g$ is also uniformly continuous on S.

[Use the inequality

$$|(f(x) + g(x)) - (f(y) + g(y))| \leqslant |f(x) - f(y)| + |g(x) - g(y)|.]$$

2. Let f be a real-valued function on X with uniform structure $\mathscr{U}$ such that for every $\epsilon > 0$ there is a uniformly continuous g satisfying $|f(x) - g(x)| \leqslant \epsilon$ everywhere on X. Show that f is uniformly continuous.

[This follows from the inequality

$$|f(x) - f(y)| \leqslant |f(x) - g(x)| + |g(x) - g(y)| + |g(y) - f(y)|.$$

See also Lemma 11.4.]

3. Show that if the real-valued uniformly continuous functions f and g are bounded, then fg is uniformly continuous.

[We have the inequality

$$|f(x)g(x) - f(y)g(y)| \leqslant |f(x) - f(y)| \cdot |g(x) - g(y)| + |f(y)| \cdot |g(x) - g(y)| + |g(y)| \cdot |f(x) - f(y)|.]$$

4. Let f be a real-valued function which is uniformly continuous on X relative to $\mathscr{U}$. Suppose that $|f(x)| \geqslant \epsilon$ for some $\epsilon > 0$ and for every $x \in X$. Show that $1/f$ is uniformly continuous on X relative to $\mathscr{U}$.

5. Show that if f and g are uniformly continuous real-valued functions, then $f \cup g = \max\{f, g\}$ and $f \cap g = \min\{f, g\}$ are uniformly continuous.

[Verify the inequality

$$|(f(x) \cup g(x)) - (f(y) \cup g(y))| \leqslant |f(x) - f(y)| + |g(x) - g(y)|.]$$

6. Show that every continuous periodic real-valued function of a real variable is uniformly continuous on the space of real numbers.

(Let $p > 0$ be a period of f. By Theorem 2 f is uniformly continuous on $[-p, p]$. If $|x_1 - x_2| < p$ we can find an integer n such that $x_i + np \in [-p, p]$ $(i = 1, 2)$.)

7. A real number $p > 0$ is called an almost period relative to the allowance $\epsilon > 0$ of the real-valued function f of a real variable if $|f(x + p) - f(x)| < \epsilon$ for every number x. The function f is *almost periodic* if for every $\epsilon > 0$ there is a length $l > 0$ such that in every interval of length l one can find a number p which is an almost period relative to ϵ. Show that every continuous almost periodic function is bounded.

(By Theorem 7.2 f is bounded on the interval $[0, l]$ where l is a length which corresponds to $\epsilon = 1$. Given any real number x, the interval $[-x, l - x]$ contains an almost period p relative to 1. Hence

$$|f(x)| \leqslant |f(x) - f(x + p)| + |f(x + p)| < 1 + |f(x + p)|.$$

Since $x + p$ belongs to $[0, l]$ we have $|f(x)| < 1 + M$ where $M > 0$ is a bound of f in $[0, l]$.)

8. Show that every continuous almost periodic function is uniformly continuous on the space of reals.

[Given $\epsilon > 0$ we determine a length l which corresponds to the allowance $\epsilon/3$. By Theorem 2 f is uniformly continuous on $[-1, 1 + l]$

and so there is a small $\delta > 0$ such that $|f(x_1) - f(x_2)| < \epsilon/3$ whenever x_1, x_2 are in $[-1, 1 + l]$ and $|x_1 - x_2| < \delta$. We prove that

$$|f(\xi_1) - f(\xi_2)| < \epsilon$$

for every pair of real numbers satisfying $|\xi_1 - \xi_2| < \delta \leqslant 1$: For let p be an almost period relative to the allowance $\epsilon/3$ in the interval $[-\xi_1, l - \xi_1]$. Then $x_1 = \xi_1 + p$ belongs to $[0, l]$ and $x_2 = \xi_2 + p$ belongs to $[-1, 1 + l]$. Hence

$$|f(\xi_1) - f(\xi_2)| \leqslant |f(\xi_1) - f(x_1)| + |f(x_1) - f(x_2)| + |f(x_2) - f(\xi_2)| < \epsilon.]$$

9. Find two metrics d_1 and d_2 compatible with the usual topology of the real numbers such that some continuous functions of a real variable are uniformly continuous with respect to d_2 but not with respect to d_1 when the range space is uniformized by its usual structure.

[Let $d_1(x, y) = |x - y|$ and $d_2(x, y) = |x^3 - y^3|$. These metrics induce the same topology on the set of reals. The function $f : x \to x^3$ is uniformly continuous with respect to d_2 but it is not uniformly continuous with respect the uniform structure generated by d_1.]

10. Find a family E of functions of a real variable and uniform structures $\mathscr{V}_1$ and $\mathscr{V}_2$ for the space of reals such that E is equicontinuous relative to $\mathscr{V}_1$ but not equicontinuous relative to $\mathscr{V}_2$.

11. Let d be a metric for the metric space X and let d_x for every $x \in X$ be the real-valued function defined by $d_x(y) = d(x, y)$ $(y \in X)$. Show that the family $\{d_x\}$ $(x \in X)$ is equicontinuous on X relative to d and the usual structure of the range space.

12. Let X be the set of all homeomorphisms of a compact plane set C into the same plane. X can be metrized as follows:

$$d(f, g) = \text{lub}\{d(f(p), g(p)) : p \in C\}.$$

Show that if $C = \{(\xi, 0) : |\xi| \leqslant 1\} \cup \{(0, \eta) : |\eta| \leqslant 1\}$, then X is separable.

[Consider the set of those homeomorphic images $f(C)$ which have the following properties: $f(C)$ is the union of finitely many closed straight line segments all of whose end points have rational coordinates. These images yield a denumerable dense subset of X.]

13. A plane set S is called a *tamely imbedded cross* if there is a homeomorphism f of the entire plane onto itself which maps the cross C onto S. Show the following: Given any tame cross $f(C)$ there is an $\epsilon > 0$

such that every other tame cross $g(C)$ satisfying $d(f(C), g(C)) < \epsilon$ intersects $f(C)$.

[Consider the restriction of the map f^{-1} to $f(C) \cup g(C)$. Since this restriction has a compact domain f^{-1} is uniformly continuous. Hence there is an $\epsilon > 0$ such that $d(p, f^{-1}(g(p))) < 1$ whenever

$$d(f(p), g(p)) < \epsilon.$$

The image $(g \circ f^{-1})(C) = f^{-1}(g(C))$ is a tame cross such that $d(i, g \circ f^{-1}) < 1$ where $i : C \to C$ is the identity map. Now Exercise 8.7 can be applied.]

14. Show that it is not possible to embed in a plane more than denumerably many disjoint tamely embedded crosses.

[By Exercise 12 and Lemmas II.12.2 and II.12.4 the set X of tamely embedded crosses is a hereditary Lindelöf space. Hence by Exercise II.12.7, every noncountable set $\mathscr{S}$ of tame crosses contains an accumulation point, say $f(C)$. By Exercise 13, crosses $g(C) \in \mathscr{S}$ which are sufficiently close to $f(C)$ intersect $f(C)$ and so $\mathscr{S}$ is not a disjoint family.]

11. The Topology of Uniform Convergence

Let S be an arbitrary nonvoid set and let $X^S = \prod X_s$ where X_s for every $s \in S$ denotes the same fixed nonvoid set X. The elements of X^S can be interpreted as functions $x : S \to X$ or as points of the product set $\prod X_s$. Since later in this section we shall assume that both S and X are topological spaces it is better to visualize these functions by their graphs. A great deal of work in modern analysis concerns the study of various topological spaces formed on sets of functions $x \in X^S$. Two such topologies were introduced earlier in Chapter I: Assuming that a topology is given on X we can introduce on X^S the product topology or the Boolean product topology. If X is uniformizable, so are the product and the Boolean product topologies of X^S. A pair of compatible uniform structures for X^S is given by the product and the Boolean product of S copies of any structure $\mathscr{V}$ which is compatible with the topology of X. If X is uniformizable and a uniform structure $\mathscr{V}$ is actually given we can introduce another uniform structure for X^S, the so-called *uniform product structure*. The topology associated with this structure is called the *topology of uniform convergence relative* to $\mathscr{V}$. The uniform product structure was first mentioned in Section I.12. We repeat the definition in terms of functions instead of the product X^S:

Definition 1. *Let $\mathscr{V}$ be a uniform structure for X and let for every $V \in \mathscr{V}$ the set U_V be defined as*

$$U_V = \{(x, y) : x, y \in X^S \quad \text{and} \quad (x(s), y(s)) \in V \quad \text{for every} \quad s \in S\}.$$

The uniform structure $\mathscr{U}$ generated by the structure base $\{U_V\}\,(V \in \mathscr{V})$ is called the uniform product structure and the topology associated with $\mathscr{U}$ is called the topology of uniform convergence relative to $\mathscr{V}$.

It is important to realize that the topology of uniform convergence depends not only on the uniform topology associated with $\mathscr{V}$ but on $\mathscr{V}$ itself. It is possible to change $\mathscr{V}$ without affecting the topology associated with $\mathscr{V}$ but changing the topology of uniform convergence. In the case of the product structure the situation is reversed: Theorem I.12.2 states that the topology associated with $\prod \mathscr{U}_s$ is always the same, namely, it is the product topology. Similarly, the topology associated with the Boolean product structure is the Boolean product of the topologies of the factor spaces.

Special instances of the topology of uniform convergence are well known from the theory of real and complex variables. For instance when we say that a sequence of real-valued functions x_n $(n = 1, 2, \ldots)$ is uniformly convergent on a set S to a limit function x, we are dealing with the topology of uniform convergence relative to the usual uniform structure of the reals. In fact "$x_n \to x$ *uniformly on* S" means that given any ϵ-neighborhood $U_\epsilon[x]$ of x there is an index $n(\epsilon)$ such that for every $n > n(\epsilon)$ we have $x_n \in U_\epsilon[x]$ or in other words $|x_n(s) - x(s)| < \epsilon$ for every real number $s \in S$. The exact relationship between convergence in the topological sense and the topology of uniform convergence will be discussed later in Section V. 9. If X is the space of reals and we are speaking about the topology of uniform convergence without specifying a uniform structure for X, we shall always mean uniform convergence relative to the usual uniform structure of the reals given, say, by the metric $d(x, y) = |x - y|$.

Let the uniform structure $\mathscr{V}$ be a pseudometric structure, say let $\mathscr{V} = \mathscr{V}_d$ where d is a pseudometric for X. Then the uniform product structure $\mathscr{U}$ has a denumerable structure base: For instance, the uniformities $U_\epsilon = \{(x, y) : d(x(s), y(s)) < \epsilon\}$ for $\epsilon = 1, \frac{1}{2}, \ldots$ form a base for $\mathscr{U}$. Thus by Theorem III.8.1, $\mathscr{U}$ is a pseudometric structure. In practice, however, it is usually not worth finding a pseudometric which generates $\mathscr{U}$ even if a pseudometric is known for $\mathscr{V}$. Instead it is more convenient to deal with another metric function called *écart*. In fact, if we introduce the nonnegative function e by setting $e(x, y) = \text{lub}\{d(x(s), y(s)) : s \in S\}$ whenever this supremum is finite, then this e defines a distance for some pairs of functions $x, y \in X^S$ which can be used to describe the structure for X^S: For if $(x, y) \in U_\epsilon$, then $e(x, y) \leqslant \epsilon$ and conversely if $e(x, y) < \epsilon$, then $(x, y) \in U_\epsilon$. Hence we have:

Lemma 1. *If $\mathscr{V}$ is a pseudometric structure for X, then the uniform product structure $\mathscr{U}$ is pseudometrizable and an écart for $\mathscr{U}$ is given by the function $e(x, y) = \text{lub}\{d(x(s), y(s)) : s \in S\}$ where d is a pseudometric for $\mathscr{V}$.*

To make things clear we define what we mean by an écart for an arbitrary structure:

Definition 2. *A nonnegative function e defined on a set $E \subseteq X \times X$ is called an écart if E is symmetric and*

(i) $e(x, x) = 0$ *for every* $x \in X$;

(ii) $e(x, y) = e(y, x)$ *for each pair* $(x, y) \in E$;

(iii) $(x, z) \in E$ *whenever* $(x, y) \in E$ *and* $(y, z) \in E$, *and in this case* $e(x, z) \leqslant e(x, y) + e(y, z)$.

The uniform structure associated with e is the structure generated by the base consisting of the sets $U_\epsilon = \{(x, y) : e(x, y) < \epsilon\}$.

Now we turn to various subspaces of X^S. If X is the space of real or complex numbers, it is clear what is meant by the boundedness of a function $x : S \to X$. Hence we can speak about the space of bounded real- or complex-valued functions topologized by the topology of uniform convergence. The notion of a bounded function can be extended to the case when X is an arbitrary set with a uniform structure $\mathscr{V}$ so we can speak about the subspace formed by the bounded functions under the topology of uniform convergence relative to $\mathscr{V}$. The only question is the definition of boundedness.

Boundedness is again a uniform notion; a function x can be bounded relative to one uniform structure while it is unbounded relative to another. A function $x : S \to X$ will be called bounded relative to the uniform structure $\mathscr{V}$ if the range of x is a "bounded set" in X. Since the range is the projection of the graph of x into the range space X this definition corresponds to the intuitive concept of a bounded function. There remains the definition of bounded sets of X relative to $\mathscr{V}$. If $\mathscr{V}$ is a pseudometric structure the obvious definition would require that the diameter of a bounded set be finite. However, every pseudometric structure can be generated by a bounded metric and so this would lead to a useless concept. The next obvious requirement would be that the diameter of a bounded set be finite relative to every pseudometric d generating the structure $\mathscr{V}$. By Theorem III.2.2 and Lemma 7.5 this definition would be in accordance with the earlier definition given for real- or complex-valued functions on S. However, this is not yet the correct definition.

The proper definition of boundedness is obtained by observing the following property of bounded sets of real numbers: If B is bounded, then for every $\epsilon > 0$ it can be decomposed into finitely many sets $B_1, \ldots, B_n$ having diameter $d(B_k) < \epsilon$. Conversely, sets satisfying this requirement are obviously bounded. Moreover, it is clear that this condition can be reformulated in terms of the uniform structure generated by the metric $d: d(x, y) = | x - y |$. Hence we can introduce the following:

Definition 3. *A set $B \subseteq X$ is called bounded with respect to the uniform structure $\mathscr{V}$ for X if for every $V \in \mathscr{V}$ there is a subdivision of B into finitely many sets B_k ($k = 1, \ldots, n$) such that*

$$(x, y) \in V \qquad \textit{for every} \qquad x, y \in B_k \qquad (k = 1, \ldots, n).$$

If X itself is bounded with respect to $\mathscr{V}$, then $\mathscr{V}$ is called a *totally bounded* or *precompact structure*. The exact relationship between precompact structures and compact spaces will be discussed in Section V.6. There we shall prove that if a space X is bounded relative to a uniform structure $\mathscr{V}$, then its diameter $d(X)$ is finite with respect to every pseudometric d satisfying $\mathscr{V}_d \leqslant \mathscr{V}$.

Now having the correct definition of boundedness we can speak about the subspace formed by the bounded functions in the topology of uniform convergence.

Lemma 2. *In the topology of uniform convergence the space of bounded functions $x : S \to X$ is a closed subspace of X^S.*

Proof. Let x be a function with the property that every neighborhood of x contains bounded functions. Then given $V \in \mathscr{V}$ there is a bounded function y in $U_W[x]$ where W is symmetric and $W \circ W \circ W \subseteq V$. Since y is bounded, the range of y is the union of finitely many sets $B_1, \ldots, B_n$ such that $(y(s), y(t)) \in W$ whenever $y(s), y(t) \in B_k$ for some index k. Since y belongs to $U_W[x]$ the sets $W[B_1], \ldots, W[B_n]$ cover the range of x. Moreover, if $x(s)$, $x(t) \in W[B_k]$, then there are points s', $t' \in S$ such that

$$(x(s), y(s')) \in W \qquad \text{and} \qquad (x(t), y(t')) \in W \qquad \text{where} \qquad y(s'), y(t') \in B_k .$$

Hence by the definition of the sets $B_1, \ldots, B_n$ we have $(y(s'), y(t')) \in W$ and so $(x(s), x(t)) \in W \circ W \circ W \subseteq V$. For a given $V \in \mathscr{V}$ we covered the range of x by finitely many sets $W[B_1], \ldots, W[B_n]$ such that if $x(s)$, $x(t) \in W[B_k]$, then $(x(s), x(t)) \in V$. Therefore the range of x is bounded. The space of bounded functions contains all of its accumulation points and so it is closed.

Further important subspaces of X^S can be defined when S is a topological space. Then we can consider the subspace formed by the continuous functions and by the bounded continuous functions on S with values in X. We prove that both of these are closed subspaces of X^S and so the bounded continuous functions form a closed subspace of the space of all continuous functions. To see all this it is sufficient to prove:

Lemma 3. *In the topology of uniform convergence the space of continuous functions $x : S \to X$ is a closed subspace of X^S.*

Proof. Suppose that every neighborhood of x contains continuous functions. We show that x is continuous at every $s \in S$. Let $V \in \mathscr{V}$ be given and let $W \circ W \circ W \subseteq V$ where W is a symmetric uniformity in $\mathscr{V}$. By hypothesis we can find a continuous y such that $(x(s), y(s)) \in W$ for every $s \in S$. Since y is continuous there is a neighborhood N_s such that $(y(s), y(t)) \in W$ for every $t \in N_s$. Hence for every $t \in N_s$ we have $(x(s), x(t)) \in W \circ W \circ W \subseteq V$ or in other words $x(t) \in V[x(s)]$. Hence x is continuous at each $s \in S$.

We can also consider the space of functions which are constant on the complement of some compact set. In general these will not form a closed subspace of X^S. If X is the set of real or complex numbers, we can consider the functions which vanish on the complement of some compact set. These are called *functions with compact support* or *functions with compact carrier*. A set $C \subseteq S$ containing all points s at which x does not vanish is called a support of x. If S is a Hausdorff space and x has a compact support, then there is a smallest compact support: By Theorem III.3.4 it is the closure of the set $\{s : x(s) \neq 0\}$.

The identically zero function has compact support ø. If S is locally compact and uniformizable, then there are also other continuous functions with compact support. For if N_s is a neighborhood of $s \in S$, then by Theorem 6.2 there is a real-valued continuous function x on S such that $x(s) = 1$ and $x(t) = 0$ for every $t \in \overline{cN_s}$. Hence N_s is a support of x. If S is locally compact at s the neighborhood N_s can be chosen compact.

The set of real- or complex-valued continuous functions with a fixed compact support is a closed subspace of X^S. However, the union of all these closed sets, that is, the set of all continuous functions having some compact support, is in general not closed in X^S. Its closure consists of those continuous functions which "*vanish at infinity*": We say that

$x : S \to X$ vanishes at infinity if for every $\epsilon > 0$ there is a compact set C in S such that $|x(s)| < \epsilon$ for every $s \notin C$. By Lemma 3 the elements of the closure are all continuous and by the definition of the topology of X^S they vanish at infinity. To see the converse first suppose that x is nonnegative and it vanishes at infinity. Then $x \cup \epsilon - \epsilon$ has compact support and is in the ϵ-neighborhood of x. The general case can be settled by considering $x^+ = x \cup 0$ and $x^- = x \cap 0$. Similarly, if x is complex valued, we can approximate its real and imaginary parts separately by functions having compact support. These subspaces are most interesting when S is a locally compact Hausdorff space.

Suppose that S is not only a topological space but a uniform structure $\mathscr{W}$ is specified for S. Then we can speak about the uniform continuity of a function $x : S \to X$ relative to the uniform structures $\mathscr{W}$ and $\mathscr{V}$. We have the following:

Lemma 4. *In the topology of uniform convergence the space of uniformly continuous functions $x : S \to X$ is a closed subspace of X^S.*

The proof is similar to those of the preceding lemmas and is left to the reader.

Finally let S be compact. Then by Lemma 7.6 every continuous function is bounded. If in addition S is uniformizable, then by Theorem 10.2 every continuous function is uniformly continuous.

We finish this section with Ascoli's theorem. This gives a sufficient condition for the compactness of a set of continuous functions in the topology of uniform convergence when S is compact. If X is a Hausdorff space, then the condition is also necessary. First we prove two lemmas:

Lemma 5. *Let S be compact and let E be an equicontinuous family in X^S. Then the topology induced on E by the topology of uniform convergence of X^S is the same as the topology induced on E by the product topology of X^S.*

Proof. Let $\mathscr{T}_u$ be the topology of uniform convergence and let $\mathscr{T}_p$ be the product topology on X^S. Since $\mathscr{T}_p \leqslant \mathscr{T}_u$ the same relation holds for the trace of $\mathscr{T}_p$ and $\mathscr{T}_u$ on E. Hence it is sufficient to prove the following: If $N_x \subseteq E$ is a neighborhood of $x \in E$ relative to the trace of $\mathscr{T}_u$, then N_x is a neighborhood of x relative to the trace of $\mathscr{T}_p$. Since N_x is a neighborhood relative to $\mathscr{T}_u$ there is a $V \in \mathscr{V}$ such that $U_V[x] \subseteq N_x$. We choose a symmetric uniformity W such that $W \circ W \circ W \subseteq V$. By the equicontinuity for every $s \in S$ there exists an open set O_s in S such that $y(O_s) \subseteq W[y(s)]$ for every $y \in E$. The family $\{O_s\}$ $(s \in S)$ being an open cover of the compact space S, there are points $s_1, \ldots, s_n$ such that the union of the sets O_{s_k} $(k = 1, \ldots, n)$ covers S.

Suppose $y \in E$ is such that $y(s_k) \in W[x(s_k)]$ for every $k = 1, ..., n$. Given any $s \in S$ we have $s \in O_{s_k}$ for a suitable k $(1 \leqslant k \leqslant n)$. By $y(O_{s_k}) \subseteq W[y(s_k)]$ and $y(s_k) \in W[x(s_k)]$ for this index k we have

$$(y(s_k), y(s)) \in W \quad \text{and} \quad (x(s_k), y(s_k)) \in W.$$

Since $x \in E$, also $x(O_{s_k}) \subseteq W[x(s_k)]$, so that $(x(s), x(s_k)) \in W$. Hence $(x(s), y(s)) \in W \circ W \circ W \subseteq U$ where s is an arbitrary element of S. In other words, $y \in U_V[x]$. We proved that if $y(s_k) \in W[x(s_k)]$ for every $k = 1, ..., n$, then $y \in U_V[x]$. Since the set of these y's is a neighborhood of x in the product topology $\mathscr{T}_p$, we see that $U_V[x]$ and so N_x are neighborhoods of x relative to $\mathscr{T}_p$. Hence $\mathscr{T}_u \leqslant \mathscr{T}_p$ on the set E.

Suppose that E is closed relative to $\mathscr{T}_p$. Since $\mathscr{T}_p \leqslant \mathscr{T}_u$, the set E is also closed relative to $\mathscr{T}_u$. In general there are sets E which are closed in $\mathscr{T}_u$ but not in $\mathscr{T}_p$. The following lemma states that for equicontinuous families this situation cannot take place:

Lemma 6. *Let S be compact. If the equicontinuous family E is closed in X^S relative to the topology of uniform convergence, then it is closed also in the product topology.*

Proof. We show that cE is open relative to the product topology of X^S. If $x \in cE$, there is a $V \in \mathscr{V}$ such that $U_V[x] \subseteq cE$. Since E is equicontinuous with respect to $\mathscr{V}$, for every $s \in S$ there is an open set O_s such that $y(O_s) \subseteq W[x(s)]$ where W is a symmetric uniformity and $W \circ W \circ W \subseteq V$. Since S is compact there are finitely many points $s_1, ..., s_n$ such that $\{O_{s_k}\}$ $(k = 1, ..., n)$ is a cover of S. Now if $y(s_k) \in W[x(s_k)]$ for $k = 1, ..., n$, then by the same reasoning as in the proof of the preceding lemma we obtain $(x(s), y(s)) \in W \circ W \circ W \subseteq V$. Therefore $y \in U_V[x] \subseteq cE$. We found a neighborhood N_x of $x \in cE$ relative to the product topology $\mathscr{T}_p$ which lies entirely in cE. Hence cE is open relative to $\mathscr{T}_p$.

Now Ascoli's theorem follows easily:

Theorem 1. *Let S be compact and let $\mathscr{V}$ be a uniform structure for X. Let E be a set in X^S such that*

(i) *the cross section $E(s) = \{x(s) : x \in E\}$ has compact closure for every $s \in S$;*

(ii) *E is equicontinuous on S relative to $\mathscr{V}$;*

(iii) *E is closed in the topology of uniform convergence on X^S.*

Then E is compact relative to the topology of uniform convergence.

Note. Conditions (i) and (ii) are also necessary. If X is a Hausdorff space relative to the uniform topology associated with $\mathscr{V}$, then (iii) is also a necessary condition.

Proof. By Theorem I.11.1 and Lemma I.11.3 the set $\prod \overline{E(s)}$ is a closed subspace of the product space $\prod X_s$. Since E is equicontinuous and closed relative to the topology of uniform convergence, by the preceding lemma it is a closed subspace of the product space $\prod \overline{E(s)}$. By Theorem III.4.2 this product is compact and so by Theorem III.3.1 the trace of the product topology of $\prod X_s$ induces a compact topology on the set E. By Lemma 5 this is the same as the topology induced on E by the topology of uniform convergence on $\prod X_s$. Hence the theorem.

Arzelà's Theorem 10.4 is a simple corollary of Ascoli's Theorem 1. For it is sufficient to show that the uniform closure of an equicontinuous family is equicontinuous. The proof of this lemma is left for the reader. (See Exercises 4 and 5 below.)

EXERCISES

1. Let X be the set of reals and let $\mathscr{V}$ be the usual uniform structure generated by the norm $|x|$. Show that the uniform product structure of X^S is generated by the écart $|x - y| = \text{lub}\{|x(s) - y(s)| : s \in S\}$.

2. Show that if S is compact and X is a normed vector space, then the norm $\|x\| = \text{lub}\{\|x(s)\| : s \in S\}$ generates the topology of uniform convergence on the set of all continuous functions $x : S \to X$.

3. Show that the Hilbert cube is compact.

4. Show that the closure of an equicontinuous family of functions defined on a compact space S is the same both in the product topology and the topology of uniform convergence.

[This follows from Lemma 6, or we have the following direct reasoning: Let $\bar{E}_p$ and $\bar{E}_u$ denote the closure of the family E with respect to the product topology and the topology of uniform convergence, respectively. We have $\bar{E}_u \subseteq \bar{E}_p$. Let $y \in \bar{E}_p$. Then $\{y\} \cup E$ is equicontinuous and so there is an O_s for every $s \in S$ such that $(y(s), y(t)) \in W$ and also $(x(s), x(t)) \in W$ for every $x \in E$ and $t \in O_s$. Let $\{O_{s_1}, ..., O_{s_n}\}$ be a cover of S. Since $y \in \bar{E}_p$ there is an $x \in E$ such that $(y(s_k), x(s_k)) \in W$ for $k = 1, ..., n$. Hence $(y(s), x(s)) \in W \circ W \circ W$ for every $s \in S$ and so $y \in \bar{E}_u$.]

5. Show that if E is equicontinuous and S is compact, then $\bar{E}_p$, the closure of E with respect to the product topology of $\prod X_s$, is also equicontinuous.

[By the preceding exercise $\bar{E}_p = \bar{E}_u$. Given $W \in \mathscr{V}$ there is an O_s such that $(x(s), x(t)) \in W$ for every $x \in E$ and $t \in O_s$. If $y \in \bar{E}_u$, there is an

$x \in E$ such that $(y(s), x(s)) \in W$ for every $s \in S$. Therefore we have $(y(s), y(t)) \in W \circ W \circ W$ for every $y \in \bar{E}_u$.]

6. Let S be compact and let E be an equicontinuous family of functions on S relative to $\mathscr{V}$. Show that the trace of the product structure and of the uniform product structure is the same structure on $E \times E$.

7. Let O be an open set in the n-dimensional Euclidean space and let E be the family of all harmonic functions $u : O \to \mathfrak{R}$ satisfying the inequality $\| u \| = \sup | u(s)| \leqslant 1$. Show that E is an equicontinuous family.

[We need the following property of harmonic functions: Let p be a point in $\mathfrak{R}^n$ and let S be the boundary of the ball $S_\rho[p]$. Then there is a continuous function $\varphi : S_\rho[p] \times S \to \mathfrak{R}$ such that for every harmonic function $u : O \to \mathfrak{R}$ whose domain O contains S and for every point q in $S_\rho[p]$ we have $u(q) = \int_S u(s)\varphi(q, s)\, ds$. The continuity of φ and compactness imply the existence of a $\delta > 0$ such that if $| p - q | < \delta$, then $| \varphi(p, s) - \varphi(q, s)| < \epsilon$ for all $s \in S$. Thus, if $| p - q | < \delta$, then for any u in E we have $| u(p) - u(q)| \leqslant \int_S | \varphi(p, s) - \varphi(q, s)|\, ds \leqslant \epsilon\mu(S)$ where $\mu(S)$ is the Euclidean volume of S.]

8. Let O be an open set in $\mathfrak{R}^n$ and let H be the set of harmonic functions $u : O \to \mathfrak{R}$. Show that H is a closed subspace of $C(O)$, the space of all continuous real-valued functions on O, in the topology of uniform convergence on compact sets.

[Harmonic functions have the following characterization: If $u(p) = \int_S u(s)\, ds$ for every $p \in O$ and for every sphere S contained in O and centered around p, then u is harmonic.]

9. Let $u_1, u_2, \ldots$ be harmonic functions on an open set O of the n-dimensional Euclidean space and let their family be uniformly bounded on compact subsets of O. Then there is a suitable subsequence which is uniformly convergent on compact subsets of O to a harmonic function.

(The set O is the union of countably many compact sets $C_1, C_2, \ldots$ and the restriction of $u_1, u_2, \ldots$ is a uniformly bounded family on each C_k. Thus by Exercise 8 and by Ascoli's theorem the uniform closure of the set of these functions is compact. The space of bounded continuous real-valued functions on C_k satisfies the first axiom of countability so compactness implies sequential compactness. Select a diagonal sequence.)

10. Suppose that the real-valued and continuous f_n $(n = 1, 2, \ldots)$ has compact support and $f_n \to f$ uniformly. Show that f has a finite maximum which is $m = \lim m_n$ where m_n is the maximum of f_n.

[Let $f_n(s_n) = m_n$ and $\mu = \operatorname{lub} f(s)$ which is finite by the uniform convergence. Choose k such that $| f_n(s) - f(s)| < \mu/8$ for every $n \geqslant k$.

There is an s such that $f(s) > 7\mu/8$ so there we have $f_n(s) > 3\mu/4$ and $f_n(s_n) = m_n > 3\mu/4$. Since $|f_n(s_n) - f_k(s_n)| < \mu/4$ for every $n \geqslant k$, we see that $f_k(s_n) \neq 0$ and so s_n belongs to the compact support of f_k. Thus the set of points s_n has an accumulation point at which the maximum μ is attained.]

12. The Algebra of Continuous Functions

In this section we shall state and prove a famous theorem of Weierstrass on the approximation of continuous functions by polynomials and an equally famous generalization of this theorem which is due to M. Stone.

Let $\mathscr{P}$ and $\mathscr{C}$ be classes of functions of a real variable, all of them defined on the same domain S. In the theory of functions of a real variable we often ask the following question: Is it possible to approximate in some sense every function of the class $\mathscr{C}$ by functions of the class $\mathscr{P}$. The oldest among these problems is the one in which S is a finite closed interval, $\mathscr{C}$ is the class of all continuous functions defined on S, $\mathscr{P}$ is the class of all polynomials in $s \in S$ whose coefficients are real numbers and approximation is understood in the uniform sense. It was believed that the answer for this question is affirmative and this was first verified by Weierstrass in 1885:

Theorem 1. *Let x be a continuous real-valued function on the finite closed interval $[a, b]$. Then for every $\epsilon > 0$ there is a polynomial p with real coefficients such that $|x(s) - p(s)| < \epsilon$ for every $s \in [a, b]$.*

Since the first proof by Weierstrass many simple proofs were given but the algebraic and topological basis of the theorem has been fully understood only by M. Stone. We shall prove two theorems of Stone from which Weierstrass' theorem and its generalizations to functions of several variables follow as simple corollaries.

Let $C(S)$ denote the class of all continuous real-valued functions x on the topological space S. This class $C(S)$ has a number of useful algebraic properties. For the absolute value $|x|$ and any constant multiple λx of a continuous real-valued function x are also continuous and if x and y are continuous, then so are $x \cup y = \max\{x, y\}$, $x \cap y = \min\{x, y\}$, xy, and $x + y$. Thus the set $C(S)$ has a rich algebraic structure: It forms a vector space over the field of reals under the operations λx and $x + y$. It is a commutative ring with respect to the operations $x + y$ and xy. Since $\lambda(xy) = (\lambda x)y = x(\lambda y)$ for every real λ and for every $x, y \in C(S)$ the set $C(S)$ is an algebra with respect to the vector operations λx and $x + y$ and the ring multiplication xy. This algebra is commutative and has a unity 1. Finally $C(S)$ is a distributive lattice under the operations $x \cup y$ and $x \cap y$. It is easy to see that $C(S)$ is in fact a vector lattice.

Theorem 2. *Let L be a sublattice of $C(S)$ where S is compact. Then the closure of L relative to the product topology is the same as its closure relative to the topology of uniform convergence.*

Proof. We must prove that if $y \in C(S)$ can be approximated by elements of L on any finite subset of the domain space S, then y can be approximated uniformly by elements of L on the entire space S. Let $\epsilon > 0$ be given. First we prove that given $u \in S$ there is a function $x_u \in L$ such that $x_u(s) < y(s) + \epsilon$ for every $s \in S$ and $y(u) - \epsilon < x_u(u)$: In fact, given any $v \in S$, by hypothesis there is an $x_{uv} \in L$ such that $|x_{uv}(u) - y(u)| < \epsilon$ and $|x_{uv}(v) - y(v)| < \epsilon$. Since x_{uv} and y are continuous functions there is an open set O_v containing v such that $|x_{uv}(s) - y(s)| < \epsilon$ for every $s \in O_v$. The family $\{O_v\}$ $(v \in S)$ is an open cover of the compact space S and so there are finitely many points $v_1, \ldots, v_m \in S$ such that S is covered by $O_{v_1} \cup \ldots \cup O_{v_m}$. Now it is easy to see that the function x_u defined by $x_u = x_{uv_1} \cap \ldots \cap x_{uv_m}$ satisfies the inequality $x_u(s) < y(s) + \epsilon$ for every $s \in S$: For given $s \in S$ we have $s \in O_{v_k}$ for some index k $(1 \leqslant k \leqslant m)$ and so $x_u(s) \leqslant x_{uv_k}(s) < y(s) + \epsilon$. The inequality $|x_{uv}(u) - y(u)| < \epsilon$ implies that $x_{uv}(u) > y(u) - \epsilon$ and so $x_u(u) > y(u) - \epsilon$. Since $x_{uv} \in L$ for every $v \in S$ and L is a lattice we have $x_u \in L$.

Next we construct a function $x \in L$ such that $|x(s) - y(s)| < \epsilon$ for every $s \in S$. We start from the inequality $x_u(u) > y(u) - \epsilon$. Since x_u and y are continuous at u there is an open set O_u containing u such that $x_u(s) > y(s) - \epsilon$ for every $s \in O_u$. The compactness of S implies that S can be covered by a finite collection of these open sets O_u $(u \in S)$, say, by $O_{u_1}, \ldots, O_{u_n}$. The function $x = x_{u_1} \cup \ldots \cup x_{u_n}$ satisfies the inequality $x(s) > y(s) - \epsilon$ for every $s \in S$. Moreover, $x_u(s) < y(s) + \epsilon$ for every $u, s \in S$, and so $x(s) < y(s) + \epsilon$ for every $s \in S$. We proved that $|x(s) - y(s)| < \epsilon$ for every $s \in S$, so x approximates y uniformly in S with an error less than $\epsilon > 0$. Since x is the maximum of finitely many functions of the lattice L, the function x itself belongs to L.

Note. In the foregoing proof we did not use the full strength of the product topology. For instead of assuming that y belongs to the weak closure of L, it is sufficient to suppose that for every $\epsilon > 0$ and $s_1, s_2 \in S$ there is an $x \in L$ such that

$$|x(s_1) - y(s_1)| < \epsilon \text{ and } |x(s_2) - y(s_2)| < \epsilon.$$

In the proof of Stone's second theorem we need the following lemma which is actually a very special case of Weierstrass' theorem:

Lemma 1. *For every $\epsilon > 0$ and for every $M > 0$ there is a polynomial p with real coefficients such that $||t| - p(t)| < \epsilon$ for every real number $t \in [-M, M]$.*

In the proof of this lemma we shall need the following result which is due to Dini:

Lemma 2. *If S is compact and (x_n) is a decreasing sequence of nonnegative continuous functions on S whose limit x is continuous, then $x_n \to x$ uniformly on S.*

Proof. For every $s \in S$ there is an index $n(s)$ and an open set O_s such that $x_{n(s)}(t) - x(t) < \epsilon$ is valid for every $t \in O_s$. The family $\{O_s\}$ $(s \in S)$ admits a finite subfamily of sets $O_{s_1}, \ldots, O_{s_m}$ whose union covers S. Hence if n is the largest of the indices $n(s_1), \ldots, n(s_m)$, then we have $x_n(t) - x(t) < \epsilon$ for every $t \in S$ and this inequality is valid also for every index larger than n. Therefore x is the uniform limit of (x_n).

Proof of Lemma 1. Let x_0 be identically 1 on $\mathfrak{R}$ and let for every $n = 1, 2, \ldots$ the function x_n be defined by the formula

$$x_n(t) = \tfrac{1}{2}(x_{n-1}(t)^2 + 1 - t^2).$$

Clearly x_n is a polynomial in t whose coefficients are reals and $0 \leqslant x_n(t) \leqslant 1$ for every $t \in [-1, 1]$. We can easily show by induction on n that $x_{n+1}(t) \leqslant x_n(t)$ for $t \in [-1, 1]$: Indeed we have $x_n - x_{n+1} = \frac{1}{2}(x_{n+1} - x_n{}^2)$ for $n = 1, 2, \ldots$. Hence $(x_n(t))$ $(-1 \leqslant t \leqslant 1)$ is a convergent sequence whose limit $x(t)$ can be computed by using the defining relation:

$$x(t) = \tfrac{1}{2} \lim x_{n-1}(t)^2 + \frac{1 - t^2}{2} = \frac{x(t)^2}{2} + \frac{1 - t^2}{2}.$$

Therefore the limit function $x(t)$ satisfies the equation $(x(t) - 1)^2 = t^2$ hence $x(t) = 1 + t$ or $x(t) = 1 - t$. Since $0 \leqslant x_n(t) \leqslant 1$ we have $0 \leqslant x(t) \leqslant 1$. This implies that $x(t) = 1 + t$ for $-1 \leqslant t \leqslant 0$ and $x(t) = 1 - t$ for $0 \leqslant t \leqslant 1$. In other words, $\lim x_n(t) = 1 - |t|$ for every $t \in [-1, 1]$. The approximating functions $x_n(t)$ are continuous on the compact set $[-1, 1]$ and their sequence is decreasing. Hence by Dini's lemma the convergence is uniform in $t \in [-1,1]$. Since the functions $x_n(t)$ are polynomials in t we proved that $|t|$ can be approximated uniformly by polynomials in $[-1, 1]$. If we replace t by t/M we obtain the same result for the interval $[-M, M]$ where $M > 0$ is arbitrary.

Theorem 3. *Let A be a subalgebra of $C(S)$ where S is compact. Then the closure of A relative to the product topology is the same as its closure relative to the topology of uniform convergence.*

Note. In practice we often use the expression "*weak topology*" to denote the product topology of X^S. In this terminology the topology of uniform convergence is called the *strong topology*.

Proof. Let $\bar{A}_p$ and $\bar{A}_u$ denote the closure of A relative to the product topology and the topology of uniform convergence, respectively. Since $\mathscr{T}_u$ is finer than $\mathscr{T}_p$ we have $\bar{A}_u \subseteq \bar{A}_p$. If we can prove that $\bar{A}_u$ is a sublattice of $C(S)$, then by Theorem 2 it will follow that $\bar{A}_u$ is closed relative to the product topology. Then $A \subseteq \bar{A}_u$ will imply that $\bar{A}_p \subseteq \bar{A}_u$ and the theorem will be proved. Therefore we wish to prove that if $x, y \in \bar{A}_u$, then $x \cup y \in \bar{A}_u$ and $x \cap y \in \bar{A}_u$. We have in general

$$x \cup y = \tfrac{1}{2}(x + y + |x - y|) \quad \text{and} \quad x \cap y = \tfrac{1}{2}(x + y - |x - y|).$$

Hence it is sufficient to prove the following:

Lemma 3. *If A is a subalgebra of $C(S)$ which is closed relative to the topology of uniform convergence, then $|x| \in A$ for every bounded $x \in A$.*

Note. In particular, if S is compact, then by Theorem 7.2 we have $|x| \in A$ for every $x \in A$.

Proof. Let $|x(s)| \leqslant M$ for every $s \in S$. By Lemma 1 there is a polynomial function p of the real variable t such that $||t| - p(t)| < \epsilon$ for every $t \in [-M, M]$. Therefore $||x(s)| - p(x(s))| = ||x(s)| - (x \circ p)(s)| < \epsilon$ for every $s \in S$. The function $x \circ p$ is a polynomial in x whose coefficients are real numbers. Hence $x \circ p \in A$ and so we proved that $|x|$ can be approximated uniformly in $s \in S$ by elements of A. Therefore $|x| \in A$ and the lemma is proved.

We say that a *family A of functions x defined on a set S separates the points of S* if to any pair of distinct points $s_1, s_2 \in S$ there is a function x in A such that $x(s_1) \neq x(s_2)$. Using this notion we have:

Theorem 4. *Let A be a subalgebra of $C(S)$ which separates the points of S. Then the closure of A relative to the product topology is either $C(S)$ or the subspace formed by those functions which vanish at a fixed point $s \in S$.*

Note. The same theorem holds also if $C(S)$ denotes the algebra of continuous complex-valued functions on S.

Proof. Let $Z \subseteq S$ be the set of those points at which every $x \in A$ vanishes. By the separation hypothesis, Z is either void or consists of a single point of S. First we prove that if $s \notin Z$ and if $t \neq s$, then there is a $u \in A$ such that $u(s) \neq 0$ and $u(t) = 0$: If x separates the points s, t and if $x(s) \neq 0$, then we can choose $u = x^2 - x(t)x$. If $x(s) = 0$, then by $s \notin Z$ we can choose $y \in A$ such that $y(s) \neq 0$. Then either $z = x + y$ or $z = x - y$ separates the points s, t and it has the property $z(s) \neq 0$. Hence again $u = z^2 - z(t)z$ satisfies the requirements. It follows that if $s_1, \ldots, s_n \notin Z$, then there are functions $x_1, \ldots, x_n \in A$ such that $x_k(s_k) = 1$ and $x_k(s_l) = 0$ for every $l \neq k$. Now let $y \in C(S)$, an $\epsilon > 0$ and a set of points $s_1, \ldots, s_n \notin Z$ be given. Then the function

$$x = y_1(s_1)\, x_1 + \ldots + y_n(s_n)\, x_n$$

has the property that $| y(s_k) - x(s_k)| < \epsilon$ for every $k = 1, \ldots, n$. Actually we have $x(s_k) = y(s_k)$ at each s_k $(k = 1, \ldots, n)$. Hence if y vanishes on Z, then $y \in \bar{A}$.

Weierstrass' Theorem 1 is an immediate consequence of the last two theorems. For let $S = [a, b]$ and let A be the class of all polynomial functions on S. Then A separates the points of S and for every $s \in [a, b]$ there is an $x \in A$ such that $x(s) \neq 0$. Hence by Theorem 4 the weak closure $\bar{A}_p$ is $C(S)$. By Theorem 3, $\bar{A}_p = \bar{A}_u$ and so the uniform closure of A is $C(S)$.

A second theorem of Weierstrass states that every continuous 2π-periodic function x can be approximated uniformly on $(-\infty, +\infty)$ by trigonometric polynomials $t(s) = \Sigma_0^n (a_k \cos ks + b_k \sin ks)$. Since x and t are periodic it is sufficient to restrict x and t to the compact domain $S = [-\pi, +\pi]$ and prove the approximation there. Here $\sin s$, $\cos s$ separate the points of S and the constant 1 never vanishes. The algebra generated by 1, $\sin s$, and $\cos s$ is the algebra of trigonometric polynomials. Hence Weierstrass' second theorem follows from the theorems of Stone.

If S is a compact set in the n-dimensional Euclidean space with points $s = (s_1, \ldots, s_n)$ and if A consists of all polynomials in $s_1, \ldots, s_n$ with real coefficients, then A separates the points of S and so every continuous function x can be approximated uniformly on S by such polynomials. This is Weierstrass' approximation theorem for functions of n real variables. This theorem has several useful generalizations; for instance, we can replace the n-dimensional Euclidean space by the product of n compact Hausdorff spaces. We obtain:

Lemma 4. *Every real-valued continuous function on the product $S_1 \times \ldots \times S_n$ of the compact Hausdorff spaces $S_1, \ldots, S_n$ can be approximated uniformly by functions x of the form $x = \Sigma_0^m x_{1k} \ldots x_{nk}$ where m is arbitrary and x_{ik} $(i = 1, \ldots, n)$ is a real-valued continuous function on S_i.*

Proof. Since S_i is a compact Hausdorff space, by Theorem III.3.6 it is normal and so by Theorem 6.2 any pair of distinct points of S_i can be separated by continuous functions. Therefore the family of those continuous functions on $S_1 \times ... \times S_n$ which depend only on one variable s_i ($i = 1, ..., n$) separate the points of $S_1 \times ... \times S_n$. Stone's theorem can be applied to the smallest algebra containing these functions.

EXERCISES

1. Let S be an arbitrary topological space and let $C_b(S)$ be the space of bounded continuous functions on S topologized by the topology of uniform convergence. Show that $C_b(S)$ is connected.

[Use Lemma 8.2 to show that $C_b(S)$ is arcwise connected. If $a, b \in C_b(S)$, then $x_r = ra + (1 - r)b$ is bounded and continuous for every r ($0 \leqslant r \leqslant 1$). The map $f : r \to x_r$ is a continuous function from the reals into $C_b(S)$ because $\| x_{r_1} - x_{r_2} \| \leqslant (\| a \| + \| b \|) | r_1 - r_2 |$.]

2. Prove that the space $C([a, b])$ of real-valued continuous functions on the closed interval $[a, b]$ is separable.

(Consider polynomials in the real variable $s \in [a, b]$ all of whose coefficients are rationals.)

3. Let S be a compact set in the n-dimensional Euclidean space. Use Stone's theorem to prove that $C(S)$ is separable.

[The algebra generated over the rationals by the functions $1, x_1, ..., x_n$ where $x_k(s) = s_k$ ($k = 1, ..., n$) contains denumerably many elements.]

4. Let S be a compact Hausdorff space having a countable base. Show that $C(S)$ is separable.

[By Exercise III.8.1 the space S is metrizable and by Theorems III.3.6 and IV.6.3 for any pair of distinct points s_1, s_2 there is a real-valued continuous function on S such that $x(s_1) \neq x(s_2)$. Let O_{s_1} and O_{s_2} be open sets such that $x(t_1) \neq x(t_2)$ for every $(t_1, t_2) \in O_{s_1} \times O_{s_2}$. The set $\{(s_1, s_2) : d(s_1, s_2) \geqslant 1/n\}$ is a compact subspace of $S \times S$. Hence it can be covered by finitely many of the sets $O_{s_1} \times O_{s_2}$ ($s_1, s_2 \in S$) and so $\{(s_1, s_2) : s_1 \neq s_2\}$ can be covered by denumerably many of these sets. The corresponding functions $x_1, x_2, ...$ separate the points of S. The algebra generated by $1, x_1, x_2, ...$ over the rationals contains denumerably many elements and is dense in $C(S)$.]

5. Show by an example that compactness is an essential requirement in the preceding exercise.

[Let $S = \{1, 2, 3, ...\}$ be topologized by the discrete topology. Then S is metrizable, separable, and locally compact. Every real-valued function is continuous on S. The functions x with $x(n) = 0$ or 1 ($n = 1, 2, 3. \ldots$) form a noncountable subset of $C(S)$ and cannot be uniformly approximated by denumerably many functions.]

6. Show by exhibiting a suitable example that in Lemma 2 the word "decreasing" is essential.

[Let $f_n(0) = f_n(2/n) = f_n(1) = 0$, let $f_n(1/n) = 1$ and let the graph of f_n be linear between these points on the closed interval [0, 1].]

7. Let S be a compact Hausdorff space and let A be an algebra of complex-valued functions on S over the field of complex numbers so that if $x \in A$ and if λ is any complex number, then $\lambda x \in A$. Suppose that A separates points and contains with each x also its complex conjugate x^*. Show that the uniform closure of A is $C(S)$ or the space of those complex-valued continuous functions on S which vanish at a fixed point of S.

(If $x \in A$, then by $x^* \in A$ the real and imaginary parts of x also belong to A. Since A is a separating algebra the real algebra spanned by the real and imaginary parts is separating and so Theorems 3 and 4 can be applied.)

NOTES

Graphs were introduced by von Neumann [1]. For applications of graphs see for example pp. 302–307 in the book of Riesz and Nagy [2] on functional analysis. The problem stated in Exercise 4.7 was first considered by Kuratowski [3]. Theorem 4.1 and the results stated in Exercises 4.11 and 4.12 are due to Whyburn [4].

Young's theorems were published first in 1908 [5]. These and also other interesting theorems can be found in his Paris address [6].

The spaces mentioned in the beginning of Section 6 are described in Urysohn [7] and in Hewitt [8]. For further results on axioms of separation in terms of continuous functions see Smirnov [9], and van Est and Freudenthal [10]. A topological space in which points can be separated by real-valued continuous functions is often called a Stone space. This terminology is due to Theorems 12.3 and 12.4 on the approximation of arbitrary real-valued continuous functions by the elements of an algebraic system of real-valued continuous functions. Theorem 6.5 was published by Dowker [11]. The content of Exercise 6.4 first appeared in [12].

There exist noncompact uniformizable spaces X such that all continuous real-valued functions on X are bounded. Simple examples were given by Mrówka [13] and Novak [14]. Among many other things the existence of such spaces is pointed out in Hewitt [15]. See also

Dieudonné [16], Doss [17], and Gál [18]. (Note: A result of Hewitt is incorrectly quoted in the last paper and as a consequence proposition (iii) on p. 1054 should be omitted.) The content of Exercise 7.11 is essentially Helly's theorem on functions of bounded variation.

In Michael's second paper [19] the following characterization is given for paracompactness: X is paracompact if and only if every open cover of X admits a closure preserving refinement. (A family $\{S_i\}$ of sets is called a closure preserving family if $\overline{\bigcup S_{i_j}} = \bigcup \bar{S}_{i_j}$ for every subfamily of $\{S_i\}$.) This result can be used to prove that the closed continuous image of a paracompact Hausdorff space is paracompact. The theorem stated in Exercise 7.1 is due to Ramanathan [20]. The content of Exercise 7.3 can be found in Hewitt's thesis [21]. Exercises 7.4, 7.7, 7.8, and 8.8 contain theorems which were proved by Klee and Utz [22]. A number of interesting theorems on the closed continuous image of a metric space were published by Vainštein [23]. Among other things the following is announced: If f is a closed continuous map of a metric space X onto a metric space Y, then the inverse image $f^{-1}(y)$ has compact boundary for every $y \in Y$.

The first articles on Peano curves were published by Peano [24] and Hilbert [25]. See also Knopp's article [26] where also continuous nowhere differentiable functions and Jordan curves with positive measure are discussed.

A real-valued function f is called continuous in the Darboux sense, or f is said to have the *Darboux property*, if the range of f is connected. In other words, f has the Darboux property if $f(X)$ is an interval. This is of course not a topological property of the function f. As a matter of fact, the domain of f need not even be a topological space. A number of articles appeared on this subject, e.g., Halperin [27] and Sierpinski [28]. The expository paper of Hopf [29] should be mentioned in connection with the remarks made at the end of Section 8.

There exist nonhomeomorphic topological spaces X_1 and X_2 such that $X_1 \times X_1$ is homeomorphic to $X_2 \times X_2$. Moreover, there exist topological spaces X, Y, and Z such that $X \times Y \sim X \times Z$ but Y and Z are not homeomorphic. Examples for this situation can be found in Fox [30].

Arzelà's theorem was published in Arzelà [31]. Of the many applications of Arzelà's theorem and of Ascoli's theorem we mention the following few: The best known is Peano's theorem on the existence of a solution of the differential equation $y' = f(x, y)$. It can be found in Birkhoff and Rota [32] or in the booklet of Kolmogoroff and Fomin on functional analysis [33]. An application to the problem of the vibrating cord is in the book of Riesz and Nagy [2]. The topological space $\mathscr{D}$

introduced by Schwartz has the Heine-Borel property, that is, a set $C \subseteq \mathscr{D}$ is compact if and only if it is closed and bounded. The proof of this proposition is also based on Ascoli's theorem. (See Schwartz [34].) Ascoli's theorem has many generalizations. A few of these one can find in Kelley's book [35], pp. 233–234. In some generalizations the topology of uniform convergence is replaced by other topologies. For instance, see Fáry [36].

Weierstrass' Theorem 12.1 was first proved in Weierstrass [37]. One of the best-known proofs is based on an interpolation formula which was discovered by Bernstein [38]. It can be found in Botts and McShane [39] or in Lorentz [40]. The least analytic proof prior to the discovery of the theorems of Stone was given by Lebesgue. A number of proofs can be found in an article by Fejér [41]. An elegant proof of Weierstrass' second theorem on approximation by trigonometric polynomials is based on the summability of Fourier series. (See for instance Titchmarsh's book on the theory of functions [42].) The summability theorem is due to Fejér [43]. Weierstrass' second theorem is a special case of an important theorem of Peter and Weyl [44].

The first theorem of Weierstrass can be generalized in another direction: Müntz' theorem states that every continuous real-valued f on $[a, b]$ can be uniformly approximated by polynomials of the form $p(s) = a_0 + a_1 s^{k_1} + ... + a_k s^{k_n}$, where $(k_1, k_2, ...)$ is a given sequence of positive integers, if and only if $\Sigma k_n^{-1} = \infty$ (See for example Kaczmarz and Steinhaus [45] or Müntz [46].)

The Stone-Weierstrass approximation theorem was published by Stone in his second memoir on Boolean algebras [47]. A number of simplifications and generalizations appeared subsequently by Dunford and Segal [48], Hewitt [49], Kakutani [50], Šilov [51], and others. A lucid exposition of the subject was given by Stone [52]. The proof of Lemma 12.1 is due to Visser.

References

1. J. von Neumann, Über adjungierte Funktionaloperatoren. *Ann. of Math.* **33**, 294–310 (1932).
2. F. Riesz and B. Sz.-Nagy, "Leçons d'Analyse Fonctionelle." Akadémia Kiadó, Budapest, 1952.
3. C. Kuratowski, On a topological problem connected with the Cantor-Bernstein theorem. *Fund. Math.* **37**, 213–216 (1950).
4. G. T. Whyburn, Open and closed mappings. *Duke Math. J.* **17**, 69–74 (1950).
5. W. H. Young, On the distinction of right and left at points of discontinuity. *Quart. J. Math.* **39**, 67–83 (1908).
6. W. H. Young, La symétrie de structure des fonctions de variables réelles. *Bull. Sci. Math.* [2] **52**, 265-280 (1928).

7. P. Urysohn, Über die Mächtigkeit der zusammenhängenden Mengen. *Math. Ann.* **94**, 262–295 (1925).
8. E. Hewitt, On two problems of Urysohn. *Ann. of Math.* [2] **47**, 503–509 (1946).
9. Yu. M. Smirnov, On the theory of completely regular spaces. *Moskov Gos. Univ. Uč. Zap.* 155 *Mat.* **5**, 137–155 (1952).
10. W. T. van Est and H. Freudenthal, Trennung durch stetige Funktionen in topologischen Raumen. *Nederl. Akad. Wetensch. Proc. Ser. A* **54**, 359–368 (1951).
11. C. H. Dowker, On countably paracompact spaces. *Canad. J. Math.* **3**, 219–224 (1951).
12. N. Vedenissoff, Sur les fonctions continues dans des espaces topologiques. *Fund. Math.* **27**, 234–238 (1936).
13. S. Mrówka, On completely regular spaces. *Fund. Math.* **41**, 105–106 (1954).
14. J. Novák, On a problem concerning completely regular sets. *Fund. Math.* **41**, 103–104 (1954).
15. E. Hewitt, Rings of real-valued continuous functions, I. *Trans. Amer. Math. Soc.* **64**, 45–99 (1948).
16. J. Dieudonné, Sur les espaces uniformes complets. *Ann. Sci. École Norm. Sup.* [3] **56**, 277–291 (1939).
17. R. Doss, On uniform spaces with a unique structure. *Amer. J. Math.* **71**, 19–23 (1949).
18. I. S. Gál, Uniformizable spaces with a unique structure. *Pacific J. Math.* **9**, 1053–1060 (1959).
19. E. Michael, Another note on paracompact spaces. *Proc. Amer. Math. Soc.* **8**, 822–828 (1957).
20. A. Ramanathan, A characterization of maximal Hausdorff spaces. *J. Indian Math. Soc.* [N.S.] **11**, 73–80 (1947).
21. E. Hewitt, A problem of set-theoretic topology. *Duke Math. J.* **10**, 309–333 (1943).
22. V. L. Klee and W. R. Utz, Some remarks on continuous transformations. *Proc. Amer. Math. Soc.* **5**, 182–184 (1954).
23. I. A. Vainštein, On closed mappings of metric spaces. *Dokl. Akad. Nauk SSSR* [N.S.] **57**, 319–321 (1947).
24. G. Peano, Sur une courbe qui remplit toute une plane. *Math. Ann.* **36**, 157–160 (1890).
25. D. Hilbert, Über die stetige Abbildung einer Linie auf ein Flächenstück. *Math. Ann.* **38**, 459–460 (1891).
26. K. Knopp, Einheitliche Erzeigung und Darstellung der Kurven von Peano, Osgood und v. Koch. *Arch. Math.* [3] **26**, 103–115 (1918).
27. I. Halperin, Discontinuous functions with the Darboux property. *Amer. Math. Monthly* **57**, 539–540 (1950).
28. W. Sierpinski, Sur une propriété de fonctions réelles quelconques. *Math. Catania* 8, 43–48 (1953).
29. H. Hopf, Vom Bolzanoschen Nullstellensatz zur algebraischen Homotopietheorie der Sphären. *Jber. Deutsch. Math. Verein.* **56**, 59–76 (1953).
30. R. H. Fox, On a problem of S. Ulam concerning Cartesian products. *Fund. Math.* **34**, 278–287 (1947).
31. C. Arzelà, Sulle serie di funzioni, I. *Mem. Accad. Sci. Bologna* **8**, 131–186 (1900),
32. G. Birkhoff and G.-C. Rota, "Ordinary Differential Equations." Ginn, Boston, 1962.
33. A. N. Kolmogoroff and S. V. Fomin, "Elements of the Theory of Functions and Functional Analysis." Vol. 1, "Metric and Normed Spaces." Graylock, New York, 1957.
34. L. Schwartz, "Théorie des distributions," Vol. I, p. 70. Hermann, Paris, 1957.

35. J. L. Kelley, "General Topology." Van Nostrand, Princeton, New Jersey, 1955.
36. I. Fáry, Un critère de compacité pour les fonctions continues. *C. R. Acad. Sci. Paris* **224**, 992–993 (1947).
37. K. T. Weierstrass, Über die analytische Darstellbarkeit sogenannter willkürlicher Funktionen reeller Argumente. *S.-B. Deutsch Akad. Wiss. Berlin Kl. Math. Phys. Tech.* 633–639, 789–805 (1885).
38. S. Bernstein, Démonstration du théorème de Weierstrass fondée sur le calcul des probabilités. *Comm. Soc. Math. Kharkoff* [2] **13**, 1–2 (1912).
39. T. A. Botts and E. J. McShane, "Real Analysis." Princeton Univ. Press, Princeton, New Jersey, 1959.
40. G. G. Lorentz, "Bernstein Polynomials." Univ. of Toronto Press, Toronto, 1953.
41. L. Fejér, Über Weierstrasssche Approximation, besonders durch Hermitesche Interpolation. *Math. Ann.* **102**, 707–725 (1930).
42. E. C. Titchmarsh, "Theory of Functions," 2nd ed. Oxford Univ. Press, London and New York, 1939.
43. L. Fejér, Untersuchungen über Fouriersche Reihen. *Math. Ann.* **58**, 51–69 (1904).
44. F. Peter and H. Weyl, Die Vollständigkeit der primitiven Darstellungen einer geschlossenen kontinuierlichen Gruppe. *Math. Ann.* **97**, 737–755 (1927).
45. H. Steinhaus, "Theorie der Orthogonalreihen," 2nd ed. Chelsea, New York, 1951.
46. Ch. Müntz, Über den Approximationssatz von Weierstrass. *Festschr. H. A. Schwarz*, pp. 303–312 (1914).
47. M. H. Stone, Applications of the theory of Boolean rings to general topology. *Trans. Amer. Math. Soc.* **41**, 375–481 (1937).
48. N. Dunford and I. E. Segal, Semigroups of operators and the Weierstrass theorem. *Bull. Amer. Math. Soc.* **52**, 911–917 (1946).
49. E. Hewitt, Certain generalizations of the Weierstrass approximation theorem. *Duke Math. J.* **14**, 419–427 (1947).
50. S. Kakutani, Concrete representations of abstract (M) spaces (a characterization of the space of continuous functions). *Ann. of Math.* [2] **42**, 994–1024 (1941).
51. G. E. Šilov, On rings of functions with uniform convergence. *Ukrain. Math. Ž.* **3**, 404–411 (1951).
52. M. H. Stone, The generalized Weierstrass approximation theorem. *Math. Mag.* **21**, 167–184, 237–254 (1948).

CHAPTER V

Theory of Convergence

Ordered convergence in its simplest form occurs when we speak of a sequence of real numbers. Using the so-called ϵ definition of a limit we can tell precisely what we mean by the convergence and the limit of sequences of real numbers but we should not be satisfied by merely being able to give an exact definition for a concept which we want to use extensively. A careful analysis of this notion was made only relatively recently by also considering more general instances of limits and eliminating the unnecessary special means used in the earlier definitions. As a result of this analysis we now see that convergence and limits are purely topological notions and as such can be described by using only topological concepts.

Another line of development in the theory of convergence centers around the problem of replacing sequences by more general mathematical notions for which convergence and limits are meaningful. Several instances of such generalized limits occur in analysis and a study of these led to new mathematical concepts which permit far-reaching generalizations of sequential convergence. On the following pages we shall introduce these new concepts and define what we mean by convergence, by adherence or accumulation points, and by limits. The classical theory of sequential convergence appears as a special case and it will illustrate the general theory.

We first introduce filters and nets. These are generalizations of sequences and as far as convergence properties are concerned they are equivalent notions. Their range of applicability, however, is different. Filters are very versatile tools in topology while nets are better adapted to arguments in analysis.

The first successful application of filters was in the theory of compact spaces. Using filters the theorem on the equivalence of the compactness and of the Bolzano-Weierstrass property can be extended from metric spaces to arbitrary topological spaces. Later, similar results were obtained by using nets instead of filters. A second basic application of filters is in the theory of complete spaces or more precisely complete uniform structures. Uniform structures are themselves filters but this is only a superficial application. It is in the definition of completeness where filters or nets are indispensible. As a generalization of Cauchy sequences we introduce Cauchy filters and define completeness in terms of these filters.

1. Filters and Nets

These two notions are set-theoretical concepts and replace sequences in the general theory of convergence. Subsequently we shall speak about the convergence and divergence of filters and of nets, their adherence points and limit points. It is possible to associate with each filter a net in such a way that they converge or diverge simultaneously, have the same accumulation points, and in the case of convergence both have identical limits. Similarly, we can replace nets by suitable filters which exhibit the same convergence properties. Hence in the theory we could restrict ourselves either only to filters or only to nets. This, however, would be as impractical as trying to forget about infinite series and to think always in terms of sequences or vice versa. When one uses filters or nets one should always choose whichever is most natural for the situation at hand.

We first define filters. In what follows X will denote a nonvoid set and for the time being we do not require the existence of any structure on X. Later when discussing convergence we shall assume that X is a topological space.

Definition 1. *A nonvoid family $\mathcal{F}$ of subsets F of a set X is called a filter in X if it satisfies the following axioms:*

(F. 1) *If $F \in \mathcal{F}$, then F is not void.*

(F. 2) *If $F_1 \in \mathcal{F}$ and $F_1 \subseteq F$, then $F \in \mathcal{F}$.*

(F. 3) *If $F_1, F_2 \in \mathcal{F}$, then $F_1 \cap F_2 \in \mathcal{F}$.*

Note. The last axiom can be replaced by a weaker one; namely:

(F. 3a) *If $F_1, F_2 \in \mathcal{F}$, then $F \subseteq F_1 \cap F_2$ for some $F \in \mathcal{F}$.*

Clearly (F.2) and (F.3a) imply (F.3).

Axiom (F.3) shows that a filter is a nonvoid family of nonvoid sets in X which has the *finite intersection property*, that is, $F_1, \ldots, F_n \in \mathcal{F}$ implies $F_1 \cap \ldots \cap F_n \neq \emptyset$. The family also has the property expressed in axiom (F.2) which can be stated in an equivalent form:

(F. 2a) *If $F_1 \in \mathcal{F}$ and if F_2 is a subset of X, then $F_1 \cup F_2 \in \mathcal{F}$.*

This form is not practical but it brings out the algebraic meaning of this axiom and of the filter itself:

Let $\mathscr{P}$ be the set of all subsets of X. Under the binary operations $P_1 \cup P_2$ and $P_1 \cap P_2$ the set $\mathscr{P}$ forms a *distributive lattice*. If a subset $\mathscr{F}$ of $\mathscr{P}$ is closed under these operations, then it is called a *sublattice*. If $F_1 \cap F_2 \in \mathscr{F}$ not only when both $F_1, F_2 \in \mathscr{F}$ but also when either F_1 or F_2 belongs to $\mathscr{F}$, then the sublattice $\mathscr{F}$ is called an *ideal* in $\mathscr{P}$. If $F_1 \cup F_2 \in \mathscr{F}$ whenever either $F_1 \in \mathscr{F}$, or $F_2 \in \mathscr{F}$, or both, then we speak about a *dual ideal* in $\mathscr{P}$. The axiom system (F.1), (F.2a), and (F.3) shows that a filter $\mathscr{F}$ is a dual ideal in $\mathscr{P}$ which does not contain the element $\emptyset$.

Usually filters are defined by giving only some of their elements from which the remaining ones can be found easily. If we discard axiom (F.2) we obtain the axioms of such a subfamily known as a *filter base:*

Definition 2. *A nonvoid family $\mathscr{B}$ of subsets B of a set X is called a filter base in X if it satisfies the following axioms:*

(Fb. 1) *If $B \in \mathscr{B}$, then B is not void.*

(Fb. 3a) *If $B_1, B_2 \in \mathscr{B}$, then $B \subseteq B_1 \cap B_2$ for some $B \in \mathscr{B}$.*

The filter $\mathscr{F}$ generated by the filter base $\mathscr{B}$ is the family of those sets $F \subseteq X$ which contain some $B \in \mathscr{B}$.

Note. Every filter $\mathscr{F}$ is a filter base and the filter generated by the filter base $\mathscr{F}$ is $\mathscr{F}$ itself.

Definition 3. *A subbase $\mathscr{S}$ for a filter in a set X is a nonvoid family of nonvoid sets $S \subseteq X$ having the finite intersection property. The filter base generated by $\mathscr{S}$ is the family $\mathscr{B}$ of all finite intersections $B = S_1 \cap \ldots \cap S_n$ of elements $S_1, \ldots, S_n$ of $\mathscr{S}$. The filter generated by $\mathscr{S}$ is the filter $\mathscr{F}$ generated by $\mathscr{B}$.*

It is convenient sometimes to speak about finer and coarser or about stronger and weaker filters and also about refinements of filters and filter bases. These concepts correspond to the notion of a subsequence. We say that a filter base $\mathscr{B}_1$ is a *refinement* of another filter base $\mathscr{B}_2$ if every $B_2 \in \mathscr{B}_2$ contains some $B_1 \in \mathscr{B}_1$. If $\mathscr{B}_1$ is a refinement of $\mathscr{B}_2$ we shall write $\mathscr{B}_1 \geqslant \mathscr{B}_2$ or $\mathscr{B}_2 \leqslant \mathscr{B}_1$. The relation $\mathscr{B}_1 \geqslant \mathscr{B}_2$ does not necessarily imply that $\mathscr{B}_1 \supseteq \mathscr{B}_2$. We define a filter $\mathscr{F}_1$ to be *finer* than a filter $\mathscr{F}_2$, or in other words, $\mathscr{F}_2$ to be *coarser* than $\mathscr{F}_1$, if $\mathscr{F}_1 \supseteq \mathscr{F}_2$. Since filters are filter bases we can also speak about $\mathscr{F}_1$ being a refinement of $\mathscr{F}_2$. It is clear from these definitions that if $\mathscr{F}_1$ is finer than $\mathscr{F}_2$, then $\mathscr{F}_1$ is a refinement of the filter base $\mathscr{F}_2$. It is simple to show that if $\mathscr{B}_1 \geqslant \mathscr{B}_2$, then $\mathscr{F}_1$, the filter generated by $\mathscr{B}_1$, is finer than $\mathscr{F}_2$, the filter generated by $\mathscr{B}_2$. A necessary and sufficient condition for $\mathscr{F}_1 = \mathscr{F}_2$ is that the generating bases $\mathscr{B}_1$ and $\mathscr{B}_2$ satisfy $\mathscr{B}_1 \leqslant \mathscr{B}_2$ and $\mathscr{B}_2 \leqslant \mathscr{B}_1$.

A filter $\mathscr{F}$ is called *countable* if it has a base consisting of at most denumerably many elements. We have:

Lemma 1. *Every countable filter has a base of the form* $\{C_n\}$ $(n = 1, 2, \ldots)$ *where* $C_1 \supseteq C_2 \supseteq \ldots$.

Proof. Let $\mathscr{B} = (B_n)$ $(n = 1, 2, \ldots)$ be a countable base for the filter $\mathscr{F}$. We consider the sets $C_n = B_1 \cap \ldots \cap B_n$ $(n = 1, 2, \ldots)$. These form a decreasing sequence $\mathscr{C}$ which is a filter base in X. The base $\mathscr{C}$ is obviously a refinement of $\mathscr{B}$ and $\mathscr{B}$ is also a refinement of $\mathscr{C}$ because by axiom (Fb.3) every finite intersection $B_1 \cap \ldots \cap B_n$ contains some set $B \in \mathscr{B}$. Hence the filters generated by $\mathscr{B}$ and $\mathscr{C}$ are identical and $\mathscr{C}$ is a decreasing base for the filter $\mathscr{F}$.

To adapt the theory of convergent sequences to this treatment we must first associate with every sequence (x_n) $(n = 1, 2, \ldots)$ of points $x_n \in X$ a filter $\mathscr{F}$ such that if X is a topological space, then $x_n \to x$ if and only if the filter $\mathscr{F}$ is also convergent and its unique limit is the point x. The filter $\mathscr{F}$ is called the elementary filter generated by the sequence (x_n). It is a countable filter and it is defined by giving a decreasing base:

Definition 4. *Let* (x_n) $(n = 1, 2, \ldots)$ *be a sequence in the set* X *and let* $\mathscr{F}$ *be the filter generated by the denumerable base* $\mathscr{B} = \{B_n\}$ $(n = 1, 2, \ldots)$ *where* $B_n = \{x_\nu : \nu \geqslant n\}$. *Then* $\mathscr{F}$ *is called the elementary filter generated by the sequence* (x_n).

Note. The elementary filter $\mathscr{F}$ is the set of all subsets F of X which contain all but finitely many elements of the sequence (x_n).

Countable filters are instances of a larger class of special filters: If a filter $\mathscr{F}$ has a base $\mathscr{B}$ of cardinality at most n, then $\mathscr{F}$ is called an *n-filter*.

The definition of a net involves directed sets. A partially ordered set D with a reflexive ordering relation $\leqslant$ is called a *directed set* if for any pair of elements $d_1, d_2 \in D$ there is a $d \in D$ with the property that $d_1 \leqslant d$ and $d_2 \leqslant d$. For example, linearly ordered sets and lattices are directed sets. If $\mathscr{F}$ is a filter in a set X, then $\mathscr{F}$ becomes a directed set by introducing the following ordering relation: $F_1 \leqslant F_2$ if and only if $F_2 \subseteq F_1$. Indeed, given $F_1, F_2 \in \mathscr{F}$ the set $F_1 \cap F_2 \in \mathscr{F}$ is an upper bound of both F_1 and F_2. This ordering relation is called *ordering by inverse inclusion*.

Definition 5. *A net or directed system on a directed set* D *with values in a set* X *is a function* x *defined on* D *whose function values* x_d *or* $x(d)$ *belong to the set* X.

We see immediately that every sequence (x_n) $(n = 1, 2, \ldots)$ of points x_n of a set X is a net in the set X and its directed set D is the set of positive integers n under the usual ordering relation. If the directed set D of a net (x_d) $(d \in D)$ is linearly ordered, we say that the net is a *linear net*. For instance, sequences are linear nets and every function of a real variable d with values x_d in a set X is a linear net. Functions of two real variables x, y need not be nets under the customary ordering "$(x_1, y_1) < (x_2, y_2)$ if $x_1 < x_2$ and $y_1 \leqslant y_2$ or if $x_1 \leqslant x_2$ and $y_1 < y_2$." For the domain of the function need not be a directed set. If the domain is an open rectangle $D = \{(x, y) : a < x < b \text{ and } c < y < d\}$, then every function on D is a net.

Now we define what we mean by the restriction of a net and by a subnet of a net:

Definition 6. *Let (x_d) $(d \in D)$ be a net and let $D_1 \subseteq D$ be a directed set. Then the restriction of the function (x_d) $(d \in D)$ to the set D_1 is called the restriction of the net to the directed set D_1.*

Every subsequence of a sequence is a restriction of the sequence to the directed set $(n_1, n_2, \ldots)$. If we insist that sequences should have $(1, 2, \ldots)$ as index set, then the definition of a subsequence should be modified as follows: (y_m) $(m = 1, 2, \ldots)$ is a subsequence of (x_n) $(n = 1, 2, \ldots)$ if there is a strictly increasing function (n_m) $(m = 1, 2, \ldots)$ such that $y_m = x_{n_m}$ for every $m = 1, 2, \ldots$. A generalization of this definition leads to subnets:

Definition 7. *Let (x_d) $(d \in D)$ and (y_i) $(i \in I)$ be nets. Suppose that d is a function which maps I into D such that:*

(i) *if $i_1 < i_2$, then $d(i_1) < d(i_2)$ and if $i_1 \leqslant i_2$ then $d(i_1) \leqslant d(i_2)$;*

(ii) *for every $d_0 \in D$ there is an $i_0 \in I$ such that $d_0 \leqslant d(i_0)$.*

If d can be chosen such that $y_i = x_{d(i)}$ for every $i \in I$, then (y_i) $(i \in I)$ is called a subnet of (x_d) $(d \in D)$.

Note. Without condition (i) every rearrangement of a sequence would be a subsequence. This is the reason for including (i). The subset $\Delta = d(I)$ is itself a directed set and by (ii) for each $d \in D$ there is a $\delta \in \Delta$ such that $d \leqslant \delta$. We say that Δ is *cofinal* with D.

EXERCISES

1. The neighborhood filter $\mathcal{N}(x)$ of a point x in a topological space is a filter. Any base $\mathcal{B}(x)$ of $\mathcal{N}(x)$ is a filter base for the filter $\mathcal{N}(x)$. For instance, the family $\mathcal{O}(x)$ of open neighborhoods of x is a filter base for $\mathcal{N}(x)$.

2. Let X be a nonvoid set and let $\mathscr{M}(x)$ be the collection of all subsets of X which contain the point x. Show that $\mathscr{M}(x)$ is a filter and there is no other filter in X which is finer than $\mathscr{M}(x)$.

3. Let X be an infinite set and let $\mathscr{F}$ be the collection of those subsets F whose complement is finite. Show that $\mathscr{F}$ is a filter in X. Try to find a filter $\mathscr{G}$ which is strictly finer than $\mathscr{F}$.

4. In the plane with polar coordinates r, φ, for every $\rho > 0$, let $F_\rho = \{(r, \varphi) : r < \rho \textit{ and } \alpha < \varphi < \beta\}$ where $0 < \alpha < \beta < 2\pi$ are fixed angles. The family $\{F_\rho\}$ $(\rho > 0)$ is a filter base in the plane for a filter $\mathscr{F}_{\alpha\beta}$ and $\mathscr{F}_{\alpha\beta} \leqslant \mathscr{F}_{\gamma\delta}$ if and only if $\alpha \leqslant \gamma < \delta \leqslant \beta$.

5. If $\mathscr{F}$ and $\mathscr{G}$ are filters in a set X, then $\mathscr{F} \cap \mathscr{G}$, the family of common elements of $\mathscr{F}$ and $\mathscr{G}$, is a filter in X. Show that $\mathscr{F} \leqslant \mathscr{G}$ if and only if $\mathscr{F} \cap \mathscr{G} = \mathscr{F}$.

6. If the family $\mathscr{O}$ of open sets of a topological space X is a filter base, then no point-pair can be separated by disjoint open sets.

7. Show that the neighborhood systems $\mathscr{N}(x)$, $\mathscr{O}(x)$, and $\mathscr{C}(x)$ form directed sets under ordering by inverse inclusion. Show the same for every base $\mathscr{B}(x)$ of the neighborhood filter $\mathscr{N}(x)$.

[Any two neighborhoods in $\mathscr{B}(x)$ have a common upper bound, namely, their intersection contains an element of $\mathscr{B}(x)$.]

8. A partially ordered set D is called an ω-directed set if every countable subset of D has an upper bound in D. Give an example to illustrate that $\mathscr{O}(x)$ need not be an ω-directed set.

(Consider the open neighborhoods $S_{1/n}[x]$ $(n = 1, 2, \ldots)$ of a point x in the plane. Then $\bigcap S_{1/n}[x] = \{x\}$ and so there is no open neighborhood of x contained in every one of these open neighborhoods.)

9. Let X be a topological space and let D be the family of all open covers $\{O_i\}$ $(i \in I)$ of X. We say that $\{Q_j\}$ is a refinement of $\{O_i\}$, in symbols, $\{O_i\} \leqslant \{Q_j\}$, if for every O_i there is a Q_j such that $Q_j \subseteq O_i$. Show that under this ordering D is a directed set. Show the same for the family of all finite covers and for the family of all locally finite covers of X.

2. Convergence of Filters, Nets, and Sequences

In this section we consider filters, nets, and sequences of points in a fixed topological space X. There are two basic notions for sequences in X: *accumulation point* and *limit point*. It is clear that a sequence can have more than one accumulation point even if X is a very simple space. The set of accumulation points is called the *adherence* of the sequence. It is less obvious that a convergent sequence can have more than one limit point because in most applications we usually deal only with Hausdorff spaces and a sequence of points in such a space can have at most one limit point. However, if X is not a Hausdorff space, a sequence can converge at the same time to several points of X. The set of all limit points is called the *limit* of the sequence. If a convergent sequence has only one limit point x, then in practice we do not distinguish between x and the set $\{x\}$ and simply say that x is the limit of the sequence. However, strictly speaking the limit is always a set in X. Now we proceed to give a precise definition of adherence and limit not only for sequences but also for arbitrary filters and nets in the spăce X.

Definition 1. *The adherence of a filter $\mathscr{F}$ in the topological space X is the closed set*

$$\text{adh}\,\mathscr{F} = \bigcap \{\bar{F} : F \in \mathscr{F}\}.$$

The elements of adh $\mathscr{F}$ *are called the adherence points of the filter* $\mathscr{F}$.

The adherence of $\mathscr{F}$ can be a void set. A point x is an adherence point of $\mathscr{F}$ if and only if $F \cap N_x \neq \emptyset$ for every set $F \in \mathscr{F}$ and for every neighborhood N_x of x. Let $\mathscr{B}$ be a base for $\mathscr{F}$. Then by $\mathscr{B} \subseteq \mathscr{F}$ we have $\bigcap \bar{F} \subseteq \bigcap \bar{B}$ and since every F contains some B we see also that $\bigcap \bar{B} \subseteq \bigcap \bar{F}$. Therefore we have:

Lemma 1. *If $\mathscr{B}$ is a base for the filter $\mathscr{F}$, then* $\text{adh}\,\mathscr{F} = \bigcap \{\bar{B} : B \in \mathscr{B}\}$.

Note. We often speak about the *adherence of a filter base* where the adherence is defined as $\text{adh}\,\mathscr{B} = \bigcap \{\bar{B} : B \in \mathscr{B}\}$. By the present lemma $\text{adh}\,\mathscr{B} = \text{adh}\,\mathscr{F}$ where $\mathscr{F}$ is the filter generated by $\mathscr{B}$.

As an illustration we determine the adherence of the elementary filter $\mathscr{F}$ generated by a sequence (x_n) $(n = 1, 2, \ldots)$: By definition $x \in \text{adh}\,\mathscr{F}$ if and only if every neighborhood N_x intersects every $B_n = \{x_\nu : \nu \geqslant n\}$, that is, if and only if every N_x contains some x_ν with $\nu \geqslant n$ where n is arbitrary. Hence adh $\mathscr{F}$ consists of those points $x \in X$ whose neighborhoods N_x contain an infinity of terms of (x_n). This is exactly how accumulation points of sequences are defined. In terms of accumulation points of sets we have: A point x is an accumulation point of (x_n) if it is an accumulation point of the set of points x_n $(n = 1, 2, \ldots)$ or if $x = x_n$ for an infinity of indices n.

Definition 2. *Let $\mathscr{F}$ be a filter in a topological space X. A point x of X is called a limit point of $\mathscr{F}$ if every neighborhood N_x of x contains some*

$F \in \mathscr{F}$. The set of all limit points is called the limit of the filter $\mathscr{F}$ and is denoted by lim $\mathscr{F}$.

A point x is a limit point of $\mathscr{F}$ if and only if $\mathscr{F}$ is finer than the neighborhood filter $\mathscr{N}(x)$ or, equivalently, if and only if $\mathscr{F}$ is a refinement of $\mathscr{N}(x)$. Obviously, we have $\lim \mathscr{F} \subseteq \operatorname{adh} \mathscr{F}$. We can also speak about the limit points of a filter base $\mathscr{B}$: We say that x is a limit point of the filter base $\mathscr{B}$ if $\mathscr{B}$ is a refinement of $\mathscr{N}(x)$. This holds if and only if each N_x contains some $B \in \mathscr{B}$. The set of limit points of $\mathscr{B}$ is called the *limit of the filter base* $\mathscr{B}$. It follows immediately that lim $\mathscr{B}$ depends only on the filter generated by $\mathscr{B}$:

Lemma 2. *The limit of a filter base $\mathscr{B}$ is identical with the limit of the filter generated by the base $\mathscr{B}$.*

One often uses the expressions "*$\mathscr{B}$ is ultimately in the set S*" or "*$\mathscr{B}$ is eventually in the set S*" to indicate the existence of a $B \in \mathscr{B}$ such that $B \subseteq S$. Using this terminology x is a limit point of $\mathscr{B}$ if and only if $\mathscr{B}$ is eventually in every neighborhood of x. We shall also use all the well-established terms of the theory of sequences: For example, if $x \in \lim \mathscr{F}$, we say that $\mathscr{F}$ converges to x; and we say that $\mathscr{F}$ is divergent or convergent according as lim $\mathscr{F}$ is void or not.

A lemma on sequences of real numbers states: A point x is an adherence point of a sequence (x_n) $(n = 1, 2, ...)$ if and only if there is a subsequence of (x_n) which is convergent to x. If we restrict ourselves to sequences, then the lemma cannot be extended to sequences on arbitrary topological spaces, but if we consider also filters on the space X, then the lemma can be generalized in a natural way:

Lemma 3. *A point x is an adherence point of a filter $\mathscr{F}$ in X if and only if there is a filter in X which is finer than $\mathscr{F}$ and is convergent to x.*

Note. The refinement of $\mathscr{F}$ need not be an elementary filter even if $\mathscr{F}$ itself is elementary.

Proof. Let x be an adherence point of $\mathscr{F}$. Then $N_x \cap F \neq \emptyset$ for every neighborhood N_x and for every $F \in \mathscr{F}$. Hence the family $\{N_x \cap F\}$ $(N_x \in \mathscr{N}(x)$ and $F \in \mathscr{F})$ is a filter in X. This filter is now a refinement of $\mathscr{F}$ and also of $\mathscr{N}(x)$ and it is convergent to x. This proves the necessity of the condition. Next we suppose that there is a filter Φ finer than $\mathscr{F}$ such that $x \in \lim \Phi$. Then $\mathscr{F} \leqslant \Phi$ and $\mathscr{N}(x) \leqslant \Phi$ which shows that $F \cap N_x \in \Phi$ for any F in $\mathscr{F}$ and N_x in $\mathscr{N}(x)$. Thus $F \cap N_x \neq \emptyset$ for every neighborhood N_x and so $x \in \bar{F}$. Since F is arbitrary we obtain $x \in \operatorname{adh} \mathscr{F}$. This proves the sufficiency of the condition.

The following result is a generalization of the well-known fact that the only accumulation point of a convergent sequence of real numbers is its limit:

Theorem 1. *If X is a Hausdorff space and $\mathscr{F}$ is a convergent filter in X, then* adh $\mathscr{F}$ *consists of a single point and* $\lim \mathscr{F} = \operatorname{adh} \mathscr{F}$.

Proof. Let x and y be distinct points in X and let O_x and O_y be disjoint open neighborhoods of x and y, respectively. If $x \in \lim \mathscr{F}$, then there is is an $F \in \mathscr{F}$ such that $F \subseteq O_x$. Therefore $F \subseteq cO_y$ and $\bar{F} \subseteq cO_y$. This shows that $y \notin \bigcap \bar{F} = \operatorname{adh} \mathscr{F}$. By $\lim \mathscr{F} \subseteq \operatorname{adh} \mathscr{F}$ we get $\lim \mathscr{F} = \operatorname{adh} \mathscr{F} = \{x\}$.

The uniqueness of the limit is a characteristic property of Hausdorff spaces:

Theorem 2. *A topological space X is a Hausdorff space if and only if every filter in X has at most one limit point.*

Proof. The necessity of the condition follows from the preceding theorem. To prove the sufficiency let us suppose that X is not a Hausdorff space. Then there are distinct points x and y in X such that $N_x \cap N_y$ is nonvoid for every $N_x \in \mathscr{N}(x)$ and $N_y \in \mathscr{N}(y)$. Then $\mathscr{F} = \{N_x \cap N_y\}$ is a filter in X. Clearly $\mathscr{F}$ is a refinement of both $\mathscr{N}(x)$ and of $\mathscr{N}(y)$ and so x and y are distinct limit points of $\mathscr{F}$. Thus $\lim \mathscr{F}$ is not a single point.

Now we discuss briefly the convergence of elementary filters and show that it is in perfect agreement with the definition of sequential convergence: $x \in \lim \mathscr{F}$ if and only if every $N_x \in \mathscr{N}(x)$ includes some $F \in \mathscr{F}$ and so x is a limit point of $\mathscr{F}$ if and only if every N_x includes a $B_n = \{x_\nu : \nu \geqslant n\}$ for some $n \geqslant 1$. Hence x is in $\lim \mathscr{F}$ if and only if every neighborhood N_x contains all but finitely many points of the sequence (x_n) $(n = 1, 2, \ldots)$.

Definition 3. *Let (x_d) $(d \in D)$ be a net in a topological space X. A point x of X is called an adherence point of (x_d) $(d \in D)$ if given $d_0 \in D$ and a neighborhood N_x there is a $d \in D$ such that $d_0 \leqslant d$ and $x_d \in N_x$. The set of all adherence points is called the adherence of the net and is denoted by* $\operatorname{adh}(x_d)$.

If for every $d_0 \in D$ there is a $d \in D$ with the property that $d_0 \leqslant d$ and $x_d \in Y$, then we say that (x_d) $(d \in D)$ *is frequently in the set* Y. Using this expression we can say that a point $x \in X$ is an adherence point of the net (x_d) $(d \in D)$ if and only if (x_d) $(d \in D)$ is frequently in every neighborhood of x. If there is a $d_0 \in D$ such that $x_d \in Y$ for every d satisfying $d_0 \leqslant d$, then we say that the net (x_d) $(d \in D)$ *is eventually or ultimately in the set* Y. A point x is called a limit point of the net if it is ultimately in every neighborhood of x:

Definition 4. *A point x of a topological space X is called a limit point of a net (x_d) $(d \in D)$ in X if for every neighborhood N_x of x there is a $d_0 \in D$ such that $x_d \in N_x$ for every d satisfying $d_0 \leqslant d$. The set of all limit points is called the limit of the net and is denoted by* $\lim(x_d)$.

A net is convergent if it has at least one limit point and is divergent if it has no limit points. We have in every case $\lim(x_d) \subseteq \operatorname{adh}(x_d)$. If X is a Hausdorff space, then every convergent net has exactly one adherence point and this is the unique limit point of the net. The converse is harder to prove and instead of giving a direct proof we are going to show the exact relationship between the convergence properties of nets and filters.

Theorem 3. *Let X be a topological space and let (x_d) $(d \in D)$ be an arbitrary net in X. Then there exists a filter $\mathscr{F}$ in X with the property that* $\operatorname{adh} \mathscr{F} = \operatorname{adh}(x_d)$ *and* $\lim \mathscr{F} = \lim(x_d)$.

Proof. For every $d \in D$ we define $B_d = \{x_\delta : d \leqslant \delta\}$. Since D is a directed set the family $\mathscr{B} = \{B_d\}$ $(d \in D)$ has the finite intersection property and so it is a filter base in X. Let $\mathscr{F}$ be the filter generated by $\mathscr{B}$. We show that $\mathscr{F}$ satisfies all requirements.

First, let $x \in \operatorname{adh}(x_d)$. Then given $d \in D$ and N_x there is a $\delta \in D$ such that $d \leqslant \delta$ and $x_\delta \in N_x$. Therefore N_x and B_d intersect for every $d \in D$ and for every N_x which means that $x \in \operatorname{adh} \mathscr{F}$. Now let us assume that $x \in \operatorname{adh} \mathscr{F}$. Let $d \in D$ and a neighborhood N_x be given. We have $N_x \cap B_d \neq \emptyset$ and so there is an $x_\delta \in N_x \cap B_d$ where δ satisfies $d \leqslant \delta$. Therefore we found a δ satisfying $d \leqslant \delta$ and such that $x_\delta \in N_x$; so $x \in \operatorname{adh}(x_d)$.

Next, let x be a limit point of (x_d) $(d \in D)$. Then for every N_x there is a $d \in D$ such that $x_\delta \in N_x$ for every δ satisfying $d \leqslant \delta$. Therefore $B_d \subseteq N_x$ and $x \in \lim \mathscr{F}$. Conversely, let x be a limit point of $\mathscr{F}$ and let N_x be a neighborhood of x. Then there is a B_d such that $B_d \subseteq N_x$ and so $x_\delta \in N_x$ for every δ satisfying $d \leqslant \delta$. In other words, $x \in \lim(x_d)$.

Theorem 4. *Let $\mathscr{F}$ be a filter in a topological space X. Then there is a net (x_d) $(d \in D)$ in X with the property that* $\operatorname{adh}(x_d) = \operatorname{adh} \mathscr{F}$ *and* $\lim(x_d) = \lim \mathscr{F}$.

Proof. Let us consider the set D of all ordered pairs $d = (x, F)$ where $x \in F \in \mathscr{F}$. We say that $d_1 \leqslant d_2$ if $F_1 \supseteq F_2$. Under this ordering D becomes a directed set and a net can be defined on D by choosing $x_d = x$ for every $d = (x, F)$. Let $x \in \lim \mathscr{F}$ and let N_x be arbitrary. Then there is an $F \in \mathscr{F}$ such that $F \subseteq N_x$. We let $d = (x, F)$. Hence if $\delta = (\xi, G)$ satisfies $d \leqslant \delta$, or in other words, if $G \subseteq F$, then $x_\delta = \xi \in G \subseteq F \subseteq N_x$ and so x is a limit point of (x_d) $(d \in D)$. Conversely, let $x \in \lim(x_d)$ and let N_x be given. Then there is a $d = (x, F)$ such that $x_\delta \in N_x$ for every δ satisfying $d \leqslant \delta$. Using this for every $\delta = (\xi, F)$ $(\xi \in F)$ we see that $x_\delta = \xi \in N_x$ for every $\xi \in F$ and so $F \subseteq N_x$. This shows that $x \in \lim \mathscr{F}$ and $\lim(x_d) = \lim \mathscr{F}$.

Now we suppose that $x \in \operatorname{adh} \mathscr{F}$ so that $N_x \cap F \neq \varnothing$ for every neighborhood N_x and for every $F \in \mathscr{F}$. Given N_x and $d = (x, F)$ we choose ξ in $N_x \cap F$ and consider $\delta = (\xi, F)$. Then $d \leqslant \delta$ and $x_\delta = \xi \in N_x$ and so $x \in \operatorname{adh}(x_d)$. On the other hand, if $x \in \operatorname{adh}(x_d)$, then given $F \in \mathscr{F}$ and N_x there is a $\delta \in D$ such that $d = (x, F) \leqslant \delta = (\xi, G)$ and $x_\delta \in N_x$. In other words, $x_\delta = \xi \in G \cap N_x \subseteq F \cap N_x$ and so F and N_x intersect for every $F \in \mathscr{F}$ and for every $N_x \in \mathscr{N}(x)$. This proves that $x \in \operatorname{adh} \mathscr{F}$ and $\operatorname{adh}(x_d) = \operatorname{adh} \mathscr{F}$.

The last two theorems show that in the study of adherence and limits we could restrict ourselves to only one of the two concepts, but as was pointed out earlier it is more convenient to use both filters and nets. Theorems 3 and 4 immediately imply the counterparts of Theorems 1 and 2 for nets:

Theorem 5. *If X is a Hausdorff space and (x_d) $(d \in D)$ is a convergent net in X, then* $\operatorname{adh}(x_d)$ *consist of a single point and* $\lim(x_d) = \operatorname{adh}(x_d)$.

Proof. Use Theorems 1 and 3.

Theorem 6. *A topological space X is a Hausdorff space if and only if every net in X has at most one limit point.*

Proof. Use Theorems 2, 3, and 4.

We close this section with a few remarks on the convergence of subnets. It is well known that every subsequence of a convergent sequence is convergent to the same limit. The corresponding result follows easily from the definition of a subnet:

If (y_i) $(i \in I)$ is a subnet of the net (x_d) $(d \in D)$, then $\lim(y_i) \supseteq \lim(x_d)$ *and* $\operatorname{adh}(y_i) \subseteq \operatorname{adh}(x_d)$. A similar proposition holds for filters: *If $\mathscr{F}_1$ is finer than $\mathscr{F}_2$, then* $\lim \mathscr{F}_1 \supseteq \lim \mathscr{F}_2$ *and* $\operatorname{adh} \mathscr{F}_1 \subseteq \operatorname{adh} \mathscr{F}_2$. This is obvious from the definitions of the concepts involved in the statement.

EXERCISES

1. Show that a filter in a pseudometric space is convergent only if for every $\epsilon > 0$ there is an $F \in \mathscr{F}$ such that its diameter $d(F) < \epsilon$.

2. Let $\mathscr{F}$ be a filter in the space of the real numbers such that for every $\epsilon > 0$ there is an $F \in \mathscr{F}$ of diameter $d(F) < \epsilon$. Show that $\mathscr{F}$ has

(i) a nonvoid adherence; (ii) a unique adherence point; (iii) a unique limit point.

[Construct by induction on $k = 1, 2, \ldots$ a sequence of closed intervals I_k of length $(\frac{2}{3})^k$ and a sequence of sets $F_k \in \mathscr{F}$ such that $F_{k+1} \subseteq F_k \subseteq I_k \subseteq I_{k-1}$ for every $k > 1$: If I_k and F_k are already given, then choose $G \in \mathscr{F}$ of length $d(G) < \frac{1}{3}(\frac{2}{3})^k$. Since $F_k \subseteq I_k$ we may choose $F_{k+1} = G \cap F_k$ and for I_{k+1} we may choose the left or the right two-thirds of I_k. By the compactness of the interval I_1 the intersection $\bigcap I_k$ determines a unique point which is an adherence point of $\mathscr{F}$.]

3. If Y is a subspace of X and if $\mathscr{F}$ is a filter in Y, then $\mathscr{F}$ is a filter base in X. Show that the adherence of $\mathscr{F}$ in X is a subset of $\bar{Y}$.

4. Show that for every filter $\mathscr{F}$ in a topological space $\lim \mathscr{F} = \bigcap \operatorname{adh} \mathscr{G}$ where the intersection is taken for every $\mathscr{G} \geqslant \mathscr{F}$.

5. Let X be an infinite set and let the topology on X be the topology of finite complements. Show that a sequence (x_n) $(n = 1, 2, \ldots)$ is convergent if and only if (i) exactly one point x occurs in (x_n) an infinite number of times, or (ii) every point occurs in (x_n) at most a finite number of times. In case (i) the point x is the unique limit and in case (ii) every point of X is a limit point.

6. Let X be infinite and let the topology be the topology of finite complements. Show that every filter in X has a nonvoid adherence. If $\mathscr{F}$ contains finite sets, then $\operatorname{adh} \mathscr{F}$ is the smallest finite set in $\mathscr{F}$ and if every F is an infinite set, then $\operatorname{adh} \mathscr{F} = X$.

7. Show that $\operatorname{adh}(x_d)$ is a closed set without using the fact that the adherence of every filter is closed.

8. Construct a net (x_d) $(d \in D)$ such that its adherence is the closed set C given in advance.

[Let D be the set of all ordered pairs $d = (n, x)$ where $n = 1, 2, \ldots$ and $x \in C$. If $n_1 \leqslant n_2$, we say that $d_1 = (n_1, x_1) \leqslant d_2 = (n_2, x_2)$ for every $x_1, x_2 \in C$. For every $d = (n, x)$ we define $x_d = x$. Since $x_d \in C$ for every $d \in D$ we have $\operatorname{adh}(x_d) \subseteq C$. Given any $x \in C$, a neighborhood N_x, and any $d \in D$ we have $x_\delta = x \in N_x$ for every $\delta = (n, x)$ and so δ can be chosen such that $d \leqslant \delta$. Hence $x \in \operatorname{adh}(x_d)$.]

3. Ultrafilters and Universal Nets

These concepts have no counterparts in the theory of sequences. It is possible to form universal nets from sequences but the nets which we obtain will not be sequences. Similarly, if we consider the elementary filters generated by sequences, we can construct some ultrafilters which are finer than these elementary filters but they are not elementary filters.

Definition 1. *An ultrafilter $\mathscr{F}$ in a set X is a filter such that there exists no filter in X which is strictly finer than $\mathscr{F}$.*

Note. In other words $\mathscr{F}$ is an ultrafilter if $\mathscr{F}$ is maximal with respect to the ordering relation $\leqslant$ defined by inclusion.

Occasionally we shall also speak about an *ultrafilter base* $\mathscr{B}$. This means that the filter generated by $\mathscr{B}$ is an ultrafilter. There is no difficulty in seeing the existence of ultrafilters in any set X: For instance, the set of all subsets containing a fixed point x is an ultrafilter. It is more interesting to show the existence of certain special types of ultrafilters:

Theorem 1. *Given any filter $\mathscr{F}$ in a set X there is an ultrafilter $\mathscr{M}$ in X which is finer than $\mathscr{F}$.*

Proof. Use Zorn's lemma: Let $\mathfrak{A}$ be the set of all filters $\mathscr{G}$ finer than $\mathscr{F}$. $\mathfrak{A} \neq \emptyset$ since $\mathscr{F} \in \mathfrak{A}$. The set $\mathfrak{A}$ is partially ordered by the ordering relation $\mathscr{G}_1 \leqslant \mathscr{G}_2$ and every linearly ordered subfamily $\mathfrak{L}$ of $\mathfrak{A}$ has an upper bound in $\mathfrak{A}$, namely, $\bigcup\{\mathscr{G} : \mathscr{G} \in \mathfrak{L}\}$ is a filter in X which is a refinement of every $\mathscr{G} \in \mathfrak{L}$. Hence Zorn's lemma applies and so there is a maximal element $\mathscr{M}$ in $\mathfrak{A}$. No other filter $\mathscr{G}$ is finer than $\mathscr{M}$ because if $\mathscr{M} \leqslant \mathscr{G}$, then $\mathscr{F} \leqslant \mathscr{G}$ and so $\mathscr{G} \in \mathfrak{A}$ and $\mathscr{G} = \mathscr{M}$.

Theorem 2. *If $\mathscr{F}$ is an ultrafilter in the set X and if $A_1 \cup \ldots \cup A_n \in \mathscr{F}$, then at least one of the sets $A_1, \ldots, A_n$ belongs to $\mathscr{F}$.*

Note. Conversely, if $\mathscr{F}$ is a filter such that for every A in X we have $A \in \mathscr{F}$ or $cA \in \mathscr{F}$, then $\mathscr{F}$ is an ultrafilter.

Proof. We can restrict ourselves to the case $n = 2$. Suppose that $A_1 \notin \mathscr{F}$ but $A_1 \cup A_2 \in \mathscr{F}$ and consider the family $\mathscr{G}$ of those sets G which have the property that $A_1 \cup G \in \mathscr{F}$. Since $A_2 \in \mathscr{G}$ the family is not void and none of its elements is the void set because $A_1 \notin \mathscr{F}$. The remaining two axioms (F.2) and (F.3) for a filter are obviously satisfied and so $\mathscr{G}$ is a filter in X. By (F.2) we have $\mathscr{F} \subseteq \mathscr{G}$ and so $\mathscr{F}$ being an ultrafilter $\mathscr{F} = \mathscr{G}$. Therefore by $A_1 \cup A_2 \in \mathscr{F}$ we have $A_2 \in \mathscr{G} = \mathscr{F}$. To prove the statement of the note let $\mathscr{F}$ be a filter satisfying the requirements and let $\mathscr{F} \subseteq \mathscr{G}$. If $G \in \mathscr{G}$, then $G \in \mathscr{F}$ or $cG \in \mathscr{F}$, but $cG \in \mathscr{F} \subseteq \mathscr{G}$ would imply that $G \cap cG = \emptyset \in \mathscr{G}$. Thus $G \in \mathscr{F}$, and so $\mathscr{G} \subseteq \mathscr{F}$. Hence $\mathscr{F} \subseteq \mathscr{G}$ implies that $\mathscr{F} = \mathscr{G}$ and so $\mathscr{F}$ is maximal.

Theorem 3. *If $\mathscr{F}$ is an ultrafilter in a topological space X, then* $\operatorname{adh} \mathscr{F} = \lim \mathscr{F}$.

Proof. This follows from Lemma 2.3: If $x \in \operatorname{adh} \mathscr{F}$, then there is a filter in X finer than $\mathscr{F}$ which is convergent and x is one of its limit points. Since $\mathscr{F}$ is an ultrafilter this filter is $\mathscr{F}$ itself and so $x \in \lim \mathscr{F}$. Therefore $\operatorname{adh} \mathscr{F} \subseteq \lim \mathscr{F}$ and so $\operatorname{adh} \mathscr{F} = \lim \mathscr{F}$.

It is possible to introduce a notion also in the theory of nets which corresponds to the notion of an ultrafilter. The simplest is to base the definition on the property of ultrafilters which is expressed in Theorem 2 and the note following it: *A net* (x_d) $(d \in D)$ *with values in a set* X *is called a universal net if for every subset* A *of* X *there is an index* $d \in D$ *such that either* $x_\delta \in A$ *whenever* $d \leqslant \delta$ *or* $x_\delta \in cA$ *whenever* $d \leqslant \delta$. In others words, (x_d) $(d \in D)$ is a universal net if for every $A \subseteq X$ it is eventually either in A or in cA.

It is easy to see that $\lim(x_d) = \operatorname{adh}(x_d)$ for every universal net on a topological space X: In fact, let $x \in \operatorname{adh}(x_d)$ and let O_x be an open neighborhood of x. As x is an adherence point the universal net cannot be eventually in cO_x and so (x_d) $(d \in D)$ is eventually in O_x. Since O_x is is arbitrary, x is a limit point. The same result can be obtained by using Theorem 2.3 and noticing that the filter associated with a universal net in Theorem 2.3 is an ultrafilter. Similarly, one can show that the net associated with an ultrafilter in Theorem 2.4 is a universal net. Every subnet of a universal net is universal. Every net has a universal subnet. We do not give detailed proofs of these results here as we do not need them in the sequel.

EXERCISES

1. Let X be the space formed by the closed interval $[0, 1]$ with the usual topology of the reals. Show that every ultrafilter $\mathscr{M}$ is convergent in X.

(The interval $I_1 = [0, 1]$ belongs to $\mathscr{M}$ and the left or right half of I_1 also belongs to $\mathscr{M}$. Construct the sequence (I_k) $(k = 1, 2, \ldots)$ of intervals such that I_{k+1} is the left or right half of I_k and $I_{k+1} \in \mathscr{M}$. Then $\bigcap I_k$ is a limit point.)

2. Show that if $\mathscr{M}$ is an ultrafilter in a set X, then $\bigcap M$ contains at most one point.

3. Show that the neighborhood filter $\mathscr{N}(x)$ is an ultrafilter if and only if x is an isolated point of X.

4. If an ultrafilter contains finite sets, then it is the family of all sets which contain a fixed point.

5. Let an infinite set be topologized by the topology of finite complements. Show that every ultrafilter is convergent.

6. Let X be a linearly ordered set and let the topology on X be the *order topology*, i.e., let O be open if it is $\emptyset$, X or if it is of the form $\{x : a < x\}$ for some $a \in X$. Show: If X has a minimal element, then every filter is convergent.

7. Show that if $\mathscr{F}$ is an ultrafilter, then the net constructed in the proof of Theorem 2.4 is a universal net.

[If $\mathscr{F}$ is an ultrafilter, then given a set A there is an $F \in \mathscr{F}$ such that $F \subseteq A$ or $F \subseteq cA$. We choose an $x \in F$ and put $d = (x, F)$. Then for every $\delta = (\xi, G)$ satisfying $d \leqslant \delta$ we have $G \subseteq F$ and so $x_\delta = \xi \in G \subseteq F$. Hence either $x_\delta \in A$ for every δ; $d \leqslant \delta$ or $x_\delta \in cA$ for every δ; $d \leqslant \delta$.]

8. Show that if (x_d) $(d \in D)$ is a universal net, then the filter constructed in the proof of Theorem 2.3 is an ultrafilter.

4. Bounds, Traces, and Products of Filters

Here is a collection of useful definitions and some practical results on filters. Further information can be found in the exercises at the end of this section.

Every family $\{\mathscr{F}_\alpha\}$ of filters $\mathscr{F}_\alpha$ in the same set X has a lower bound in X, e.g., the filter consisting of the single set $F = X$ is a lower bound for any filter in X. Among the lower bounds there is a largest called the greatest lower bound of the family and denoted by glb$\{\mathscr{F}_\alpha\}$. It is the intersection of the families $\mathscr{F}_\alpha$ and for this reason it is also called the *intersection filter* and can be denoted by $\bigcap \mathscr{F}_\alpha$.

A family $\{\mathscr{F}_\alpha\}$ need not have an upper bound. A finite family $\{\mathscr{F}_1, ..., \mathscr{F}_n\}$ has an upper bound if and only if $F_1 \cap ... \cap F_n \neq \emptyset$ for all choices of $F_1 \in \mathscr{F}_1, ..., F_n \in \mathscr{F}_n$. If this condition is satisfied, then the sets $F_1 \cap ... \cap F_n$ form a filter which is the least upper bound of the family. An infinite family $\{\mathscr{F}_\alpha\}$ has an upper bound if and only if every finite subfamily has an upper bound. If $\{\mathscr{F}_\alpha\}$ has an upper bound, then there is a *least upper bound*: It is the filter consisting of the sets $\bigcap F_\alpha$ where $F_\alpha = X$ for all but finitely many indices α. The least upper bound is denoted by lub$\{\mathscr{F}_\alpha\}$. The family $\bigcup \mathscr{F}_\alpha$ is a subbase for lub$\{\mathscr{F}_\alpha\}$ provided the latter exists. If any two filters of $\{\mathscr{F}_\alpha\}$ are comparable, then lub$\{\mathscr{F}_\alpha\}$ exists and it is $\bigcup \mathscr{F}_\alpha$.

Suppose that the cardinality of $\{\alpha\}$ is at most n. If in addition every $\mathscr{F}_\alpha$ is an n-filter and if lub$\{\mathscr{F}_\alpha\}$ exists, then it is an n-filter. In particular, the least upper bound of a countable family of countable filters is itself a countable filter. An increasing sequence of elementary filters $\mathscr{F}_1 \leqslant \mathscr{F}_2 \leqslant ...$ can always be majorized by an elementary filter. The diagonal process applied to the sequences generating the filters $\mathscr{F}_n$ leads to a sequence which generates an upper bound for $\{\mathscr{F}_n\}$. This upper bound, however, need not be the least upper bound of $\{\mathscr{F}_n\}$.

Let $\mathscr{B}$ be a filter base in X and let Y be a subset of X. The family $\mathscr{B}_Y = \{B \cap Y\} (B \in \mathscr{B})$ is called the *trace* of $\mathscr{B}$ in Y. The trace is a filter base in Y if and only if $B \cap Y$ is nonvoid for every $B \in \mathscr{B}$. If $\mathscr{B}_Y$ is a filter base in Y, then the filter generated by $\mathscr{B}_Y$ is $\mathscr{F}_Y = \{F \cap Y\} (F \in \mathscr{F})$ where $\mathscr{F}$ is the filter generated by $\mathscr{B}$: For it is obvious that $\mathscr{B}_Y \subseteq \mathscr{F}_Y$ and so the filter $\mathscr{F}_Y$ is finer than the filter generated by $\mathscr{B}_Y$. On the other hand, if $F \in \mathscr{F}$, then there is a $B \in \mathscr{B}$ such that $B \subseteq F$ and so $B \cap Y \in \mathscr{B}_Y$. This shows that $F \cap Y$ belongs to the filter generated by $\mathscr{B}_Y$. Hence this filter is exactly $\mathscr{F}_Y$. If $\mathscr{F}_Y$ is a filter then it is called the filter *induced* on Y by $\mathscr{F}$. If $\mathscr{M}$ is an ultrafilter in X, then $\mathscr{M}_Y$ is a filter in Y if and only if $Y \in \mathscr{M}$. For either Y or cY must belong to $\mathscr{M}$. If the induced filter $\mathscr{M}_Y$ exists, then it is an ultrafilter in Y. Indeed, if A lies in Y, then A or $c_X A$ belongs to $\mathscr{M}$ and so by $Y \in \mathscr{M}$ we have $A \in \mathscr{M}_Y$ or $c_Y A \in \mathscr{M}_Y$.

Let X be a product set and let a filter $\mathscr{F}_s$ be given in each of its factors X_s. Then the set of all products $\prod F_s$, where $F_s \in \mathscr{F}_s$ and $F_s = X_s$ for all but finitely many indices, is a filter base $\mathscr{B}$ in X. The filter generated by $\mathscr{B}$ is called the *product* of the filters $\mathscr{F}_s$ and is denoted by $\prod \mathscr{F}_s$. It is the coarsest filter in X such that its projection in X_s is $\mathscr{F}_s$ for every s. If $\mathscr{B}_s$ is a base for $\mathscr{F}_s$, then the set of all products $\prod B_s$, where $B_s = X_s$ for all but finitely many indices and $B_s \in \mathscr{B}_s$ for the rest, is a base for $\prod \mathscr{F}_s$. If the index set is finite, then the definition is simpler: The set of all products $\prod B_s$ is a base for the product filter. It is easy to verify that the neighborhood filter $\mathscr{N}(x)$ of a point $x = (x_s) \in X$ in the product topology is the product of the neighborhood filters $\mathscr{N}(x_s)$ of the points x_s in X_s.

Now let X be a topological space. If $\mathscr{B}$ is a filter base in X, then the the family $\overline{\mathscr{B}} = \{\bar{B}\} (B \in \mathscr{B})$ is also a filter base. If $\mathscr{B}_1$ and $\mathscr{B}_2$ generate the same filter, then $\overline{\mathscr{B}}_1$ and $\overline{\mathscr{B}}_2$ do also. The filter associated with the filter base $\overline{\mathscr{F}}$ is called the *closure* of the filter $\mathscr{F}$. We use the same symbol $\overline{\mathscr{F}}$ to denote the base $\{\bar{F}\} (F \in \mathscr{F})$ and the closure itself. In general $\lim \overline{\mathscr{F}} \subseteq \lim \mathscr{F}$ because $\mathscr{F}$ is finer than $\overline{\mathscr{F}}$.

Lemma 1. *We have* $\lim \overline{\mathscr{F}} = \lim \mathscr{F}$ *for every filter* $\mathscr{F}$ *in* X *if and only if* X *is a* (T_3) *space.*

Proof. First let X be a (T_3) space and let $x \in X$. Then given O_x there is an open set Q_x having the property that $x \in Q_x \subseteq \bar{Q}_x \subseteq O_x$. If $x \in \lim \mathscr{F}$, then there is an $F \in \mathscr{F}$ such that $F \subseteq Q_x$ and so $\bar{F} \subseteq \bar{Q}_x \subseteq O_x$. Therefore every open neighborhood O_x contains a set $\bar{F} \in \overline{\mathscr{F}}$ and so

$x \in \lim \overline{\mathscr{F}}$. Next suppose that X is not a (T_3) space. Then there is a point x and an open set O_x containing x with the following property: If Q_x is an open set containing x, then $\bar{Q}_x$ is not contained in O_x. We choose for $\mathscr{F}$ the filter $\mathscr{N}(x)$ of all neighborhoods of x. Then $\{\bar{Q}_x\}(Q_x \in \mathscr{N}(x))$ is a base for $\overline{\mathscr{F}}$ and no element of this base is contained in O_x. Thus $x \notin \lim \overline{\mathscr{N}(x)}$ but clearly $x \in \lim \mathscr{N}(x)$.

EXERCISES

1. A uniform structure for a set X is a filter in $X \times X$. Considered as filters show that every family of uniform structures for a fixed set X has a least upper bound which is also a uniform structure for X.

2. Considered as filters show that every family of uniform structures for a fixed set has a greatest lower bound which satisfies axioms (U.1)–(U.5).

3. Let Y be a subspace of X and let $\mathscr{F}_Y$ be a filter in Y. Then $\mathscr{F}_Y$ is a filter base in X for a filter $\mathscr{F}_X$. Show that $Y \cap \lim \mathscr{F}_X = \lim \mathscr{F}_Y$ and $Y \cap \operatorname{adh} \mathscr{F}_X = \operatorname{adh} \mathscr{F}_Y$.

4. Show that the product $\prod \mathscr{U}_\kappa$ of the uniform structures $\mathscr{U}_\kappa$ for the sets X_κ is a uniform structure for the product set $\prod X_\kappa$.

5. Let $\mathscr{T}_i$ $(i \in I)$ be a topology on the fixed set X and let $\mathscr{N}_i(x)$ be the neighborhood filter of a point $x \in X$ relative to $\mathscr{T}_i$. Show that $\mathscr{N}(x) = \operatorname{lub}\{\mathscr{N}_i(x)\}$ where $\mathscr{N}(x)$ denotes the neighborhood filter of x relative to $\mathscr{T} = \operatorname{lub}\{\mathscr{T}_i\}$.

[A set O_x is an open neighborhood of x relative to $\mathscr{T}$ if and only if $O_x = O_x^{i_1} \cap \dots \cap O_x^{i_n}$ for suitable indices $i_1, \dots, i_n$ and open neighborhoods $O_x^{i_1} \in \mathscr{O}_{i_1}, \dots, O_x^{i_n} \in \mathscr{O}_{i_n}$. Hence $N_x \in \mathscr{N}(x)$ if and only if there are indices $i_1, \dots, i_n$ and neighborhoods $N_x^{i_1} \in \mathscr{N}_{i_1}(x), \dots, N_x^{i_n} \in \mathscr{N}_{i_n}(x)$ such that $N_x = N_x^{i_1} \cap \dots \cap N_x^{i_n}$. This shows that $N_x \in \mathscr{N}(x)$ if and only if $N_x \in \operatorname{lub}\{\mathscr{N}_i(x)\}$.]

5. Applications of Filters and Nets to Compactness

Filters and nets are important tools in topology. In the sequel we use them when discussing compactness and completeness. Among the many applications of filters these are the most fundamental and incidentally the first where filters were used.

For metric spaces compactness and the Bolzano-Weierstrass property are equivalent and so X is compact if and only if every sequence (x_n) of points $x_n \in X$ has an adherence point, or in other words X is compact if and only if every elementary filter in X has a nonvoid adherence. Using filters this proposition can be extended to arbitrary topological spaces:

Theorem 1. *A topological space X is compact if and only if every filter $\mathscr{F}$ in X has a nonvoid adherence.*

In view of Theorems 2.3 and 2.4 this result is equivalent to the following:

Corollary. *X is compact if and only if every net in X has a nonvoid adherence.*

Proof. The necessity follows from Theorem III.1.1. For if $\mathscr{F} = \{F\}$ is a filter in X, then the family $\{\bar{F}\}$ $(F \in \mathscr{F})$ has the finite intersection property. Hence if X is compact, then $\bigcap \{\bar{F} : F \in \mathscr{F}\} = \operatorname{adh} \mathscr{F}$ is not void. The sufficiency can be proved indirectly: Suppose that X is not compact. Then by Theorem III.1.1 there is a family $\mathscr{B}$ of closed sets having the finite intersection property but with total intersection void. Hence $\mathscr{B}$ is a filter base in X whose adherence is void.

If every filter has a nonvoid adherence in X, then by Theorem 3.3 every ultrafilter has a nonvoid limit. Conversely, if every ultrafilter in X is convergent, then every filter in X has a nonvoid adherence because by Theorem 3.1. a filter $\mathscr{F}$ can be majorized by an ultrafilter $\mathscr{M}$ and so $\lim \mathscr{M} \subseteq \operatorname{adh} \mathscr{F}$. Hence using the preceding theorem we obtain another compactness criterion:

Theorem 2. *A topological space X is compact if and only if every ultrafilter in X is convergent.*

This condition can be used for instance to prove Tychonoff's theorem on compactness: We need a few results on filters in product spaces. Let $\mathscr{F} = \{F\}$ be a filter in the product $X = \prod X_s$. We denote by F_s the projection of the set F into the coordinate space X_s. Then we have:

Lemma 1. *The family $\mathscr{F}_s = \{F_s\}$ of all projections F_s $(F \in \mathscr{F})$ is a filter in X_s which is called the projection of $\mathscr{F}$ into X_s.*

Proof. First $F' \cap F'' \supseteq F$ implies that $F_s' \cap F_s'' \supseteq F_s$ and so $\mathscr{F}_s$ is a filter base. Next if A_s is a set in X_s which contains a projection F_s of a set $F \in \mathscr{F}$, then the cylinder $A = \{x : x_s \in A_s\}$ contains F and so it belongs to $\mathscr{F}$. This shows that A_s itself is a projection of a set $A \in \mathscr{F}$ and so $\mathscr{F}_s$ is a filter.

Lemma 2. *If $\mathscr{F} \leqslant \mathscr{G}$, then $\mathscr{F}_s \leqslant \mathscr{G}_s$ for every index s.*

Proof. If $F_s \in \mathscr{F}_s$ is the projection of $F \in \mathscr{F}$, then $F \in \mathscr{G}$ and so $F_s \in \mathscr{G}_s$. Hence $\mathscr{F}_s \leqslant \mathscr{G}_s$.

Lemma 3. *If $\mathscr{M}$ is an ultrafilter in the product set X, then the projection of $\mathscr{M}$ into each of the factors X_s is an ultrafilter.*

Proof. Let $\mathscr{F}$ be a filter in $X = \prod X_s$ and let $\mathscr{F}_s$ denote the projection of $\mathscr{F}$ into X_s. Suppose that we can find some filter $\mathscr{G}_s$ in X_s which is strictly finer than $\mathscr{F}_s$. Then there is a set G_s in $\mathscr{G}_s$ such that $F_s \nsubseteq G_s$ for every $F_s \in \mathscr{F}_s$. We introduce the nonvoid sets $\varphi(F) = \{x : x \in F$ *and* $x_s \in G_s\}$. The family $\mathscr{B} = \{\varphi(F)\}\,(F \in \mathscr{F})$ is a filter base in X which is not coarser than $\mathscr{F}$ because $\varphi(F) \subseteq F$ for every $F \in \mathscr{F}$. Moreover, if $F \in \mathscr{F}$, then $\varphi(F)_s \subseteq G_s$ while $F_s \nsubseteq G_s$ for every $F \in \mathscr{F}$. Therefore no element of $\mathscr{F}$ is contained in $\varphi(F)$ and so $\mathscr{B}$ is strictly finer than $\mathscr{F}$. Hence $\mathscr{F}$ is not an ultrafilter.

The preceding lemma does not have a converse. One can easily find a filter $\mathscr{F}$ in a product set X such that $\mathscr{F}$ itself is not a ultrafilter but every projection $\mathscr{F}_s$ is an ultrafilter: We can choose $\mathscr{F}$ as the product of a family of ultrafilters $\mathscr{M}_s$ in the factors X_s. For instance, $\mathscr{M}_s$ can be the family of all sets in X_s which contain a fixed point x_s.

Lemma 4. *If $\mathscr{F}$ is a filter in a product space X, then* $(\text{adh}\,\mathscr{F})_s \subseteq \text{adh}\,\mathscr{F}_s$ *and* $(\lim \mathscr{F})_s \subseteq \lim \mathscr{F}_s$ *for every index s.*

Proof. Suppose that $x \in \text{adh}\,\mathscr{F}$ so that $x_s \in (\text{adh}\,\mathscr{F})_s$. If N_{x_s} is a neighborhood of x_s in X_s, then $N_x = \{\xi : \xi_s \in N_{x_s}\}$ is a neighborhood of x in X. Since x is an adherence point of $\mathscr{F}$ we have $N_x \cap F \neq \emptyset$ for every $F \in \mathscr{F}$ and so $N_{x_s} \cap F_s \neq \emptyset$ for every $F_s \in \mathscr{F}_s$. Thus $x_s \in \text{adh}\,\mathscr{F}_s$ and $(\text{adh}\,\mathscr{F})_s \subseteq \text{adh}\,\mathscr{F}_s$. If, moreover, $x \in \lim \mathscr{F}$, i.e., if $x_s \in (\lim \mathscr{F})_s$, then there is an $F \in \mathscr{F}$ such that $F \subseteq N_x$. Therefore $F_s \subseteq N_{x_s}$ and this shows that $x_s \in \lim \mathscr{F}_s$.

The second part of this lemma implies that the projections of a convergent filter $\mathscr{F}$ are convergent filters in the factor spaces. The converse is also correct:

Lemma 5. *If every projection $\mathscr{F}_s$ of a filter $\mathscr{F}$ in a product space is convergent, then $\mathscr{F}$ is a convergent filter.*

Proof. For each index s we choose a point x_s in $\lim \mathscr{F}_s$. Let $x \in X$ be the point whose sth coordinate is x_s. We show that $x \in \lim \mathscr{F}$. For let $O = \prod O_s$ be an open set in X which contains x. By the definition of the product topology $O_s = X_s$ for all but finitely many indices. For each s there is an $F^s \in \mathscr{F}$ such that $F^s{}_s \subseteq O_s$ because $x_s \in O_s$ and $x_s \in \lim \mathscr{F}_s$. We may choose $F^s = X$ for all but finitely many indices. Then $F = \bigcap F^s$ is a finite intersection and so $F \in \mathscr{F}$. Moreover, $F \subseteq \prod F^s{}_s \subseteq \prod O_s = O$. Hence $x \in \lim \mathscr{F}$.

The proof actually shows more than is stated in the lemma; namely, we proved that $x \in \lim \mathscr{F}$ provided $x_s \in \lim \mathscr{F}_s$ for every s. Therefore $\lim \mathscr{F} \supseteq \prod \lim \mathscr{F}_s$. If we combine this with the second half of Lemma 4 we obtain:

Lemma 6. *If $\mathscr{F}$ is a filter in a product space, then* $\lim \mathscr{F} = \prod \lim \mathscr{F}_s$.

It is interesting that the first half of Lemma 4 does not have a true converse even if the number of factors in the product space is finite. Hence in general adh $\mathscr{F}$ and $\prod$ adh $\mathscr{F}_s$ are different sets. For example, we may choose $X_1 = X_2$ to be the set of reals under the discrete topology. We consider the family $\mathscr{F}$ of all sets F in $X = X_1 \times X_2$ having the property that the lines $\{(\xi_1, x_2) : \xi_1 \in X_1\}$ and $\{(x_1, \xi_2) : \xi_2 \in X_2\}$ are subsets of F for all but finitely many values of $x_1 \in X_1$ and $x_2 \in X_2$. The family $\mathscr{F}$ is clearly a filter in X and its projections are $\mathscr{F}_1 = \{X_1\}$ and $\mathscr{F}_2 = \{X_2\}$. Hence adh $\mathscr{F}_1 = X_1$ and adh $\mathscr{F}_2 = X_2$. The filter $\mathscr{F}$, however, does not have any adherence point because the set $\{(x_1, x_2)\}$ consisting of the single point (x_1, x_2) is a neighborhood of (x_1, x_2) and its complement is an element of $\mathscr{F}$.

The theorem on the compactness of a product space is an immediate consequence of Lemmas 3 and 5 and Theorem 2: Let $\mathscr{M}$ be an ultrafilter in the product space $X = \prod X_s$ of compact factor spaces X_s. Then by Lemma 3 the projection $\mathscr{M}_s$ is an ultrafilter in X_s and so by Theorem 2 it is convergent. Lemma 5 implies that $\mathscr{M}$ is convergent. Since every ultrafilter $\mathscr{M}$ is convergent, X is compact by Theorem 2. The axiom of choice is needed, namely, it was used in the proof of Theorem 3.1 which is used to establish Theorem 2.

Countable compactness and the Lindelöf property can also be characterized in terms of filters. We introduce first an order classification of filters according to their intersection properties and the cardinalities of their bases: Let $\boldsymbol{m}$ and $\boldsymbol{n}$ denote infinite cardinals satisfying $\boldsymbol{m} \leqslant \boldsymbol{n}$. A family of sets is said to have the $\boldsymbol{m}$-intersection property if every subset of cardinality at most $\boldsymbol{m}$ has a nonvoid intersection. By an $(\boldsymbol{m}, \boldsymbol{n})$-*filter* in a set X we understand a filter $\mathscr{F}$ in X such that $\mathscr{F}$ has the $\boldsymbol{m}$-intersection property and $\mathscr{F}$ has a base $\mathscr{B}$ with card $\mathscr{B} \leqslant \boldsymbol{n}$. A filter having the $\boldsymbol{m}$-intersection property is called an $(\boldsymbol{m}, \infty)$-*filter* and a filter having a base with cardinality at most $\boldsymbol{n}$ is a $(1, \boldsymbol{n})$-*filter* or an $\boldsymbol{n}$-*filter*. A $(1, \omega)$-filter is usually called a *countable filter*. A $(1, \infty)$-*filter* is a filter in the usual sense.

Theorem 3. *A topological space X is countably compact if and only if every countable filter in X has a nonvoid adherence.*

Proof. Let X be countably compact and let $\mathscr{B}$ with card $\mathscr{B} \leqslant \omega$ be a filter base for a countable filter $\mathscr{F}$ in X. Then the family $\{\bar{B}\}$ $(B \in \mathscr{B})$ has the finite intersection property and so by the same reasoning as in the proof of Theorem III.1.1 we see that it has the ω-intersection property. Since card $\mathscr{B} \leqslant \omega$ this implies that $\bigcap \bar{B} = \operatorname{adh} \mathscr{F}$ is not void. Conversely, if X is not countably compact, then by the same theorem there is a family of closed sets $\mathscr{B}$ with card $\mathscr{B} \leqslant \omega$ and having the finite intersection property but with total intersection void. Thus $\mathscr{B}$ is a filter base for a countable filter $\mathscr{F}$ and adh $\mathscr{F} = \varnothing$.

Similarly one can prove that X is a Lindelöf space if and only if every (ω, ∞)-filter in X has a nonvoid adherence.

Another necessary and sufficient condition for countable compactness was given in Theorem III.1.3. Combining these we obtain:

Every countable filter in a topological space X has a nonvoid adherence if and only if every infinite sequence (x_n) of points $x_n \in X$ has an accumulation point.

This result can also be verified directly without reference to countable compactness.

EXERCISES

1. Prove by using filters: If X_1 and X_2 are compact subspaces of a space X, then $X_1 \cup X_2$ is compact.

2. Show by using filters that the topology of finite complements is compact.

3. Use filters to prove that an order topology is compact if and only if the linearly ordered set X has a minimal element.

4. Let $m < n$ be infinite cardinals, let X be a set of cardinality n, and let $\mathscr{F}$ be the family of all subsets F of X whose complement has cardinality at most m. Show that $\mathscr{F}$ is an (m, n)-filter.

5. It is possible to prove without using the axiom of choice that the product of finitely many compact spaces is compact. Prove this result by using filters and avoiding the axiom of choice.

[Let $\mathscr{F}$ be a filter in $X_1 \times X_2$ and let $\mathscr{F}_1$ be the projection of $\mathscr{F}$ into the compact space X_1. Let $x_1 \in \operatorname{adh} \mathscr{F}_1$ and let $\mathscr{B}_2$ be the family of all sets

$$B_2 = B_2(F, O_{x_1}) = \{\xi_2 : (\xi_1, \xi_2) \in F \quad \textit{and} \quad \xi_1 \in O_{x_1}\}$$

where $F \in \mathscr{F}$ and O_{x_1} is an open neighborhood of x_1 in X_1. Then $\mathscr{B}_2$ is a filter base in X_2. If X_2 is compact, then $\mathscr{B}_2$ has an adherence point x_2. It follows that $O_{x_2} \cap B_2 \neq \emptyset$ for every open neighborhood O_{x_2} of x_2. Hence $(O_{x_1} \times O_{x_2}) \cap F \neq \emptyset$ for every O_{x_1} in X_1 and O_{x_2} in X_2. In other words, $(x_1, x_2) \in \operatorname{adh} \mathscr{F}$ and $X_1 \times X_2$ is compact.]

6. Show the existence of a sequence (x^n) in a product space such that every projection of (x^n) has an adherence point but $\operatorname{adh}(x^n)$ is void.

[Let $X_1 = X_2 = \{0, 1, 2, ...\}$ and let the topology on X_1 and X_2 be discrete. Consider the following sequence of points $x^n \in X = X_1 \times X_2$:

$$((1, 0),\quad (0, 2),\quad (3, 0),\quad (0, 4), ...).$$

Every set F_m in the projection $\mathscr{F}_m$ $(m = 1, 2)$ of the elementary filter $\mathscr{F}$ generated by (x^n) contains 0 and so $0 \in \operatorname{adh} \mathscr{F}_m$. If $x = (x_1, x_2) \in X$ is a possible adherence point of $\mathscr{F}$, then both for $m = 1$ and $m = 2$ there are arbitrarily high indices n such that $x_m = x_m{}^n$. Since $x_m{}^n$ is 0 or n it follows that $x_1 = x_2 = 0$. The point $x = (0, 0)$ is not an adherence point of $\mathscr{F}$ because there is no x^n such that $x_1{}^n = x_2{}^n = 0$.]

7. If X is not compact, then the complements of compact sets form a filter base $\mathscr{B}$. Show that $\operatorname{adh} \mathscr{B}$ is the set of those points of X which do not have compact neighborhoods.

[Clearly $\operatorname{adh} \mathscr{B} = \bigcap \overline{cC}$ where the intersection is taken for every compact set in X. If $x \in C^i$ for some compact set C, then $cC \in \mathscr{B}$ and and $x \notin \overline{cC}$. Hence $x \notin \operatorname{adh} \mathscr{B}$ and so $\operatorname{adh} \mathscr{B}$ is a subset of the set of those points which have no compact neighborhoods. Conversely, if x has no compact neighborhood, then for every compact set $x \in c(C^i) = C^e \cup C^b = \overline{cC}$. Therefore $x \in \bigcap \overline{cC} = \operatorname{adh} \mathscr{B}$. As a corollary we obtain the following: *X is locally compact if* $\operatorname{adh} \mathscr{B}$ *is void* where $\mathscr{B}$ is described above.]

6. Cauchy Filters and Complete Spaces

The importance of Cauchy sequences is well known from the theory of one real variable. If (x_n) is a convergent sequence of real numbers, then it is a Cauchy sequence, that is, $| x_m - x_n | \to 0$ as $m, n \to \infty$ simultaneously. A theorem of Cauchy states that, conversely, every one of these sequences has a limit in the space of the real numbers. The limit need not be rational even if all elements of the sequence are rational. The notion of a Cauchy sequence can be immediately extended to metric spaces : (x_n) is a Cauchy sequence relative to the metric d if

$d(x_m, x_n) \to 0$ as $m, n \to \infty$ simultaneously. Every convergent sequence of points in a metric space is a Cauchy sequence.

If every Cauchy sequence is convergent, then the metric space is called *complete* relative to the given metric. The reals form a complete space under the metric $d(x, y) = |x-y|$ and the space formed by the rationals is not complete. Completeness is not a topological invariant, a change in the metric can destroy completeness while the topology is preserved. For example, the topology generated on the set of reals by the metric

$$d(x, y) = \left| \frac{x}{1+|x|} - \frac{y}{1+|y|} \right|$$

is the usual topology but the space is not complete relative to d because $(1, 2, 3, \ldots)$ is a Cauchy sequence without a limit. The theory of completeness can be extended to uniform structures by considering filters instead of sequences.

Definition 1. *A Cauchy filter $\mathscr{F}$ relative to a uniform structure $\mathscr{U}$ for X is a filter in X such that for every $U \in \mathscr{U}$ there is an $F \in \mathscr{F}$ satisfying $F \times F \subseteq U$.*

Note. If a family $\mathscr{F}$ is such that for every $U \in \mathscr{U}$ there is a nonvoid set $F \in \mathscr{F}$ satisfying $F \times F \subseteq U$, we say that "$\mathscr{F}$ contains arbitrarily small sets." A filter $\mathscr{F}$ is a Cauchy filter if it contains arbitrarily small sets.

An elementary filter $\mathscr{F}$ generated by a sequence (x_n) is a Cauchy filter if and only if for every $U \in \mathscr{U}$ we can find an integer $n(U)$ such that $(x_m, x_n) \in U$ for every $m, n \geqslant n(U)$. If $\mathscr{U}$ is generated by the metric d, then $\mathscr{F}$ is a Cauchy filter relative to $\mathscr{U}$ if and only if $\mathscr{F}$ contains sets of arbitrarily small diameter. An elementary filter generated by a sequence (x_n) is a Cauchy filter if and only if for every $\epsilon > 0$ there is an $n(\epsilon)$ with the property that $d(x_m, x_n) < \epsilon$ for every $m, n \geqslant n(\epsilon)$. More briefly, $\mathscr{F}$ generated by (x_n) is a Cauchy filter provided $d(x_m, x_n) \to 0$ as $m, n \to \infty$.

The familiar results on Cauchy sequences can all be generalized to arbitrary Cauchy filters:

If $\mathscr{F}$ is a Cauchy filter and if $\mathscr{F} \leqslant \mathscr{G}$, then $\mathscr{G}$ is a Cauchy filter.

If $\mathscr{F}$ is a Cauchy filter relative to $\mathscr{U}$ and if $\mathscr{V} \leqslant \mathscr{U}$, then $\mathscr{F}$ is a Cauchy filter relative to $\mathscr{V}$.

Less formally we can say: "The finer the uniform structure the fewer Cauchy filters exist on it."

If $\mathscr{F}$ or $\mathscr{G}$ is a Cauchy filter and if $F \cap G \neq \emptyset$ for every $F \in \mathscr{F}$, $G \in \mathscr{G}$, then $\mathscr{F} \cup \mathscr{G} = \{F \cap G\}$ $(F \in \mathscr{F}, G \in \mathscr{G})$ is a Cauchy filter.

Lemma 1. *If $\mathscr{F}$ is a Cauchy filter, then* adh $\mathscr{F}$ = lim $\mathscr{F}$.

Proof. Let $U \in \mathscr{U}$ and let V be a uniformity satisfying $V \circ V \subseteq U$. Since $\mathscr{F}$ is a Cauchy filter we can choose an $F \in \mathscr{F}$ such that $F \times F \subseteq V$. If $x \in \text{adh}\, \mathscr{F}$, then $V[x]$ being a neighborhood of x, the intersection $V[x] \cap F$ is nonvoid, and so $(x, a) \in V$ for some $a \in F$. By $F \times F \subseteq V$ we have $(a, b) \in V$ for every $b \in F$ and so $(x, b) \in V \circ V \subseteq U$ for every $b \in F$. Therefore given $U[x]$ with x in adh $\mathscr{F}$ there is an F satisfying $F \subseteq U[x]$. This shows that $x \in \lim \mathscr{F}$ and adh $\mathscr{F} \subseteq \lim \mathscr{F}$.

Lemma 2. *If* lim $\mathscr{F}$ *is not void, then $\mathscr{F}$ is a Cauchy filter relative to every uniform structure $\mathscr{U}$ compatible with the topology of the space.*

Proof. Given $U \in \mathscr{U}$ we determine a symmetric uniformity V such that $V \circ V \subseteq U$. If x is an arbitrary element of lim $\mathscr{F}$, then there is an $F \in \mathscr{F}$ satisfying $F \subseteq V[x]$. Hence $(x, a) \in V$ for every $a \in F$ and so by the symmetry $(a, b) \in V \circ V$ for any pair of points, $a, b \in F$. Since $V \circ V \subseteq U$ this shows that $F \times F \subseteq U$ and so $\mathscr{F}$ is a Cauchy filter.

Definition 2. *A uniform structure $\mathscr{U}$ for a set X is called complete if every Cauchy filter relative to $\mathscr{U}$ is convergent to some point of the set X.*

A pseudometric d on a set X is called complete if the uniform structure generated by d is a complete structure for X.

It is customary to speak about *complete spaces* instead of complete structures although completeness is a property of the structure $\mathscr{U}$ and in general is not determined by the uniform topology associated with $\mathscr{U}$. Occasionally we follow this practice but it is understood that we always mean complete structures or complete metrics.

It is necessary to show that for metric structures the earlier definition of completeness by Cauchy sequences coincides with the present one. This is the content of the following:

Theorem 1. *A pseudometric d is complete if and only if every Cauchy sequence is convergent relative to the uniform topology induced by d.*

Proof. The necessity is obvious. To prove the sufficiency let $\mathscr{F}$ be a Cauchy filter relative to the pseudometric d for the set X. Then for every $n = 1, 2, \ldots$ there is a set $F_n \in \mathscr{F}$ such that its diameter $d(F_n) < 1/n$. We define a sequence (x_n) $(n = 1, 2, \ldots)$ by choosing for every n the point x_n in the set $F_1 \cap \ldots \cap F_n$. Then $d(x_m, x_n) < 1/m$ for every $m \leqslant n$ and so (x_n) is a Cauchy sequence. By hypothesis (x_n) has a limit x in X. We show that $x \in \lim \mathscr{F}$. In fact, given $\epsilon > 0$, let n be so large that $1/n < \epsilon/2$. Since x is a limit point of (x_n) we can choose n such that $x_n \in S_{\epsilon/2}[x]$ and so $d(F_n) < 1/n$ and $x_n \in F_n$ imply that $F_n \subseteq S_\epsilon[x]$. So every Cauchy filter $\mathscr{F}$ is convergent and $\mathscr{U}_d$ is a complete structure.

Theorem 2. *If $\mathscr{U}$ is a complete structure for the set X, then a filter $\mathscr{F}$ is convergent in the uniform topology associated with $\mathscr{U}$ if and only if it is a Cauchy filter.*

The necessity is a consequence of Lemma 2 and the sufficiency is obvious from the definition of completeness. The completeness of a structure is a very important property. For the preceding theorem implies that in complete structures it is possible to decide on the convergence of a filter without actually determining any of its limit points.

Lemma 3. *If $\mathscr{U}$ is a complete structure for X and if the subset Y is closed relative to the uniform topology associated with $\mathscr{U}$, then the trace of $\mathscr{U}$ is a complete structure for Y.*

Briefly and less accurately: Every closed subspace of a complete space is complete.

Proof. If $\mathscr{F}$ is a Cauchy filter in Y, then $\mathscr{F}$ is a filter base for a Cauchy filter in X. Since X is complete, $\mathscr{F}$ has a limit point in X. However, $\bar{F} \subseteq Y$ for every $F \in \mathscr{F}$ and so adh $\mathscr{F} = \bigcap \bar{F} \subseteq Y$. Therefore $\mathscr{F}$ has a limit point in Y and $\mathscr{F}$ is a convergent filter in Y.

If the uniform space associated with the structure $\mathscr{U}$ is a Hausdorff space, then we have the following converse:

Lemma 4. *Let $\mathscr{U}$ be a separated uniform structure for the set X. Then the completeness of the trace of $\mathscr{U}$ on $Y \times Y$ implies that Y is a closed subset of X relative to the uniform topology associated with the structure $\mathscr{U}$.*

Briefly we may say: If X is a Hausdorff space, then every complete subspace is closed.

Proof. Let $y \in \bar{Y}$ and consider the sets $V[y] \cap Y$ where $V \in \mathscr{U}$. Since $y \in \bar{Y}$ these sets are not void and so $\mathscr{B} = \{V[y] \cap Y\}$ $(V \in \mathscr{U})$ is a filter base in Y. Given $U \in \mathscr{U}$ we have $V[y] \times V[y] \subseteq V \circ V \subseteq U$ provided V is a symmetric uniformity such that $V \circ V \subseteq U$. It follows that $(V[y] \cap Y) \times (V[y] \cap Y) \subseteq U \cap (Y \times Y)$ and so $\mathscr{B}$ is a base for a Cauchy filter in Y. Hence $\mathscr{B}$ has a limit point η in Y. Since $\mathscr{B}$ is a filter base in X, the point η is a limit point of $\mathscr{B}$ also in X. The point y is obviously a limit point of $\mathscr{B}$ in X and X being a Hausdorff space it is the only limit point. Therefore $\eta = y$ and so $y \in Y$. This proves that $\bar{Y} = Y$.

A useful completeness criterion is given by the following:

Lemma 5. *Let S be a dense subspace of the uniform space $\bar{S}$ with uniform structure $\bar{\mathscr{U}}$. If every Cauchy filter in S containing arbitrarily small open sets converges to some point of $\bar{S}$, then $\bar{S}$ is complete.*

Proof. Let $\mathscr{F}$ be a Cauchy filter in $\bar{S}$. Consider the family $\mathscr{B} = \{U[F] \cap S\}$ $(U \in \overline{\mathscr{U}}$ and $F \in \mathscr{F})$. Since S is dense in $\bar{S}$ every $U[F] \cap S$ is nonvoid and $(U[F_1] \cap S) \cap (U[F_2] \cap S) = U[F_1 \cap F_2] \cap S$. Hence $\mathscr{B}$ is a filter base in S. $\mathscr{B}$ is a base for a Cauchy filter because if $F \times F \subseteq V$ where V is symmetric, then $V[F] \times V[F] \subseteq V \circ V \circ V$: In fact, if $x \in V[F]$ and $y \in V[F]$, then $(x, a) \in V$ and $(b, y) \in V$ for some $a, b \in F$. Thus by $(a, b) \in F \times F \subseteq V$ we have $(x, y) = (x, a) \circ (a, b) \circ (b, y) \in V \circ V \circ V$. We prove that $\mathscr{B}$ contains arbitrarily small open sets: First, the interior of any $V \in \overline{\mathscr{U}}$ taken in the product space $\bar{S} \times \bar{S}$ is also an element of $\overline{\mathscr{U}}$ because for instance $W \subseteq V^i$ provided $W \circ W \circ W \subseteq V$. Given U choose $V \in \overline{\mathscr{U}}$ such that $V \circ V \circ V \subseteq U$ and choose $F \in \mathscr{F}$ such that $F \times F \subseteq V$. Then $O = V^i[F] \cap S$ is open in S and $O \times O \subseteq U$. By hypothesis, $\mathscr{B}$ as a Cauchy filter base in $\bar{S}$ is convergent to some point $x \in \bar{S}$. We show that $x \in \operatorname{adh} \mathscr{F}$ which is equivalent to saying that $x \in \lim \mathscr{F}$. Given an open neighborhood O_x of x, there is a $U \in \overline{\mathscr{U}}$ such that $U[x] \subseteq O_x$. We determine the symmetric $V \in U$ such that $V \circ V \subseteq U$. Since $x \in \operatorname{adh} \mathscr{B}$ and $V[F] \cap S$ is an element of $\mathscr{B}$ we have $V[x] \cap V[F] \cap \neq \varnothing$ for every $F \in \mathscr{F}$. Hence there is an $s \in S$ and an $a \in F$ such that $(x, s) \in V$ and $(s, a) \in V$. Therefore $(x, a) \subseteq V \circ V \subseteq U$ and $a \in U[x] \cap F$. This shows that $U[x]$ and F intersect for every $U \in \mathscr{U}$ and for every $F \in \mathscr{F}$. Thus $x \in \operatorname{adh} \mathscr{F}$.

Theorem 3. *If the uniform topology associated with a structure $\mathscr{U}$ for a set X is compact, then $\mathscr{U}$ is a complete structure.*

Proof. Since X is compact, by Theorem 5.1 every filter in X has a nonvoid adherence and so by Lemma 1 every Cauchy filter is convergent in X.

The connection between compact spaces and complete structures can be given by introducing the notion of a *precompact structure*:

Definition 3. *A uniform structure $\mathscr{U}$ for a set X is called precompact if for every $U \in \mathscr{U}$ there are finitely many points $x_1, \ldots, x_n \in X$ such that $U[x_1] \cup \ldots \cup U[x_n] = X$.*

The expression "precompact structure" was introduced with regards to the following result:

Theorem 4. *A uniformizable space X is compact if and only if there is a precompact and complete structure for X which is compatible with its topology.*

Proof. The necessity is immediate: For every $U \in \mathscr{U}$ the family $\{U[x]^i\}$ $(x \in X)$ is an open cover of X and so by the compactness of X there is a finite subcover. Hence $\mathscr{U}$ is precompact. The sufficiency of the condition is a consequence of Theorem 5.2 and the following:

Lemma 6. *If $\mathscr{U}$ is precompact, then every ultrafilter is a Cauchy filter with respect to $\mathscr{U}$.*

Proof. Let $\mathscr{M}$ be an ultrafilter in X and let $U \in \mathscr{U}$. We determine a symmetric $V \in \mathscr{U}$ such that $V \circ V \subseteq U$. Since $\mathscr{U}$ is precompact there are finitely many points $x_1, \ldots, x_n$ such that $V[x_1] \cup \ldots \cup V[x_n] = X$. Then by Theorem 3.2 we have $V[x_k] \in \mathscr{M}$ for some k and so for $M = V[x_k] \in \mathscr{M}$ we have $M \times M \subseteq U$.

Definition 4. *A uniform structure $\mathscr{U}$ for a set X is called totally bounded if for every $U \in \mathscr{U}$ there are finitely many sets $B_1, \ldots, B_n$ in X such that $B_k \times B_k \subseteq U$ and $\bigcup B_k = X$.*

Note. One can suppose that $B_1, \ldots, B_n$ are open sets.

Theorem 5. *A uniform structure $\mathscr{U}$ is precompact if and only if it is totally bounded.*

Proof. If $\bigcup B_k = X$ and $B_k \times B_k \subseteq U$, then $U[x_k] \supseteq B_k$ for any $x_k \in B_k$ and so $\bigcup U[x_k] = X$. If $\mathscr{U}$ is precompact and $U \in \mathscr{U}$, then there is a symmetric $V \in \mathscr{U}$ satisfying $V \circ V \circ V \circ V \subseteq U$ and there are finitely many points $x_1, \ldots, x_n$ such that $\bigcup V[x_k] = X$. Then $V[x_k] \times V[x_k] \subseteq V \circ V$ and so $B_k = (V \circ V)[x_k]^i$ $(k = 1, \ldots, n)$ satisfies the requirements.

Every uniform structure $\mathscr{U}$ for a set X can be extended to an enlarged set $\bar{X}$ such that the new structure is complete. Among these extensions there is a smallest which is called the completion of $\mathscr{U}$ or the completion of X with respect to $\mathscr{U}$. A detailed discussion in the special case of metric structures follows and the general case is studied in Section 9.

EXERCISES

1. Show: If $\mathscr{F}$ is a Cauchy filter and $\mathscr{F} \leqslant \mathscr{G}$, then $\lim \mathscr{F} = \lim \mathscr{G}$. (It follows that $\mathscr{G}$ is a Cauchy filter, so $\lim \mathscr{F} = \operatorname{adh} \mathscr{F}$ and $\lim \mathscr{G} = \operatorname{adh} \mathscr{G}$. By $\mathscr{F} \leqslant \mathscr{G}$ we have $\operatorname{adh} \mathscr{F} \supseteq \operatorname{adh} \mathscr{G}$ and $\lim \mathscr{F} \subseteq \lim \mathscr{G}$.)

2. Let $x_n = \Sigma_{k=1}^{n} a_k 2^{-k}$ where $(a_1, a_2, \ldots)$ is a nonperiodic sequence of 0's and 1's. Show that (x_n) is a Cauchy sequence with respect to the usual metric of the reals but it has no rational limit.

3. Find the Cauchy sequences when X is the set of rationals and the metric is

$$d(x, y) = \left| \frac{1}{x - a} - \frac{1}{y - a} \right|,$$

where a is irrational.

(A sequence is a Cauchy sequence with respect to d if and only if it is a Cauchy sequence in the usual sense and its limit is not a, or if $|x_n| \to \infty$ as $n \to \infty$.)

4. Show that the function d defined by

$$d(x, y) = \frac{|x - y|}{\sqrt{(x^2 + 1)(y^2 + 1)}}$$

is a metric for the reals. Determine the Cauchy sequences.

[Notice that

$$d(x, y) = \left| \frac{1}{x - i} - \frac{1}{y - i} \right|$$

and use the triangle inequality for the complex numbers. A sequence (x_n) is a Cauchy sequence with respect to d if it is a Cauchy sequence in the usual sense or if $|x_n| \to \infty$ as $n \to \infty$.]

5. Show that $\mathscr{U}$ is precompact if and only if for every $U \in \mathscr{U}$ there are finitely many sets $B_1, ..., B_n$ in X such that $B_k \times B_k \subseteq U$ and $\bigcup B_k = X$. (One can suppose that $B_1, ..., B_n$ are open sets.)

6. Show that the space of the real numbers is complete with respect to the metric $|x - y|$.

(Apply the result stated in Exercise 2.2 or use Theorem 1 and show that every Cauchy sequence is convergent.)

7. Show that if $\mathscr{F}$ is a Cauchy filter in $\prod X_s$ with respect to $\prod \mathscr{U}_s$, then $\mathscr{F}_s$ is a Cauchy filter in X_s with respect to $\mathscr{U}_s$.

8. Suppose that the metric d for the set X has the property that every closed set of finite diameter is compact. Show that d is a complete.

[Let $\mathscr{F}$ be a Cauchy filter and let $F \in \mathscr{F}$ be determined such that $d(F) \leqslant 1$. Then it follows that $d(\bar{F}) \leqslant 1$. The family $\mathscr{B} = \{B : B \subseteq F$ *and* $B \in \mathscr{F}\}$ is a base for $\mathscr{F}$ and adh $\mathscr{F} = \bigcap \bar{B}$. Since $\bar{F}$ is compact and $\bar{B} \subseteq \bar{F}$ for every $B \in \mathscr{B}$ we see that adh $\mathscr{F}$ is not void.]

9. Let $X = [0, 1)$ be metrized by

$$d(x, y) = \left| \frac{1}{1 - x} - \frac{1}{1 - y} \right|.$$

Prove that X is complete with respect to d.

10. Prove that if $\mathscr{U}$ is precompact, then so is its trace on any subset $Y \times Y$ of $X \times X$.

(Use the precompactness criterion given in Exercise 5.)

11. Let $\mathscr{U}_s$ $(s \in S)$ be a precompact structure for the set X_s $(s \in S)$. Show that $\mathscr{U} = \prod \mathscr{U}_s$ is a precompact structure for $X = \prod X_s$.

(If $U \in \mathscr{U}$, then there are finitely many indices $s_1, \ldots, s_m \in S$ and uniformities $U_{s_1} \in \mathscr{U}_{s_1}, \ldots, U_{s_m} \in \mathscr{U}_{s_m}$ such that $U \supseteq \prod U_s$ where $U_s = U_{s_i}$ for every $s = s_i$ and $U_s = X_s \times X_s$ otherwise. Choose $x^1_{s_i}, \ldots, x^n_{s_i} \in X_{s_i}$ such that $\bigcup U[x^k_{s_i}] = X_{s_i}$. Here n depends on the index s_i. Choose a fixed point x_s in X_s for every $s \neq s_1, \ldots, s_m$ and consider those finitely many points $x \in X$ whose sth coordinate is x_s for every s different from $s_1, \ldots, s_m$ and is one of the $x^k_{s_i}$ points for every $i = 1, \ldots, m$. U evaluated on this finite set gives X.)

12. Show that a product structure $\mathscr{U} = \prod \mathscr{U}_s$ is complete if and only if every factor $\mathscr{U}_s$ is complete.

[Exercise 7 and Lemma 5.5 provide the sufficiency. The necessity follows from Lemma 5.6 and the following remark: If $\mathscr{F}_s$ is a Cauchy filter in X_s and a_σ is fixed in X_σ for each $\sigma \neq s$, then $\{x : x_s \in F_s \textit{ and } x_\sigma = a_\sigma\}$ $(F_s \in \mathscr{F}_s)$ is the base for a Cauchy filter in X.]

13. Use the results given in the preceding two exercises to show that the product of compact uniformizable spaces is compact.

14. Carefully examine the preceding proof of Tychonoff's theorem for uniformizable spaces and list the points where the axiom of choice is used.

15. Let X be a (T_3) space and let A be a dense subset of X such that every filter $\mathscr{F}$ in A is a filter base in X with nonvoid adherence. Show that X is compact.

(Modify the proof of Lemma 5.)

7. Completion of Metric Structures

This section serves a double purpose. First it acquaints the reader with the notion of completion of metric spaces, and secondly it forms a background for the general theory of completion and compactification of uniform spaces. We define what is meant by the completion, describe the process of completion by using Cauchy sequences, and show that the completion of a metric space is not determined by the topology but depends on the given metric.

A metric space is complete with respect to a metric d if every Cauchy sequence has a limit. This means that $d(x_m, x_n) \to 0$ as $m, n \to \infty$ implies the existence of a point x such that $d(x_m, x) \to 0$ as $m \to \infty$. Two metric spaces X and $\tilde{X}$ with metrics d and $\tilde{d}$ are called *isometric* if there is a one-to-one correspondence $x \leftrightarrow \tilde{x}$ between the elements of X and $\tilde{X}$ which preserves the distance between points, that is, $d(x, y) = \tilde{d}(\tilde{x}, \tilde{y})$ for any two corresponding point-pairs $x, \tilde{x}$ and $y, \tilde{y}$.

Isometric spaces are homeomorphic. This follows from the fact that the isometric image of a ball $S_\epsilon[x] = \{y : d(x, y) < \epsilon\}$ is the ball $S_\epsilon[\tilde{x}] = \{\tilde{y} : \tilde{d}(\tilde{x}, \tilde{y}) < \epsilon\}$ where $\tilde{x}$ is the point corresponding to the center x.

Definition 1. *A metric space $\bar{X}$ with metric δ is called the completion of the metric space X with metric d if $\bar{X}$ is complete with respect to the metric δ and if there is a dense set in $\bar{X}$ which is isometric to X.*

Note. Since X and the dense subspace are homeomorphic, X can be considered as a dense subspace of the completion $\bar{X}$ and the metric d can be considered as the restriction of δ to X.

Theorem 1. *Every metric space X has a completion with respect to any metric d.*

Proof. We consider the class of all Cauchy sequences with respect to the metric d and we introduce an equivalence relation by saying that $(x_n) \sim (y_n)$ if $d(x_n, y_n) \to 0$ as $n \to \infty$. The relation $\sim$ is obviously reflexive and symmetric and its transitivity follows from the triangle inequality. Let $\bar{X}$ be the set of equivalence classes into which the family of Cauchy sequences is divided by the relation $\sim$. The elements of $\bar{X}$ will be denoted by $\xi, \eta, \zeta, \ldots$.

We can introduce a metric δ on $\bar{X}$ by setting

$$\delta(\xi, \eta) = \lim_{n \to \infty} d(x_n , y_n)$$

where (x_n) and (y_n) denote arbitrary representative sequences of the classes ξ and η, respectively. First we show that the limit exists and is independent of the choice of the representative sequences. This is an easy consequence of the triangle inequality: For any four points a, b, c, d in X we have

$$| d(a, b) - d(c, d) | \leqslant d(a, c) + d(b, d).$$

Applying this inequality to the points x_m, y_m, x_n, y_n of the Cauchy sequences (x_n) and (y_n) we see that $d(x_n, y_n)$ is a Cauchy sequence of real numbers and so it is convergent. This proves the existence of the limit. If $(x_n), (s_n) \in \xi$, then $d(x_n, s_n) \to 0$ and similarly if $(y_n), (t_n) \in \eta$, then $d(y_n, t_n) \to 0$. Hence by the foregoing inequality

$$| d(x_n , y_n) - d(s_n , t_n) | \leqslant d(x_n , s_n) + d(y_n , t_n)$$

and so $\lim_{n\to\infty} d(x_n, y_n) = \lim_{n\to\infty} d(s_n, t_n)$. Therefore $\delta(\xi, \eta)$ is independent of the choice of the representative sequences.

It is clear that $\delta(\xi, \xi) = 0$ and $\delta(\xi, \eta) = \delta(\eta, \xi)$ for every $\xi, \eta \in \bar{X}$. The triangle inequality follows from the corresponding inequality for the metric d on X: By the latter we have

$$d(x_n, z_n) \leqslant d(x_n, y_n) + d(y_n, z_n)$$

for $n = 1, 2, \ldots$ and so taking the limit on the right and on the left-hand sides we obtain $\delta(\xi,\zeta) \leqslant \delta(\xi, \eta) + \delta(\eta, \zeta)$. We proved that δ is a pseudometric on $\bar{X}$. $\delta(\xi, \eta) = 0$ if and only if $d(x_n, y_n) \to 0$ as $n \to \infty$ or equivalently if and only if $(x_n) \sim (y_n)$, that is, if and only if $\xi = \eta$. Therefore δ is a proper metric and so $\bar{X}$ is a Hausdorff space.

Every sequence of the form $(x, x, \ldots)$ where $x \in X$ is a Cauchy sequence and so it determines an equivalence class in $\bar{X}$ which we denote by $\tilde{x}$. Obviously $\tilde{x} = \tilde{y}$ if and only if $x = y$ so that $x \leftrightarrow \tilde{x}$ $(x \in X)$ is a one-to-one correspondence between X and the set $\tilde{X}$ of all points $\tilde{x}$ in $\bar{X}$. The distance $\delta(\tilde{x}, \tilde{y})$ can be computed by choosing $(x, x, \ldots)$ and $(y, y, \ldots)$ as representatives and it follows that $\delta(\tilde{x}, \tilde{y}) = d(x, y)$. This shows that the one-to-one correspondence $x \leftrightarrow \tilde{x}$ $(x \in X)$ is an isometry.

We show that $\tilde{X}$ is a dense subset of the metric space $\bar{X}$: Let $\epsilon > 0$ and $\xi \in \bar{X}$ be given. We shall find a point $\tilde{x} \in \tilde{X}$ in the ϵ-neighborhood $S_\epsilon[\xi]$ of ξ. Let $(x_1, x_2, \ldots)$ be any representative sequence of the class ξ. Since this is a Cauchy sequence we can choose an index m so large that $d(x_m, x_n) \leqslant \epsilon/2$ for every $n \geqslant m$. We fix m and define $\tilde{x}$ by the sequence $(x_m, x_m, \ldots)$. Then

$$\delta(\tilde{x}, \xi) = \lim_{n \to \infty} d(x_m, x_n) \leqslant \frac{\epsilon}{2}$$

and so $\tilde{x} \in S_\epsilon[\xi]$. Therefore $\tilde{X}$ is dense in $\bar{X}$.

The proof will be finished by showing that $\bar{X}$ is complete. First we present some preliminary remarks on Cauchy sequences: We can select from any Cauchy sequence a subsequence (ξ_p) such that $\delta(\xi_p, \xi_q) < 1/3p$ for every $p \leqslant q$. If a subsequence of a Cauchy sequence is convergent, then the original sequence is convergent to the same limit. Hence in the proof of the completeness we may restrict ourselves to Cauchy sequences satisfying $\delta(\xi_p, \xi_q) < 1/3p$ for every $p \leqslant q$. Let p be a positive integer. By omitting a finite number of terms from the beginning of a Cauchy sequence we obtain a subsequence (x_m) such that $d(x_m, x_n) < 1/3p$ for every $m, n \geqslant 1$. The resulting sequence is a Cauchy sequence which is equivalent to the original one.

Now let (ξ_p) be a Cauchy sequence in $\bar{X}$ such that $\delta(\xi_p, \xi_q) < 1/3p$ for every $p \leqslant q$. By the last remark we can choose for every index p a representative sequence (x_{mp}) for ξ_p such that $d(x_{mp}, x_{np}) < 1/3p$ for every $m, n \geqslant 1$.

Since

$$\delta(\xi_p, \xi_q) = \lim_{n\to\infty} d(x_{np}, x_{nq}) < \frac{1}{3p}$$

there is an r depending on p and q for which $d(x_{rp}, x_{rq}) < 1/3p$. However, $d(x_{mp}, x_{np}) < 1/3p$ and $d(x_{mq}, x_{nq}) < 1/3q \leqslant 1/3p$ for every $m, n \geqslant 1$ and so $d(x_{mp}, x_{nq}) \leqslant d(x_{mp}, x_{rp}) + d(x_{rp}, x_{rq}) + d(x_{rq}, x_{nq}) < 1/p$ for every $m, n \geqslant 1$ and $q \geqslant p \geqslant 1$. We define $\xi \in \bar{X}$ by the representative sequence (x_m) where $x_m = x_{1m}$ for every $m \geqslant 1$. Then

$$\delta(\xi_p, \xi) = \lim_{m\to\infty} d(x_{mp}, x_{1m}) \leqslant \frac{1}{p}$$

because for $m \geqslant p \geqslant 1$ we have $d(x_{mp}, x_{1m}) < 1/p$. Consequently, $\delta(\xi_p, \xi) \to 0$ as $p \to \infty$ and ξ is the limit of the Cauchy sequence (ξ_p). Hence $\bar{X}$ is a complete space.

If X is complete with respect to a metric d, then every Cauchy sequence (x_n) is convergent to some limit $x \in X$ and so it is equivalent to a sequence of the form $(x, x, ...)$. Hence every equivalence class is a class $\tilde{x}$ and $\bar{X} = \tilde{X}$. Therefore the completion of a complete space X is isomorphic to the given space so that the space and its completion can be considered as identical.

The completion of a metric space X depends on the metric d which is used to define the Cauchy sequences, their equivalence, and the distance between the equivalence classes. For instance, let X be the space of real numbers and consider the metrics

$$d_1(x, y) = |x - y| \quad \text{and} \quad d_2(x, y) = \left| \frac{x}{1 + |x|} - \frac{y}{1 + |y|} \right|.$$

The first metric leads to Cauchy sequences in the usual sense which are known to be convergent. Hence X and its completion with respect to d_1 are identical. In the second metric there are equivalence classes which do not correspond to convergent sequences. For instance, $(1, 2, 3, ...)$ is a Cauchy sequence with respect to d_2 but it is not convergent to any real number x because

$$\lim_{n\to\infty} \delta(x, n) = 1 - \frac{x}{1 + |x|} > 0.$$

It is easy to show that either (x_n) is convergent in the ordinary sense or $x_n \to +\infty$ or $x_n \to -\infty$. Hence only two new equivalence classes occur, and representatives for them are $(1, 2, 3, ...)$ and $(-1, -2, -3, ...)$, respectively. The occurence of new equivalence classes does not necessarily imply that $\bar{X}$ and X are not isometric or homeomorphic. To see that this is not the case we must actually exhibit a topological property in which X and $\bar{X}$ differ. Compactness or equivalently the Bolzano-Weierstrass property can serve this purpose. The reader should verify for himself that every infinite subset of $\bar{X}$ has an accumulation point in $\bar{X}$. This is a very easy task because the function $\xi = x/(1 + |x|)$ occurring in the definition of the metric d_2 is a homeomorphism between the real line and the open interval $(-1, 1)$. The completion $\bar{X}$ of X with respect to d_2 is homeomorphic to the compact interval $[-1, 1]$.

Let $X_1, \ldots, X_k$ be metric spaces with metrics $d_1, \ldots, d_k$ and let the Cartesian product $X = X_1 \times \ldots \times X_k$ be metrized by the distance $d = d_1 + \ldots + d_k$:

$$d((x_1, \ldots, x_k), (y_1, \ldots, y_k)) = d_1(x_1, y_1) + \ldots + d_k(x_k, y_k).$$

Let $\bar{X}_1, \ldots, \bar{X}_k$ and $\bar{X}$ denote the completion of the spaces $X_1, \ldots, X_k$, and X, respectively. The extended metrics on $\bar{X}_1, \ldots, \bar{X}_k$, and $\bar{X}$ we denote by $\delta_1, \ldots, \delta_k$, and δ. Then it is simple to show that $\delta = \delta_1 + \ldots + \delta_k$ and $\bar{X}$ is isometric with the product $\bar{X}_1 \times \ldots \times \bar{X}_k$ metrized by $\delta = \delta_1 + \ldots + \delta_k$. Briefly:

The completion of a finite Cartesian product is the Cartesian product of the completion of its factors.

In the proof we may restrict ourselves to the case of two factors. Then $((x_{1n}, x_{2n}))$ is a Cauchy sequence in $X_1 \times X_2$ if and only if (x_{1n}) and (x_{2n}) are Cauchy sequences, and $((x_{1n}, x_{2n})) \sim ((y_{1n}, y_{2n}))$ if and only if $(x_{1n}) \sim (y_{1n})$ and $(x_{2n}) \sim (y_{2n})$. Hence the correspondence $((x_{1n}, x_{2n})) \leftrightarrow ((x_{1n}), (x_{2n}))$ leads to a one-to-one correspondence between the elements of $\bar{X}$ and of $\bar{X}_1 \times \bar{X}_2$. The metric on $\bar{X}$ is given by

$$\delta(\xi, \eta) = \lim_{n\to\infty} d((x_{1n}, x_{2n}), (y_{1n}, y_{2n}))$$

and so using $d = d_1 + d_2$ we see that

$$\delta(\xi, \eta) = \lim_{n\to\infty} d_1(x_{1n}, y_{1n}) + \lim_{n\to\infty} d_2(x_{2n}, y_{2n}).$$

Hence $\delta(\xi, \eta) = \delta_1(\xi_1, \eta_1) + \delta_2(\xi_2, \eta_2)$.

EXERCISES

1. Find the completion of the rationals when

$$d(x, y) = \left| \frac{1}{x - a} - \frac{1}{y - a} \right|$$

and a is irrational.

[Cauchy sequences with respect to d are sequences which converge in the usual sense to a limit different from a or tend to ∞. If (x_n) is a Cauchy sequence with respect to d, then it is convergent to its ordinary limit or to ∞. Hence $\bar{X}$ is in a one-to-one correspondence with the set of reals different from a and the symbol ∞. The distance in $\bar{X}$ is given by

$$\delta(\xi, \eta) = \left| \frac{1}{\xi - a} - \frac{1}{\eta - a} \right|$$

where ξ and η denote the usual limits of $(x_n) \in \xi$ and $(y_n) \in \eta$, respectively, and $\delta(\xi, \infty) = 1/|\xi - a|$. The space $\bar{X}$ is homeomorphic to the ordinary reals.]

2. The function d given by

$$d(x, y) = \left| \frac{1}{1 + x^2} - \frac{1}{1 + y^2} \right|$$

is a metric for the set $\{x : 1 \leqslant x < +\infty\}$. Determine the completion of d.

[Bounded Cauchy sequences are Cauchy sequences in the usual sense and they are equivalent if and only if they have the same limit. A nonbounded sequence has the Cauchy property if and only if $x_n \to +\infty$ and all these sequences are equivalent. Hence $\bar{X}$ can be identified with the set $\{x : 1 \leqslant x \leqslant \infty\}$ where ∞ corresponds to the nonbounded Cauchy sequences. Since $\delta(x, \infty) = 1/(1 + x^2)$, the space $\bar{X}$ is homeomorphic to the extended real number system.]

3. Show that the completion of the reals with respect to the metric

$$d(x, y) = \left| \frac{1}{x + i} - \frac{1}{y + i} \right|$$

is homeomorphic to the circle.

[The bounded Cauchy sequences correspond to the reals and all nonbounded Cauchy sequences are equivalent. If ∞ denotes the point corresponding to the class of the nonbounded Cauchy sequences, then $\delta(x,\infty) = 1/|x + i|$. Since $\delta(x, \infty) \to 0$ as $|x| \to \infty$, the space $\bar{X}$ is a circle.]

4. The function $d(u, v) = |1/u - 1/v|$ is a metric for the set of nonzero complex numbers. Show that the completion of this space is homeomorphic to the plane.

(First show that $\bar{X}$ is homeomorphic to the punctured sphere whose north pole corresponds to the nonbounded Cauchy sequences.)

8. Baire's Category Theorem, the Principles of Uniform Boundedness and of the Condensation of Singularities

A set A of a topological space X is called *nowhere dense* if $\bar{A}^i = \varnothing$. Thus if A is nowhere dense, then $N_x \cap c\bar{A} \neq \varnothing$ for every point x of X and for any neighborhood N_x. It follows that cA is dense in X. However, cA can be dense in X without A being nowhere dense, e.g., both the rationals and irrationals form dense subsets of the space of reals.

Lemma 1. *If A is nowhere dense, then $\bar{A}$ is not the entire space.*

Proof. Since X is open and closed we have $\bar{X}^i = X$ while A being nowhere dense $\bar{A}^i = \varnothing$.

Lemma 2. *The union of finitely many nowhere dense sets is nowhere dense.*

Proof. It is sufficient to consider the case of two nowhere dense sets, say, A and B. For simplicity let $O = \overline{A \cup B}^i$ so that $O \subseteq \bar{A} \cup \bar{B}$. Thus $O \cap c\bar{B}$ is an open subset of $\bar{A}$ and so by $\bar{A}^i = \varnothing$ it must be empty. Hence $O \subseteq \bar{B}$ and by $\bar{B}^i = \varnothing$ it follows that $O = \varnothing$.

Definition 1. *A set A of a topological space is called a set of first category if it can be represented as a countable union of nowhere dense sets. Any other set is called a set of the second category.*

There are many results in real-variable theory and in functional analysis the proof of which depends on these concepts. For instance, the *existence of a continuous nowhere differentiable function* can be most easily demonstrated by an argument involving categories. Similarly, the concept of category comes up in one of the proofs showing the *existence of a continuous periodic function whose Fourier series diverges on a dense subset of the real line.* The basis of these proofs is Baire's category theorem:

Theorem 1. *If a topological space X can be metrized such that it becomes complete, then X is of the second category, i.e., it is not expressible as a countable union of nowhere dense subsets.*

Proof. Let d be a complete metric and let $A_1, A_2, \ldots$ be nowhere dense subsets of X. Our object is to show that $\bigcup A_n$ is not the whole space. To this end we construct a sequence of points $x_1, x_2, \ldots$ and a nested sequence of open balls $O_1, O_2, \ldots$ centered around them by induction as follows: Since A_1 is nowhere dense, by Lemma 1 $c\bar{A}_1$ is not void and so we can choose a point x_1 and an open ball O_1 of diameter $d(O_1) < \frac{1}{2}$ with

x_1 as center such that $x_1 \in O_1 \subseteq c\bar{A}_1$. Suppose $x_1, ..., x_{n-1}$ and $O_1, ..., O_{n-1}$ are already determined such that $x_k \in O_k$, $d(O_k) < 2^{-k}$, and $O_1 \supseteq \bar{O}_2 \supseteq ... \supseteq \bar{O}_{n-2} \supseteq O_{n-2} \supseteq \bar{O}_{n-1}$. Since A_n is nowhere dense and the open set O_{n-1} is not void, $O_{n-1} \nsubseteq \bar{A}_n$ proving that $O_{n-1} \cap c\bar{A}_n$ is not empty. We can thus choose x_n and an open ball O_n with center at x_n such that $x_n \in \bar{O}_n \subseteq O_{n-1} \cap c\bar{A}_n$ and $d(O_n) < \frac{1}{2}d(O_{n-1})$. Clearly $d(x_m, x_n) \leqslant 2^{-m}$ for any $m \leqslant n$ and so $x_1, x_2, ...$ is a Cauchy sequence whose limit we denote by x. We claim that $x \notin A_n$ for every $n = 1, 2, ...$. Indeed, $x_m \in O_n$ for $m \geqslant n$, so $x \in \bar{O}_n$ proving that $x \in O_{n-1}$. Thus by $O_{n-1} \subseteq c\bar{A}_{n-1}$ we have $x \notin A_{n-1}$ for $n > 1$.

As an application of Baire's category theorem we prove the *existence of continuous nowhere differentiable functions*. We let X be the vector space of real-valued continuous functions x on the interval $[-1, 1]$ satisfying $x(-1) = x(1)$ and we topologize X by the metric $d(x, y) = \| x - y \|$ where $\| x \| = \text{lub} | x(s) |$. Since the uniform limit of continuous functions is itself continuous it follows that X is complete with respect to the metric d. We let A_n denote the set of those elements of X for which there is some point s at which $| x(s + h) - x(s) | \leqslant | h | n$ for every h satisfying both $-1 \leqslant s + h \leqslant 1$ and $| h | \leqslant 1/n$. If x is differentiable at some point s, then it belongs to A_n for a sufficiently high value of n. Thus it is enough to show that $\bigcup A_n$ is a proper subset of X. Moreover, by Theorem 1 the space X is of second category so it is sufficient to demonstrate that each A_n is nowhere dense in X.

We first prove that A_n is a closed set by showing that its complement $X - A_n$ is open. This can be done by using a compactness argument: If $x \notin A_n$, then for any $s \in [-1, 1]$ there is an h_s such that $| h_s | \leqslant 1/n$ and $| x(s + h_s) - x(s) | > | h_s | n$. Due to the strict inequality and the continuity of x we thus have

$$| x(\sigma + h_s) - x(\sigma) | > | h_s | n + 2\epsilon_s$$

for some $\epsilon_s > 0$ and for every σ in a suitable open neighborhood O_s of s. Let $s_1, ..., s_m$ be chosen such that the corresponding neighborhoods O_{s_k} cover $[-1, 1]$ and let ϵ be the smallest one among the ϵ_{s_k}'s. Now, if $\| x - y \| < \epsilon$, then by the above inequality we have

$$| y(\sigma + h_{s_k}) - y(\sigma) | > | h_{s_k} | n$$

for every $\sigma \in O_{s_k}$ proving that $y \notin A_n$ and A_n is closed.

We complete the proof by showing that the interior of A_n is void and consequently A_n is nowhere dense: Given $x \in A_n$ and $\epsilon > 0$, by the Weierstrass approximation theorem (Theorem IV.12.1) there is a polynomial p such that $\| x - p \| < \epsilon/2$. The derivative of p is bounded on $[-1, 1]$ and so by the mean value theorem there is a constant M such that

$$| p(s + h) - p(s) | \leqslant | h | M$$

for every s and h satisfying $| h | \leqslant 1/n$. Let q be a "sawtooth" function on $[-1, +1]$ such that $\| q \| < \epsilon/2$ and the slope of the linear segments are greater than $M + n$. Then $p + q \notin A_n$ though $p + q$ belongs to the ϵ-neighborhood of x. This completes the proof of the existence of a continuous function x on $[-1, 1]$ satisfying $x(-1) = x(1)$ and such that x is nowhere differentiable. By continuing x periodically one obtains a bounded, continuous, and nowhere differentiable function on the real line.

In order to discuss the principle of uniform boundedness we must first study some of the fundamental concepts of functional analysis.

Definition 2. *A vector space X over the reals $\mathfrak{R}$ or the field of complex numbers $\mathfrak{C}$ is called a normed vector space if there is a nonnegative real valued function $\| \cdot \| : X \to \mathfrak{R}$, called its norm such that*

(Nv. 1) $\| x \| = 0$ *if and only if x is the zero vector θ;*

(Nv. 2) $\| \lambda x \| = | \lambda | \cdot \| x \|$ *for every x in X and every field element λ;*

(Nv. 3) $\| x_1 + x_2 \| \leqslant \| x_1 \| + \| x_2 \|$ *for every $x_1 , x_2 \in X$.*

Note. One can speak about normed vector spaces also if the field is not $\mathfrak{R}$ or $\mathfrak{C}$ but any other normed field, e.g., the rationals Q or the field of p-adic numbers for some prime p.

Every normed vector space carries a natural topology derived from the metric $d(x_1 , x_2) = \| x_1 - x_2 \|$. This topology is such that vector addition and multiplication of vectors by scalars λ are continuous functions mapping $X \times X$ and $\mathfrak{R} \times X$ or $\mathfrak{C} \times X$ onto X. A vector space X over $\mathfrak{R}$ or $\mathfrak{C}$ which is endowed with a topology such that both of these maps are continuous is called a *topological vector space.*

A linear transformation $\alpha : X \to Y$ of one vector space X into another Y over the same field is a function α such that $\alpha(x_1 + x_2) = \alpha(x_1) + \alpha(x_2)$ and $\alpha(\lambda x) = \lambda\alpha(x)$ for all choices of x_1 , x_2 , x, and λ. In other words, a linear transformation is a *vector space homomorphism.* If both X and Y carry topologies, then we can speak about the continuity of a linear transformation α. The field of complex numbers $\mathfrak{C}$ can be interpreted as a normed vector space over the reals $\mathfrak{R}$ or over the field $\mathfrak{C}$, and similarly $\mathfrak{R}$ can be considered as a normed vector space over itself. If the range space Y is $\mathfrak{C}$, then any, not necessarily linear, map $\alpha : X \to \mathfrak{C}$ is called a *functional.* In particular, α might be a *linear functional*, a *real functional*, a *continuous functional*, or any combination of these such as, for instance, a real continuous linear functional.

If X is a vector space over some field, A is a subset of X, and $x \in X$, we let

$$A + x = x + A = \{a + x : a \in A\}$$

and similarly, if λ is a field element, we let

$$\lambda A = \{\lambda a : a \in A\}.$$

More generally one can define $A + B$, $A - B$, and ΛA for subsets A, B, and Λ.

Lemma 3. *If X and Y are normed vector spaces over the same normed field F and if the linear transformation $\alpha : X \to Y$ is continuous at some point of X, then it is continuous everywhere.*

Note. The same holds if X and Y are topological vector spaces but the proof is somewhat longer as one has to prove that $A + x$ is open for any open set A and $x \in X$.

Proof. It is sufficient to prove that a linear transformation α is continuous at a point x if and only if it is continuous at the zero vector θ. In view of $\alpha(\theta) = 0$, this statement follows immediately from the identity

$$\| \alpha(\xi) - \alpha(x) \| = \| \alpha(\xi - x) \|.$$

Let the map $\alpha : X \to Y$ be such that $\| \alpha(\lambda x) \| = | \lambda | \cdot \| \alpha(x) \|$ although α need not be necessarily linear. If α is continuous at θ, then there is a $\delta > 0$ such that $\| \alpha(x) \| \leqslant 1$ for every x satisfying $\| x \| \leqslant \delta$ and so $\| \alpha(x) \| \leqslant \delta^{-1} \| x \|$ for any x in X. Conversely, if there exists a positive constant M such that $\| \alpha(x) \| \leqslant M \| x \|$ for every x in X, then α is continuous. The smallest possible value of M is called the *norm* of α and is denoted by $\| \alpha \|$. We have

$$\| \alpha \| = \operatorname*{lub}_{x \neq \theta} \frac{\| \alpha(x) \|}{\| x \|} = \operatorname*{lub}_{\|x\| \leqslant 1} \| \alpha(x) \| = \operatorname*{lub}_{\|x\| = 1} \| \alpha(x) \| .$$

If α is not continuous at θ we let $\| \alpha \| = +\infty$. Thus by the foregoing lemma we have:

Lemma 4. *A linear transformation $\alpha : X \to Y$ is continuous if and only if $\| \alpha \|$ is finite.*

The principle of uniform boundedness concerns sequences of real nonnegative functionals α_n having the same domain space X and satisfying $\alpha_n(\lambda x) = |\lambda| \alpha_n(x)$. If the set of norms $\|\alpha_1\|, \|\alpha_2\|, \ldots$ is bounded, then so is the set of values $\alpha_1(x), \alpha_2(x), \ldots$ for each fixed x. (Although in general the upper bound depends on the choice of x.) The principle states that in certain cases the converse is true:

Theorem 2. *Let X be a normed vector space and let $\alpha_1, \alpha_2, \ldots$ be nonnegative continuous functionals on X such that*

$$\alpha_n(\lambda x) = |\lambda| \alpha_n(x)$$

and

$$\alpha_n(x + y) \leqslant \alpha_n(x) + \alpha_n(y).$$

Then the boundedness of the set of values $\alpha_1(x), \alpha_2(x), \ldots$ for each fixed element x of a second category subset of X implies the boundedness of the set of norms $\|\alpha_1\|, \|\alpha_2\|, \ldots$.

Corollary. *If X is complete and the set of values $\alpha_1(x), \alpha_2(x), \ldots$ is bounded for every x in X, then the set of norms $\|\alpha_1\|, \|\alpha_2\|, \ldots$ is bounded.*

Note. A complete normed vector space is usually called a *Banach space.*

Proof. Consider the sets

$$C_k = \{x : \alpha_n(x) \leqslant k \quad \text{for every} \quad n = 1, 2, \ldots\}.$$

Since α_n is continuous, the sets C_k $(k = 1, 2, \ldots)$ are closed and according to the hypothesis their union is of second category. Thus at least one of them is dense somewhere, i.e., $C_k^i \neq \emptyset$ for some index k and so there is a ball $S_\epsilon[x]$ lying in C_k. If ϵ is sufficiently small, C_k also contains the closure of $S_\epsilon[x]$. Hence if y is an arbitrary element of X having norm $\|y\| < \epsilon$, then by the definition of C_k we have

$$\alpha_n(y) \leqslant \alpha_n(y + x) + \alpha_n(-x) \leqslant k + B$$

where B denotes an upper bound of the numbers $\alpha_1(x), \alpha_2(x), \ldots$. Thus for any y satisfying $\|y\| \leqslant 1$ we have

$$\alpha_n(y) \leqslant \frac{1}{\epsilon}(k + B) = M$$

and so $\|\alpha_n\| \leqslant M$ for $n = 1, 2, \ldots$.

The *principle of condensation of singularities* can be formulated as follows:

Theorem 3. *Let X be a normed vector space and let α_{mn} $(m, n = 1, 2, \ldots)$ be nonnegative functionals on X such that*

$$\alpha_{mn}(\lambda x) = |\lambda| \alpha_{mn}(x);$$
$$\alpha_{mn}(x + y) \leqslant \alpha_{mn}(x) + \alpha_{mn}(y);$$

and the set of norms

$$\| \alpha_{m1} \|, \quad \| \alpha_{m2} \|, \ldots$$

is unbounded for every $m = 1, 2, \ldots$. Then the set of points $x \in X$ for which the set of values

$$\alpha_{m1}(x), \quad \alpha_{m2}(x), \ldots$$

is bounded for some $m = 1, 2, \ldots$ is a set of the first category.

Corollary. *If X is complete, then there is a subset S of second category in X such that for every x in S and for every $m = 1, 2, \ldots$ the set of values*

$$\alpha_{m1}(x), \quad \alpha_{m2}(x), \ldots$$

is bounded.

Proof. Let H_m be the set of those elements x in X for which the set of values $\alpha_{m1}(x)$, $\alpha_{m2}(x)$, ... is bounded. Then H_m is of first category because otherwise the principle of uniform boundedness (Theorem 2) would imply the boundedness of the set of norms $\| \alpha_{m1} \|, \| \alpha_{m2} \|, \ldots$. Therefore $H = \bigcup H_m$ is of the first category.

As an application let us discuss the convergence of Fourier series of continuous functions. Denote by X the vector space of real-valued, continuous, and 2π-periodic functions on the real line. By the periodicity every element of X is a bounded function and so we can define $\| x \| = \text{lub}|x(s)|$. Since the uniform limit of 2π-periodic and continuous functions is also 2π-periodic and continuous it follows that X is complete with respect to the norm $\| \cdot \|$. For x in X and for some real number t we let

$$s_n(x; t) = \frac{1}{2\pi} \int_0^{2\pi} D_n(s - t)x(s)\, ds$$

where D_n is the *Dirichlet kernel*, i.e.,

$$D_n(t) = \frac{\sin(n + \frac{1}{2})t}{\sin(t/2)}.$$

It is well known that $s_n(x; t)$ is the value of the nth partial sum of the Fourier series of the function x at t.

For fixed t the map $x \to s_n(x; t)$ defines a real linear functional on X which is continuous because

$$| s_n(x; t) | \leqslant \frac{1}{2\pi} \left(\int_0^{2\pi} | D_n(s - t) | \, ds \right) \| x \| = L_n \| x \| .$$

This inequality also shows that the norm of the linear functional $s_n(\cdot; t)$ does not exceed L_n. The step function sign $D_n(s - t)$ can be uniformly approximated by elements of X. This implies that the norm of $s_n(\cdot; t)$ is at least L_n. Therefore

$$\| s_n(\cdot; t) \| = L_n = \frac{1}{2\pi} \int_0^{2\pi} | D_n(s) | \, ds .$$

The norm L_n is called the nth *Lebesgue constant.*

It is easy to show that the set of norms $L_1, L_2, \ldots$ is not bounded. Indeed, if

$$\alpha_k = \frac{4k + 1}{4n + 2} \pi \qquad \text{and} \qquad \beta_k = \frac{4k + 3}{4n + 2} \pi,$$

then

$$L_n > \frac{1}{2\pi} \sum_{k=0}^{2n} \int_{\alpha_k}^{\beta_k} | D_n(s) | \, ds \geqslant \frac{\sqrt{2}}{4\pi} \sum_{k=0}^{2n} \int_{\alpha_k}^{\beta_k} \frac{dt}{| \sin(t/2) |}$$

and so

$$L_n > \frac{\sqrt{2}}{\pi} \sum_{k=0}^{2n} \frac{1}{4k + 3} .$$

Hence by the principle of uniform boundedness (Theorem 2) there is an x in X such that the set of numbers $s_1(x; t), s_2(x; t), \ldots$ is not bounded. Thus we have proved: *There exists a 2π-periodic and continuous function whose Fourier series diverges at a given point t.*

Consider an arbitrary denumerable set of real numbers $t_1, t_2, \ldots$ and define

$$\alpha_{mn}(x) = s_n(x; t_m)$$

so that $\alpha_{mn}(x)$ is the value of the nth partial sum of the Fourier series of x at t_m. We proved already the existence of a function x in X such that the set of functional values $\alpha_{m1}(x), \alpha_{m2}(x), \ldots$ is not bounded. Therefore by the principle of condensation of singularities (Theorem 3) we obtain: *Given any denumerable set of points there exists a 2π-periodic and continuous function whose Fourier series diverges at the points of the given set.*

Before leaving this subject we shall mention a few more useful facts about vector spaces. We recall that in a real vector space a set A is called *convex* if A contains with any two points a_1, a_2 the segment consisting of all points $\lambda_1 a_1 + \lambda_2 a_2$ where $0 \leqslant \lambda_1, \lambda_2 \leqslant 1$, and $\lambda_1 + \lambda_2 = 1$. In other words, A is convex if $\lambda A + (1 - \lambda)A \subseteq A$ for every λ satisfying $0 \leqslant \lambda \leqslant 1$. Notice also that $O + A$ is open if O is open. For $O + a$ is open and $O + A = \bigcup \{O + a : a \in A\}$.

Theorem 4. *If A is a convex set in a normed real vector space, then $\bar{A}$ and A^i are convex sets.*

Proof. Let $\bar{a}_1, \bar{a}_2 \in \bar{A}$. Then for every $\epsilon > 0$ there are points $a_1, a_2 \in A$ such that $\|\bar{a}_1 - a_1\| < \epsilon/2$ and $\|\bar{a}_2 - a_2\| < \epsilon/2$. Hence

$$\|(\lambda_1\bar{a}_1 + \lambda_2\bar{a}_2) - (\lambda_1 a_1 + \lambda_2 a_2)\| \leqslant |\lambda_1| \cdot \|\bar{a}_1 - a_1\| + |\lambda_2| \cdot \|\bar{a}_2 - a_2\| < \epsilon$$

and so $\lambda_1\bar{a}_1 + \lambda_2\bar{a}_2 \in \bar{A}$. Next suppose that $a_1, a_2 \in A^i$. Then we can find an $\epsilon > 0$ such that $S_\epsilon[a_1] \subseteq A$ and $S_\epsilon[a_2] \subseteq A$. If $0 < \lambda < 1$, then $\lambda S_\epsilon[a_1]$ and $(1 - \lambda)S_\epsilon[a_2]$ are open sets so that

$$O = \lambda S_\epsilon[a_1] + (1 - \lambda)\, S_\epsilon[a_2]$$

is open. Since A is convex and $S_\epsilon[a_1]$, $S_\epsilon[a_2] \subseteq A$, the set O is contained in A. Therefore O is an open set in A which contains $\lambda a_1 + (1 - \lambda)a_2$ and so $\lambda a_1 + (1 - \lambda)a_2$ belongs to the interior of A.

Theorem 5. *The finite-dimensional linear subspaces of a normed real or complex vector space are closed sets.*

Proof. We apply induction on the algebraic dimension of the linear subspace L. Points are closed sets so if L is zero dimensional, then the theorem holds. Now suppose that the linear subspace L_0 is a closed set in the vector space X and suppose that x is a point not in L_0. Let L denote the linear subspace generated by L_0 and $x : L = \{\lambda x + y : y \in L_0\}$. If z is a point in the closure of L, then for every $m = 1, 2, \ldots$ there exist scalars λ_m and points $y_m \in L_0$ such that $\|\lambda_m x + y_m - z\| < 1/m$. Hence by the triangle inequality $\|(\lambda_m - \lambda_n)x + (y_m - y_n)\| < 1/m + 1/n$. Since L_0 is a closed set and $x \notin L_0$ there is a constant $k > 0$ such that $k|\lambda| \leqslant \|\lambda x + y\|$ for every λ and $y \in L_0$. Therefore $|\lambda_m - \lambda_n| < (1/m + 1/n)/k$ for $m, n = 1, 2, \ldots$ and so (λ_m) $(m = 1, 2, \ldots)$ is a convergent sequence of real or complex numbers. If λ is the limit of (λ_m), then $\|\lambda x + y_m - z\| \to 0$ as $m \to \infty$ and so $\lambda x + L_0$ being a closed set $z \in \lambda x + L_0 \subseteq L$. Therefore L is also a closed linear subspace.

EXERCISES

1. Let a_{mn} $(m, n = 1, 2, ...)$ be complex numbers such that for every convergent sequence of complex numbers $x = (x_1, x_2, ...)$ the sum

$$\alpha_m(x) = \sum_{n=1}^{\infty} a_{mn}x_n$$

exists. Prove that if the transformed sequence $(\alpha_1(x), \alpha_2(x), ...)$ is convergent for every convergent $(x_1, x_2, ...)$, then there is a positive bound B such that

$$\sum_{n=1}^{\infty} | a_{mn} | \leqslant B$$

for every $m = 1, 2, ...$.

2. Let X be a normed vector space with norm $\| x \|$ and let X^n be the set of all ordered n-tuples $(x_1, ..., x_n)$ where $x_1, ..., x_n \in X$. Show that $\max\{\| x_1 \|, ..., \| x_n \|\}$ and $\| x_1 \| + ... + \| x_n \|$ define norms on X^n.

3. Show that

$$(x, y) = \sum_{i,k} \frac{x_i y_k}{i + k + 1}$$

is an *inner product* for the n-dimensional real vector space, i.e.,

(i) $(x, x) \geqslant 0$ and $(x, x) = 0$ only if $x = \theta$;

(ii) $(x, y) = (y, x)$;

(iii) $(\lambda x, y) = \lambda(x, y)$;

(iv) $(x, y + z) = (x, y) + (x, z)$.

[The only hard part is to prove that $(x, x) > 0$ for every $x \neq \theta$. This follows from

$$(x, x) = \int_0^1 \Big(\sum_{k=1}^{n} x_k t^k\Big)^2 dt.]$$

4. Prove that every open ball in a normed vector space over $\mathfrak{R}$ is a convex set.

(The sets $A + x$ and λA for $\lambda > 0$ are convex if and only if A is convex. Hence it is sufficient to show that the unit ball $S_1[\theta]$ is convex. Now if $\| x_1 \| < 1$ and $\| x_2 \| < 1$, then $\| \lambda_1 x_1 + \lambda_2 x_2 \| < \lambda_1 + \lambda_2 = 1$.)

5. Let A be a convex set and let $A \subseteq B \subseteq \bar{A}$. Give an example to show that B need not be convex.

6. Let S be a set in the complex plane and let (p_n) $(n = 1, 2, ...)$ be a sequence of polynomials which is uniformly convergent in the usual sense on S to a limit function p. Show that if the sequence of degrees of the p_n's is bounded, then p is a polynomial.

(First suppose that S is bounded, $|s| \leqslant r$ for every $s \in S$. Consider the complex vector space X formed by the family of all bounded complex-valued functions x on S and introduce $\| x \| = \text{lub}\{|x(s)| : s \in S\}$. The polynomials of degree at most m form a finite-dimensional subspace L_m. By Theorem 5, L_m is a closed set and so $p \in L_m$ for a sufficiently high value of m. Now let $r \to \infty$ to cover the general case. The conclusion is correct even if the convergence is not uniform.)

7. A vector space X over $\mathfrak{C}$ is called a *unitary space* if a complex-valued function (x, y) is given on $X \times X$ such that

(i) $(x, x) \geqslant 0$ and $(x, x) = 0$ only if $x = \theta$;

(ii) $(x, y) = \overline{(y, x)}$;

(iii) $(\lambda x, y) = \lambda(x, y)$;

(iv) $(x, y + z) = (x, y) + (x, z)$.

Show that a unitary space X is a normed vector space with norm $\| x \| = \sqrt{(x, x)}$.

[First one proves *Schwarz's inequality* $|(x, y)| \leqslant \| x \| \cdot \| y \|$ by using the inequality $0 \leqslant (x - \lambda y, x - \lambda y)$ with $\lambda = (x, y)/(y, y)$. Axiom (Nv. 3) follows from this inequality. In fact, $\| x + y \|^2 = (x + y, x + y) = (x, x) + (y, y) + (x, y) + (y, x) \leqslant \| x \|^2 + \| y \|^2 + 2|(x, y)| \leqslant (\| x \| + \| y \|)^2$.]

8. Give a necessary and sufficient condition that two norms $\| x \|_1$ and $\| x \|_2$ generate the same topology on a vector space X and use this criterion to show that the norms given in Exercises 2 and 3 generate the same topology on the n-dimensional real vector space.

(The topology generated by $\| x \|_1$ is finer than the topology generated by $\| x \|_2$ if for every $\epsilon_2 > 0$ there is an $\epsilon_1 > 0$ such that $S^2_{\epsilon_2}[\theta] \supseteq S^1_{\epsilon_1}[\theta]$.)

9. Completions and Compactifications

It was proved in Section 7 that every metric space X can be embedded isometrically in a complete metric space $\bar{X}$ such that X is dense in $\bar{X}$ relative to the topology induced on $\bar{X}$ by the extended metric. In the present section we generalize this result to arbitrary uniform structures and show that the completion is unique in a certain sense. For the sake of simplicity we shall speak about pairs $X, \mathscr{U}$ instead of sets X with uniform structures $\mathscr{U}$.

Definition 1. *The pairs $X, \mathscr{U}$ and $Y, \mathscr{V}$ are called uniformly isomorphic if there is a one-to-one correspondence $f: X \to Y$ between the elements of X and Y such that the map*

$$(x_1, x_2) \in X \times X \to (f(x_1), f(x_2)) \in Y \times Y$$

yields a one-to-one correspondence between the elements of $\mathscr{U}$ and $\mathscr{V}$. Any map f having this property is called a uniform isomorphism of $X, \mathscr{U}$ onto $Y, \mathscr{V}$.

It is clear that uniform isomorphism of pairs $X, \mathscr{U}$ is a reflexive, symmetric, and transitive relation and so as far as the theory of uniform structures is concerned uniformly isomorphic pairs $X, \mathscr{U}$ can be considered identical.

Definition 2. *A pair $\bar{X}, \overline{\mathscr{U}}$ is called a completion of the pair $X, \mathscr{U}$ if $\overline{\mathscr{U}}$ is a complete uniform structure for $\bar{X}$ and if there is a set $\tilde{X}$ in $\bar{X}$ such that*

(i) *$\tilde{X}$ is dense in $\bar{X}$ relative to the uniform topology induced on $\bar{X}$* by *$\overline{\mathscr{U}}$, and*

(ii) *the pair $X, \mathscr{U}$ is uniformly isomorphic to $\tilde{X}$ and the relativization of $\overline{\mathscr{U}}$ to X.*

It is easy to see that if the metric space $\bar{X}$ with metric δ is the completion of the space X having metric d in the sense described in Definition 7.1 then $\bar{X}, \mathscr{U}_\delta$ is a completion of $X, \mathscr{U}_d$ in the above sense.

Theorem 1. *Every pair $X, \mathscr{U}$ has a completion $\bar{X}, \overline{\mathscr{U}}$.*

Proof. We shall consider all those filters $\mathscr{X}$ in X which contain arbitrarily small open sets. Thus if $U \in \mathscr{U}$ is given then there is some open set O in $\mathscr{X}$ such that $O \times O \subseteq U$ and so every $\mathscr{X}$ is a Cauchy filter. We call two such filters $\mathscr{X}_1$ and $\mathscr{X}_2$ equivalent if they contain the same open sets, i.e., if $\mathscr{X}_1 \cap \mathscr{O} = \mathscr{X}_2 \cap \mathscr{O}$ where $\mathscr{O}$ is the family of open sets of the uniform topology associated with $\mathscr{U}$. Since every $\mathscr{X}$ of a given class contains the same open sets we may speak about the open sets of the class itself. The equivalence classes will be denoted in general by $\xi, \eta, \ldots$ but we shall also study some particular equivalence classes associated with points $x, y, \ldots$ of X and these we shall denote by $\tilde{x}, \tilde{y}, \ldots$. Namely, we let $\tilde{x}$ denote the class containing the neighborhood filter $\mathscr{N}(x)$ of x in the uniform topology of X. We define $\bar{X}$ to be the set of all equivalence classes ξ and we let $\tilde{X}$ denote the subset of special equivalence classes $\tilde{x}$ where $x \in X$.

In order to define $\overline{\mathscr{U}}$ we consider the set $\mathscr{U}_B$ of those symmetric uniformities $U \in \mathscr{U}$ which are open subsets of $X \times X$ and for every $U \in \mathscr{U}_B$ we let

$$\bar{U} = \{(\xi, \eta) : O \times Q \subseteq U \quad \text{for some} \quad O \in \xi \quad \text{and} \quad Q \in \eta\}.$$

We prove that the family $\overline{\mathscr{U}}_B$ of these sets $\bar{U}$ is a base for a uniform structure $\overline{\mathscr{U}}$. First let U_1 and U_2 be open uniformities and let $(\xi, \eta) \in \bar{U}_1 \cap \bar{U}_2$. Then $O_i \times Q_i \subseteq U_i$ for some $O_i \in \xi$ and $Q_i \in \eta$ where $i = 1, 2$ and so $(O_1 \cap O_2) \times (Q_1 \cap Q_2) \subseteq U_1 \cap U_2$ proving that $(\xi, \eta) \in \overline{U_1 \cap U_2}$ and $\bar{U}_1 \cap \bar{U}_2 \subseteq \overline{U_1 \cap U_2}$. Therefore by $\overline{U_1 \cap U_2} \subseteq \bar{U}_i$ $(i = 1, 2)$ we have $\overline{U_1 \cap U_2} = \bar{U}_1 \cap \bar{U}_2$ and so $\bar{U}_1 \cap \bar{U}_2 \in \overline{\mathscr{U}}_B$. Since we considered only symmetric open uniformities the sets $\bar{U}$ are also symmetric and so only the last axiom (Ub.6) remains to be proved. If $U \in \mathscr{U}$ and if V is a symmetric uniformity such that $V \circ V \circ V \subseteq U$ then $V \subseteq U^i$ because $V[x] \times V[y] \subseteq V \circ V \circ V \subseteq U$ for any $(x, y) \in V$. This shows that the family $\mathscr{U}_B$ of symmetric open uniformities is a base for $\mathscr{U}$. Thus in order to prove (Ub.6) it will be sufficient to show that $V \circ V \subseteq U$ implies $\bar{V} \circ \bar{V} \subseteq \bar{U}$. Indeed, if $(\xi_1, \xi_2) \in \bar{V}$ and $(\xi_2, \xi_3) \in \bar{V}$ then $O_1 \times O_2 \subseteq V$ and $Q_2 \times Q_3 \subseteq V$ for suitable $O_1 \in \xi_1$; $O_2, Q_2 \in \xi_2$; and $O_3 \in \xi_3$. Thus if y is any point in $O_2 \cap Q_2$ then $(x_1, y) \in V$ for every $x_1 \in O_1$ and $(y, x_3) \in V$ for every $x_3 \in O_3$ which shows that $O_1 \times O_3 \subseteq V \circ V \subseteq U$. Since $O_i \in \xi_i$ $(i = 1, 3)$ we see that $(\xi_1, \xi_3) \in \bar{U}$ and so $\bar{V} \circ \bar{V} \subseteq \bar{U}$. We proved that $\overline{\mathscr{U}}_B$ is a structure base for a uniform structure $\overline{\mathscr{U}}$ for $\bar{X}$.

The trace of $\overline{\mathscr{U}}$ on $\tilde{X}$ is the natural image of the original structure $\mathscr{U}$ because the trace of each $\bar{U} \in \overline{\mathscr{U}}_B$ is the uniformity $U \in \mathscr{U}_B$ used to construct $\bar{U}$: Indeed, if $(x, y) \in U$ then U being open we have $O_x \times O_y \subseteq O$ for suitable open sets $O_x \in \tilde{x}$, $O_y \in \tilde{y}$ proving that $(\tilde{x}, \tilde{y}) \in \bar{U}$. Conversely, if $(\tilde{x}, \tilde{y}) \in \bar{U}$ then $O_x \times O_y \subseteq U$ for suitable $O_x \in \tilde{x}$ and $O_y \in \tilde{y}$ and so by $x \in O_x$, $y \in O_y$ we have $(x, y) \in U$.

In order to prove that X is dense in $\bar{X}$ let a point $\xi \in \bar{X}$ and a neighborhood $\bar{U}[\xi]$ be given. Since $\xi \in \bar{U}[\xi]$ there is an open set $O \in \xi$ such that $O \times O \subseteq U$. If we choose x in O then $O \in \tilde{x}$ and so by $O \in \xi$ and $O \times O \subseteq U$ we have $(\tilde{x}, \xi) \in \bar{U}$ or in other words $\tilde{x} \in \bar{U}[\xi]$. Hence $\tilde{X}$ is dense in $\bar{X}$. If we prove that every filter $\mathscr{X}$ containing arbitrarily small open sets converges to some point of $\bar{X}$ then the completeness of $\overline{\mathscr{U}}$ will immediately follow from Lemma 6.5. We claim that the equivalence class ξ containing $\mathscr{X}$ is in $\lim \mathscr{X}$ and so $\mathscr{X}$ is indeed convergent. For any U in $\mathscr{U}_B$ there is an $O \in \xi$ such that $O \times O \subseteq U$

and we have $O \in \tilde{y}$ for any $y \in O$. Therefore $O \subseteq \bar{U}[\xi]$ proving that every neighborhood of ξ contains elements of the filter $\mathscr{X}$, i.e., $\xi \in \lim \mathscr{X}$. Therefore $\bar{\mathscr{U}}$ is complete and the proof of the existence of $\bar{X}$, $\bar{\mathscr{U}}$ is finished.

Now we turn to the question of *separated completions* of separated structures. First suppose that X is a topological space in which separation axiom (T_0) does not necessarily hold. Then an equivalence relation R can be introduced on X by letting $x\,R\,y$ provided x and y cannot be separated by open sets, i.e., if every open set O of X either contains both x and y or it does not contain any one of them. We notice that the corresponding quotient space X/R will be a (T_0) space. This remark will be used in the following situation:

If $\mathscr{U}$ is a not necessarily separated uniform structure for some set X then an equivalence relation R can be introduced on X by letting $x\,R\,y$ provided $(x, y) \in U$ for every $U \in \mathscr{U}$. The natural map $f : X \to X/R$ induces a map of $X \times X$ onto $(X/R) \times (X/R)$. The images U/R of the uniformities U of $\mathscr{U}$ form a uniform structure $\mathscr{U}/R$ for X/R and the uniform topology associated with $\mathscr{U}/R$ is the quotient topology $\mathscr{T}/R$ where $\mathscr{T}$ denotes the uniform topology generated on X by $\mathscr{U}$. Moreover $x\,R\,y$ if and only if x and y cannot be separated by open sets. Thus by the above remark X/R is a (T_0) space and so by Theorem II.9.1 $\mathscr{U}/R$ is a separated structure, called the *separated structure associated with* $\mathscr{U}$.

Let $f : X \to Y$ be a surjective map and let $\mathscr{F}$ be a filter in Y. Then the sets $f^{-1}(F)$ $(F \in \mathscr{F})$ form a filter base for a filter in X which is called the *inverse image* of $\mathscr{F}$ under the map f and is denoted by $f^{-1}(\mathscr{F})$. Now suppose that $f^{-1}(\mathscr{F})$ is convergent to x and f is continuous at x. Then for every neighborhood N_y of $y = f(x)$ there is a neighborhood N_x such that $f(N_x) \subseteq N_y$ and by $x \in \lim f^{-1}(\mathscr{F})$ there is an $f^{-1}(F)$ contained in N_x. Thus for each N_y there is an F in $\mathscr{F}$ such that $F \subseteq N_y$. Therefore we have:

Theorem 2. *If the surjective map $f : X \to Y$ is continuous then* $f(\lim f^{-1}(\mathscr{F})) \subseteq \lim \mathscr{F}$.

Note. More generally, if f is continuous at x and if $x \in \lim f^{-1}(\mathscr{F})$ then $f(x) \in \lim \mathscr{F}$.

Let us return to the case of a uniform structure $\mathscr{U}$ and the equivalence relation R defined on X by $\mathscr{U}$. We let $Y = X/R$ and f be the natural map of X onto X/R. Then it is easy to check that the inverse image of every Cauchy filter in X/R is a Cauchy filter in X. Thus if $\mathscr{U}$ is complete then by the foregoing theorem every Cauchy filter lying in X/R is convergent and so the uniform structure $\mathscr{U}/R$ is complete. Hence we proved:

Theorem 3. *If $\mathscr{U}$ is a complete uniform structure then so is the separated uniform structure associated with $\mathscr{U}$.*

Let $\bar{X}$, $\bar{\mathscr{U}}$ be a completion of X, $\mathscr{U}$ and let $\tilde{X}$ be a subset of $\bar{X}$ such that $\tilde{X}$ endowed with the restriction $\tilde{\mathscr{U}}$ of $\bar{\mathscr{U}}$ is uniformly isomorphic to X, $\mathscr{U}$. Let $\bar{X}/R$, $\bar{\mathscr{U}}/R$ be the separated uniform structure associated with $\bar{X}$, $\bar{\mathscr{U}}$. If X, $\mathscr{U}$ is separated then so is its uniformly isomorphic image $\tilde{X}$, $\tilde{\mathscr{U}}$, so each $\tilde{x}$ of $\tilde{X}$ forms an equivalence class all by itself and consequently $\tilde{X}$, $\tilde{\mathscr{U}}$ is uniformly isomorphic to $\tilde{X}/R$, $\tilde{\mathscr{U}}/R$. This implies that X, $\mathscr{U}$ and $\tilde{X}/R$, $\tilde{\mathscr{U}}/R$ are uniformly isomorphic. Moreover $\tilde{X}/R$ is dense in $\bar{X}/R$ because $\tilde{X}$ is dense in $\bar{X}$. Hence $\bar{X}/R$, $\bar{\mathscr{U}}/R$ is a completion of X, $\mathscr{U}$ and $\bar{\mathscr{U}}/R$ is a separated structure. We proved:

Theorem 4. *Every separated uniform structure X, $\mathscr{U}$ has separated completions.*

Now let X, $\mathscr{U}$ and Y, $\mathscr{V}$ be uniform structures, let $\bar{X}$, $\bar{\mathscr{U}}$ and $\bar{Y}$, $\bar{\mathscr{V}}$ be their completions, and let $\tilde{X}$ and $\tilde{Y}$ denote the uniformly isomorphic images of X and Y, respectively. If $\xi \in \bar{X}$ then $\tilde{X}$ being dense in $\bar{X}$ the sets $N_\xi = \bar{N}_\xi \cap \tilde{X}$ where $\bar{N}_\xi \in \mathscr{N}(\xi)$ form a filter base in $\tilde{X}$. If f is a map of X into Y then f can be interpreted as a map $\tilde{f}$ of $\tilde{X}$ into $\tilde{Y}$. The sets $\tilde{f}(N_\xi)$ form a base for a filter in $\bar{Y}$ which we denote by $\tilde{f}(\mathscr{N}(\xi))$. If in addition f is uniformly continuous with respect to the uniform structures $\mathscr{U}$ and $\mathscr{V}$ then $\tilde{f}(\mathscr{N}(\xi))$ is a Cauchy filter and so $\bar{Y}$, $\bar{\mathscr{V}}$ being complete it is convergent. If $\bar{\mathscr{V}}$ is a separated structure then $\lim \tilde{f}(\mathscr{N}(\xi))$ consists of a single point η which we shall call the image of ξ under the extended map and we shall write $\eta = \bar{f}(\xi)$. This process determines an extension of the map $f : X \to Y$ to a map $\bar{f} : \bar{X} \to \bar{Y}$.

We claim that the extension $\bar{f}$ is uniformly continuous with respect to the uniform structures $\bar{\mathscr{U}}$ and $\bar{\mathscr{V}}$. In fact given $\bar{V}$ we determine a symmetric $\bar{U}$ and $\bar{W}$ such that

$$\tilde{f}((\bar{U} \circ \bar{U} \circ \bar{U}) \cap (\tilde{X} \times \tilde{X})) \subseteq \bar{W} \cap (\tilde{Y} \times \tilde{Y})$$

and $\bar{W} \circ \bar{W} \circ \bar{W} \subseteq \bar{V}$. Now if $(\xi_1, \xi_2) \in \bar{U}$ then by the definition of η_i $(i = 1,2)$ there is an N_{ξ_i} such that $\tilde{f}(N_{\xi_i}) \subseteq \bar{W}[\eta_i]$. Thus $\tilde{X}$ being dense in $\bar{X}$ we can find $\tilde{x}_1, \tilde{x}_2 \in \tilde{X}$ such that $(\tilde{x}_i, \xi_i) \in \bar{U}$ and $(\tilde{y}_i, \eta_i) \in \bar{W}$. Then we have $(\tilde{x}_1, \tilde{x}_2) \in \bar{U} \circ \bar{U} \circ \bar{U}$ and so $(\tilde{y}_1, \tilde{y}_2) \in \bar{W}$. Hence $(\eta_1, \eta_2) \in \bar{W} \circ \bar{W} \circ \bar{W} \subseteq \bar{V}$ follows from the hypothesis $(\xi_1, \xi_2) \in \bar{U}$ and this shows that $\bar{f}$ is uniformly continuous. Therefore we have:

Theorem 5. *Let $f : X \to Y$ be uniformly continuous with respect to a uniform structure $\mathscr{U}$ for X and a separated structure $\mathscr{V}$ for Y. Let $\bar{X}$, $\bar{\mathscr{U}}$ be a completion of X, $\mathscr{U}$ and let $\bar{Y}$, $\bar{\mathscr{V}}$ be a separated completion of Y, $\mathscr{V}$.*

Then f admits a unique extension $\bar{f} : \bar{X} \to \bar{Y}$ *which is uniformly continuous with respect to* $\bar{\mathcal{U}}$ *and* $\bar{\mathcal{V}}$

Note. The uniqueness is an easy consequence of the continuity of $\bar{f}$.

Theorem 6. *Any two separated completions of a separated uniform structure* $X, \mathcal{U}$ *are uniformly isomorphic.*

Proof. Let $\bar{X}, \bar{\mathcal{U}}$ and $\bar{Y}, \bar{\mathcal{V}}$ be separated completions of $X, \mathcal{U}$ and let $\tilde{X}$ and $\tilde{Y}$ be the uniformly isomorphic images of X in $\bar{X}$ and $\bar{Y}$, respectively. The uniform isomorphisms $\tilde{X} \to X$ and $X \to \tilde{Y}$ yield a uniform isomorphism $f : \tilde{X} \to \tilde{Y}$. Thus f is a one-to-one map of $\tilde{X}$ onto $\tilde{Y}$ such that both f and $g = f^{-1}$ are uniformly continuous. Hence by Theorem 5 these maps can be extended to uniformly continuous maps $\bar{f} : \bar{X} \to \bar{Y}$ and $\bar{g} : \bar{Y} \to \bar{X}$. The composition $\bar{f} \circ \bar{g} : \bar{X} \to \bar{X}$ is uniformly continuous and on $\tilde{X}$ it is the identity map i. Since the identity map $\bar{\imath} : \bar{X} \to \bar{X}$ is an obvious uniformly continuous extension of i to $\bar{X}$ it follows that $\bar{f} \circ \bar{g} = \bar{\imath}$. Therefore $\bar{f} : \bar{X} \to \bar{Y}$ is a surjective injection such that both $\bar{f}$ and its inverse $\bar{g}$ are uniformly continuous. In other words $\bar{f}$ and $\bar{g}$ are uniform isomorphisms.

Theorem 7. *If* $\mathcal{U}$ *is a precompact structure then its completions are compact.*

Proof. In view of Theorem 6.4 it is sufficient to prove that $\bar{\mathcal{U}}$ is precompact. Given $\bar{U}$ in $\bar{\mathcal{U}}$ choose a symmetric $\bar{V}$ such that $\bar{V} \circ \bar{V} \subseteq \bar{U}$ and consider $V = \bar{V} \cap (\tilde{X} \times \tilde{X})$. Since $X, \mathcal{U}$ is precompact there are finitely many points $\tilde{x}_1, ..., \tilde{x}_n$ in $\tilde{X}$ such that the union of the sets $V[\tilde{x}_i]$ $(i = 1, ..., n)$ covers $\tilde{X}$. Now if $\xi \in \bar{X}$ then X being dense in $\bar{X}$ we can find $\tilde{x}$ in $\tilde{X}$ such that $(\xi, \tilde{x}) \in \bar{V}$. Since there is an $\tilde{x}_i$ such that $(x_i, \tilde{x}) \in \bar{V}$ we have $(\xi, x_i) \subseteq \bar{V} \circ \bar{V} \subseteq \bar{U}$, and this proves that the union of the sets $\bar{U}[x_i]$ $(i = 1, ..., n)$ is $\bar{X}$.

Definition 3. *If* $X, \mathcal{U}$ *is precompact then* $\bar{X}, \bar{\mathcal{U}}$ *is called a compactification of* $X, \mathcal{U}$.

By Theorem III.3.8 the uniform structure $\bar{\mathcal{U}}$ is completely determined by the topology of $\bar{X}$. However this uniform structure plays an important role because it appears explicitely in the uniform isomorphism of X onto $\tilde{X}$. In general it is possible to embed the space X into the compact space $\bar{X}$ as a dense subset in many different ways and two such embeddings are considered equivalent if and only if the natural one-to-one correspondence between these images $\tilde{X}$ is a uniform isomorphism with respect to the traces of the unique structure of $\bar{X}$.

We finish this section with the construction of three important uniform structures, two of which lead to compactifications while the third is generally not precompact. We start from a topological space X and a family $\mathfrak{F}$ of real valued, continuous functions $f : X \to \mathfrak{R}$. Every $f \in \mathfrak{F}$ defines a uniform structure for X, namely the inverse image $f^{-1}(\mathscr{U})$ of the usual structure $\mathscr{U}$ of the reals under the map f. (See p. 226.) We define $\mathscr{U}_{\mathfrak{F}} = \text{lub}\{f^{-1}(\mathscr{U}) : f \in \mathfrak{F}\}$ and so by Lemma IV.10.2 $\mathscr{U}_{\mathfrak{F}}$ is the weakest uniform structure on X which makes the functions f uniformly continuous when the range space $\mathfrak{R}$ is uniformized by its usual uniformity $\mathscr{U}$. Since every f is continuous the uniform topology associated with $\mathscr{U}_{\mathfrak{F}}$ is not stronger than the topology originally given on X and equality might occur only if X is uniformizable.

Lemma 1. *If for every point x and for every neighborhood N_x there is an f in $\mathfrak{F}$ such that*

$$\{\xi : |f(\xi) - f(x)| < \epsilon\} \subseteq N_x$$

for some $\epsilon > 0$ then the uniform topology associated with $\mathscr{U}_{\mathfrak{F}}$ coincides with the topology originally given on X.

Note. A necessary and sufficient condition can be obtained by considering finite intersections $\bigcap \{\xi : |f(\xi) - f(x)| < \epsilon\}$.

Proof. Every neighborhood N_x contains some uniform neighborhood, namely $U_f^{\epsilon}[x] \subseteq N_x$ where

$$U_f^{\epsilon} = \{(\xi, \eta) : |f(\xi) - f(\eta)| < \epsilon\}.$$

Lemma 2. *The inverse image of a precompact structure is precompact.*

Proof. Let $f : X \to Y$ where X is a set and let $\mathscr{U}$ be a precompact uniform structure for Y. By the precompactness of $\mathscr{U}$ for every $U \in \mathscr{U}$ there is a finite set B in Y such that $U[B] = Y$. If A is a finite set such that $f(A) = B$ then $f^{-1}(U)[A] = X$. Since the sets $f^{-1}(U)$ $(U \in \mathscr{U})$ form a structure base for the inverse image $f^{-1}(\mathscr{U})$ we see that $f^{-1}(\mathscr{U})$ is precompact.

Lemma 3. *Let $\mathscr{U}_S$ be a subbase for a uniform structure $\mathscr{U}$ such that for every $S \in \mathscr{U}_S$ there is a finite set A satisfying $S[A] = X$. Then $\mathscr{U}$ is a precompact structure.*

Proof. If $U \in \mathscr{U}$ then there are uniformities $S_1, ..., S_m$ in $\mathscr{U}_S$ such that $U \subseteq S_1 \cap ... \cap S_m$. The object is to find a finite set A such that

$U[A] = X$. By hypothesis there exist finite sets A_k such that $S_k[A_k] = X$ for every $k = 1, \dots, m$. Hence if $A = A_1 \cup \dots \cup A_m$ then $S_k[A] = X$ for $k = 1, \dots, m$ and so

$$U[A] \supseteq (S_1 \cap \dots \cap S_m)[A] = S_1[A] \cap \dots \cap S_m[A] = X.$$

Lemma 4. *The least upper bound of precompact structures is precompact.*

Proof. Let $\mathcal{U}_i$ $(i \in I)$ be precompact structures for X and let $\mathcal{U} = \text{lub}\{\mathcal{U}_i\}$. A subbase for $\mathcal{U}$ consists of all finite intersections $S = U_1 \cap \dots \cap U_m$ where $U_1, \dots, U_m \in \bigcup \mathcal{U}_i$. Since each $\mathcal{U}_i$ is precompact there are finite sets $A_1, \dots, A_m$ such that $U_k[A_k] = X$ for $k = 1, \dots, m$. Hence for $A = A_1 \cup \dots \cup A_m$ we have

$$S[A] = (U_1 \cap \dots \cap U_m)[A] = U_1[A] \cap \dots \cap U_m[A] = X.$$

By Lemma 3 the least upper bound $\mathcal{U}$ is precompact.

Let X be a uniformizable space and let $\mathfrak{F}$ be the family of all bounded, real-valued continuous functions on X. By Lemma 2 every $f^{-1}(\mathcal{U})$ is a precompact structure for X and so by Lemma 4 $\text{lub}\{f^{-1}(\mathcal{U}) : f \in \mathfrak{F}\}$ is also precompact. We denote this particular uniform structure of the uniformizable space X by $\mathcal{U}_{SC}$ and call it the *Stone-Čech structure* of X. If O_x is an open neighborhood of $x \in X$ then by Theorem IV.6.2 there is an f in $\mathfrak{F}$ such that $0 \leqslant f(\xi) \leqslant 1$ everywhere, $f(\xi) = 0$ for every $\xi \in cO$ and $f(x) = 0$. Thus $U_f^\epsilon[x] = \{\xi : |f(\xi)| < \epsilon\}$ is contained in O_x for any positive $\epsilon < 1$. Thus by Lemma 1 the Stone-Čech structure $\mathcal{U}_{SC}$ is compatible with the topology of X. Clearly $\mathcal{U}_{SC}$ is the weakest uniform structure compatible with the topology of the uniformizable space X such that every bounded, real-valued, and continuous function f is uniformly continuous with respect to $\mathcal{U}_{SC}$ and any uniform structure $\mathcal{V}$ compatible with the topology of the reals. [Indeed since $\overline{f(X)}$ is compact by Theorem III.3.8 the particular nature of $\mathcal{V}$ is irrelevant.] The completion of $X, \mathcal{U}_{SC}$ is a compactification $\bar{X}, \bar{\mathcal{U}}_{SC}$ such that every real-valued, bounded, and continuous function f has a uniformly continuous extension to $\bar{X}$. (Since $\bar{X}$ is compact continuity and uniform continuity are equivalent notions.)

The following proposition shows that among all precompact structures compatible with the topology of the uniform space X there is a strongest one, namely the Stone-Čech structure $\mathcal{U}_{SC}$:

Lemma 5. *If $\mathcal{V}$ is a precompact structure for X then $\mathcal{V} = \text{lub}\{f^{-1}(\mathcal{U})\}$ where $\mathcal{U}$ is the usual structure of the reals and f varies over the family $\mathfrak{F}$*

of those real-valued functions on X which are uniformly continuous with respect to $\mathscr{V}$ and $\mathscr{U}$.

Proof. Let $\bar{f}$ denote the extension of the function f to the compactification $\bar{X}, \bar{\mathscr{V}}$. A function $\bar{f}$ on $\bar{X}$ is the extension of some $f \in \mathfrak{F}$ if and only if $\bar{f}$ is continuous. Since $\bar{X}$ is uniformizable by Theorem IV.6.2 and by Lemma 1 lub$\{\bar{f}^{-1}(\mathscr{U})\}$ is compatible with the topology of $\bar{X}$. However there is only one uniform structure which is compatible with a compact topology and so lub$\{\bar{f}^{-1}(\mathscr{U})\}$ is the extension $\bar{\mathscr{V}}$ of $\tilde{\mathscr{V}}$. Since the restriction of $\bar{f}^{-1}(\mathscr{U})$ is $f^{-1}(\mathscr{U})$ we see that $\mathscr{V} =$ lub$\{f^{-1}(\mathscr{U})\}$.

If X is a locally compact Hausdorff space then by Theorem III.5.3 it is uniformizable. We let $\mathfrak{F}$ be the family of all continuous functions $f: X \to \mathfrak{R}$ having compact support and define $\mathscr{U}_A =$ lub$\{f^{-1}(\mathscr{U})\}$ where $\mathscr{U}$ is the usual structure of the reals. Given a point x and an open neighborhood O_x by Theorem IV.6.2 there is a real-valued, continuous function f such that $f(x) = 1$ and $f(\xi) = 0$ for every $\xi \in cO_x$. If $\bar{O}_x$ is compact then $f \in \mathfrak{F}$ and $\{\xi : |f(\xi) - f(x)| < 1\} \subseteq O_x$. Hence by Lemma 1 $\mathscr{U}_A =$ lub$\{f^{-1}(\mathscr{U})\}$ is compatible with the topology of X. Moreover, if a uniform structure $\mathscr{V}$ is compatible with the topology of X then every $f \in \mathfrak{F}$ is uniformly continuous with respect to $\mathscr{V}$ and the usual structure of the reals. Therefore by Lemma IV.10.2 we have $f^{-1}(\mathscr{U}) \leqslant \mathscr{V}$ and so $\mathscr{U}_A \leqslant \mathscr{V}$. This proves that the precompact structure $\mathscr{U}_A$ is the weakest uniform structure compatible with the topology of the locally compact Hausdorff space X. It is called the *Alexandroff structure* of X and the corresponding completion is the *Alexandroff compactification* or the *one-point compactification* of X. One can prove that if X is uniformizable but not locally compact then there is no weakest uniform structure among those which are compatible with the topology of X.

If X is a compact Hausdorff space then X with its unique uniform structure is its own Alexandroff compactification. If the locally compact Hausdorff space X is not compact then its one-point compactification can be constructed very easily as follows: We let $\bar{X} = X \cup \{\infty\}$ and define O to be an open set of $\bar{X}$ if O is an open set in X or if cO is a compact set in X.

Finally let X be a uniformizable space and let $\mathfrak{F}$ be the family of all real-valued, continuous functions f on the product space $X \times X$. Then the *Weil structure* or *universal uniform structure* $\mathscr{U}_W$ is defined to be $\mathscr{U}_W =$ lub $f^{-1}(\mathscr{U})$ where $\mathscr{U}$ is the usual structure of the reals and f varies over $\mathfrak{F}$. The importance of $\mathscr{U}_W$ is due to the fact that it is the strongest uniform structure compatible with the topology of X.

This can be seen immediately from the following remarkable property of $\mathscr{U}_W$:

Theorem 8. *If $Y, \mathscr{V}$ is arbitrary and if $\varphi : X \to Y$ is continuous then φ is uniformly continuous with respect to $\mathscr{U}_W$ and $\mathscr{V}$.*

Proof. By Theorem III.8.2 we have $\mathscr{V} = \text{lub}\{\mathscr{V}_d\}$ where d runs through all pseudometrics satisfying $\mathscr{V}_d \leqslant \mathscr{V}$. If $V \in \mathscr{V}$ then by Lemma III.8.1 there is a d such that $V \in \mathscr{V}_d \leqslant \mathscr{V}$ and so for some positive ϵ we have

$$V \supseteq \{(\eta_1, \eta_2) : d(\eta_1, \eta_2) < \epsilon\} = V_\epsilon .$$

Let $f : X \times X \to \mathfrak{R}$ be defined by the following rule:

$$f(x_1, x_2) = d(\varphi(x_1), \varphi(x_2)).$$

Then $U = \{(\xi_1, \xi_2) : f(\xi_1, \xi_2) < \epsilon\}$ belongs to $\mathscr{U}_W$ and $\varphi(U) \subseteq V_\epsilon \subseteq V$. Thus $\varphi : X \to Y$ is uniformly continuous.

EXERCISES

1. Given a prime number p for every integer m let $|m|_p = p^{-\nu}$ where p^ν is the highest power of p which divides m. For rationals m/n let $|m/n|_p = |m|_p/|n|_p$. Denote by $\mathfrak{R}_p$ the completion of the rationals Q with respect to the metric $d(x, y) = |x - y|_p$. Prove that addition and multiplication give uniformly continuous mappings of $Q \times Q$ onto Q. Extend the definition of addition and multiplication to the completion $\mathfrak{R}_p$. By determining the possible ideals of $\mathfrak{R}_p$ show that $\mathfrak{R}_p$ is a field, called the field of *p-adic numbers*. Show that $\mathfrak{R}_p$ is locally compact.

2. Let S be a nonvoid set, let X be a topological space, and let X^S be the set of all functions $x : S \to X$. Prove the existence of a unique topology on X^S such that any filter $\mathscr{F}$ in X^S is convergent if and only if its projections $\mathscr{F}_s$ $(s \in S)$ are convergent in X.

3. Let S be a nonvoid set, let $X, \mathscr{V}$ be a uniform space, and let X^S be the set of all functions $x : S \to X$. Prove that a filter $\mathscr{F}$ in X^S is uniformly convergent with respect to $\mathscr{V}$ if and only if $\mathscr{F}$ is convergent in the topology of uniform convergence.

[The filter $\mathscr{F}$ is said to converge uniformly to x if for each uniformity $V \in \mathscr{V}$ there is an F in $\mathscr{F}$ such that $(x(s), y(s)) \in V$ for every $y \in F$ and for every $s \in S$.]

4. Prove that the universal uniform structure of a paracompact Hausdorff space is complete.

NOTES

Filter bases and filters occur in the publications of several early topologists, e.g., in 1913 Carathéodory constructed prime ends by describing a base for the neighborhood filters of these ideal points of a compactification. (See [1] and [2].) As far as I know ultrafilters were first introduced by Henri Cartan in 1937 [3] but they also appear in a paper of Wallman which was published approximately at the same time [4]. It is hard to decide who was the one who first considered nets because for instance infinite sequences and limits taken along various paths were considered for a long time. However it is plain that a general theory of convergence based on the use of directed sets and nets was first published by E. H. Moore and H. L. Smith in 1922 [5]. Subnets were introduced by Kelley in 1950 [6].

Cauchy filters are straightforward generalizations of Cauchy sequences which were indeed first isolated by Cauchy more than a century ago. Theorem 7.1 on the completion of metric spaces is due to Hausdorff [7]. The proof is an extension of the method used to construct the real numbers from the rationals by using Cauchy sequences. First and second category sets were introduced by Baire in 1899 [8] in a long paper which contains also the proof of Theorem 8.1. The principle of uniform boundedness and the principle of condensation of singularities were published in 1927 [9] but special cases of these principles were known and used much earlier and in the early Twenties several mathematicians knew the principle of uniform boundedness essentially in its present form. As I learned from Professor Zygmund, first Banach and Steinhaus planned to publish a proof based on the use of infinite series and sequences but soon discovered the proofs given in Section 8. The old methods were found again by me who needed them to obtain an extension of these principles. (See [10] and [11].) The principle of uniform boundedness can be further extended, e.g., it is included in the *principle of equicontinuity* [12]. Precompact structures can be interpreted as a type of binary relations on the power set of X. These binary relations, called *proximity relations* were introduced by V. A. Efremovič [13] and their equivalence with compactifications was proved by Smirnov [14, 15]. A readable account on this subject including a direct proof of the equivalence of proximity relations and precompact structures can be found in [16].

The equivalence of filters and nets was first shown by R. G. Bartle [17] who proved that for every filter (net) there exists a net (filter) with the same limit. He also proved the corresponding proposition with respect to adherence. In Section 2 we proved this equivalence in a stronger form by showing the existence of a net (filter) such that both its limit and adherence are the same as that of a given filter (net).

REFERENCES

1. C. Carathéodory, Über die Begrenzung einfach zusammenhängender Gebiete. *Math. Ann.* **73**, 323–370 (1913).
2. I. S. Gál, Conformally invariant metrics and uniform structures. *Nederl. Akad. Wetensch. Proc. Ser. A* **63**, 218–244 (1960).
3. H. Cartan, Théorie des filtres, Filtres et ultrafiltres. *C. R. Acad. Sci. Paris* **205**, 595–598 and 777–779 (1937).
4. H. Wallman, Lattices and topological spaces. *Ann. of Math.* **39**, 112–126 (1938).
5. E. H. Moore and H. L. Smith, A general theory of limits. *Amer. J. Math.* **44**, 102–121 (1922).
6. J. L. Kelley, Convergence in topology. *Duke Math. J.* **17**, 277–283 (1950).
7. F. Hausdorff, "Grundzüge der Mengenlehre." Veit, Leipzig, 1914.
8. R. Baire, Sur les fonctions de variables réelles. *Ann. di Mat.* (3) **3**, 1–123 (1899).
9. St. Banach and H. Steinhaus, Sur le principe de la condensation de singularités. *Fund. Math.* **9**, 50–61 (1927).
10. I. S. Gál, On sequences of operations in complete vector spaces. *Amer. Math. Monthly* **60**, 527–538 (1953).
11. I. S. Gál, The principle of condensation of singularities. *Duke Math. J.* **20**, 27–36 (1953).
12. N. Dunford and J. T. Schwartz, "Linear Operators," Vol. I. Wiley (Interscience), New York, 1958.
13. V. A. Efremovič, The geometry of proximity. *Mat. Sb.* [N.S.] **31** (73), 189–200 (1952).
14. Yu. Smirnov, On proximity spaces in the sense of V. A. Efremovič. *Dokl. Akad. Nauk SSSR* [N.S.] **84**, 895–898 (1952).
15. Yu. Smirnov, On proximity spaces. *Mat. Sb.* [N.S.] **31** (73), 543–574 (1952).
16. I. S. Gál, Proximity relations and precompact structures. *Nederl. Akad. Wetensch. Proc. Ser. A* **62**, 304–326 (1959).
17. R. G. Bartle, Nets and filters in topology. *Amer. Math. Monthly* **62**, 551–557 (1955).

Author Index

Numbers in italics indicate the pages on which the complete references are given.

Subject Index

K

L

M

N

O

Mathematics

FUNCTIONAL ANALYSIS (Second Corrected Edition), George Bachman and Lawrence Narici. Excellent treatment of subject geared toward students with background in linear algebra, advanced calculus, physics and engineering. Text covers introduction to inner-product spaces, normed, metric spaces, and topological spaces; complete orthonormal sets, the Hahn-Banach Theorem and its consequences, and many other related subjects. 1966 ed. 544pp. $6^1/_8$ x $9^1/_4$. 0-486-40251-7

DIFFERENTIAL MANIFOLDS, Antoni A. Kosinski. Introductory text for advanced undergraduates and graduate students presents systematic study of the topological structure of smooth manifolds, starting with elements of theory and concluding with method of surgery. 1993 edition. 288pp. $5^3/_8$ x $8^1/_2$. 0-486-46244-7

VECTOR AND TENSOR ANALYSIS WITH APPLICATIONS, A. I. Borisenko and I. E. Tarapov. Concise introduction. Worked-out problems, solutions, exercises. 257pp. $5^5/_8$ x $8^1/_4$. 0-486-63833-2

AN INTRODUCTION TO ORDINARY DIFFERENTIAL EQUATIONS, Earl A. Coddington. A thorough and systematic first course in elementary differential equations for undergraduates in mathematics and science, with many exercises and problems (with answers). Index. 304pp. $5^3/_8$ x $8^1/_2$. 0-486-65942-9

FOURIER SERIES AND ORTHOGONAL FUNCTIONS, Harry F. Davis. An incisive text combining theory and practical example to introduce Fourier series, orthogonal functions and applications of the Fourier method to boundary-value problems. 570 exercises. Answers and notes. 416pp. $5^3/_8$ x $8^1/_2$. 0-486-65973-9

COMPUTABILITY AND UNSOLVABILITY, Martin Davis. Classic graduate-level introduction to theory of computability, usually referred to as theory of recurrent functions. New preface and appendix. 288pp. $5^3/_8$ x $8^1/_2$. 0-486-61471-9

AN INTRODUCTION TO MATHEMATICAL ANALYSIS, Robert A. Rankin. Dealing chiefly with functions of a single real variable, this text by a distinguished educator introduces limits, continuity, differentiability, integration, convergence of infinite series, double series, and infinite products. 1963 edition. 624pp. $5^3/_8$ x $8^1/_2$. 0-486-46251-X

METHODS OF NUMERICAL INTEGRATION (SECOND EDITION), Philip J. Davis and Philip Rabinowitz. Requiring only a background in calculus, this text covers approximate integration over finite and infinite intervals, error analysis, approximate integration in two or more dimensions, and automatic integration. 1984 edition. 624pp. $5^3/_8$ x $8^1/_2$. 0-486-45339-1

INTRODUCTION TO LINEAR ALGEBRA AND DIFFERENTIAL EQUATIONS, John W. Dettman. Excellent text covers complex numbers, determinants, orthonormal bases, Laplace transforms, much more. Exercises with solutions. Undergraduate level. 416pp. $5^3/_8$ x $8^1/_2$. 0-486-65191-6

RIEMANN'S ZETA FUNCTION, H. M. Edwards. Superb, high-level study of landmark 1859 publication entitled "On the Number of Primes Less Than a Given Magnitude" traces developments in mathematical theory that it inspired. xiv+315pp. $5^3/_8$ x $8^1/_2$. 0-486-41740-9

INTRODUCTORY REAL ANALYSIS, A.N. Kolmogorov, S. V. Fomin. Translated by Richard A. Silverman. Self-contained, evenly paced introduction to real and functional analysis. Some 350 problems. 403pp. $5^3/_8 \times 8^1/_2$. 0-486-61226-0

APPLIED ANALYSIS, Cornelius Lanczos. Classic work on analysis and design of finite processes for approximating solution of analytical problems. Algebraic equations, matrices, harmonic analysis, quadrature methods, much more. 559pp. $5^3/_8 \times 8^1/_2$. 0-486-65656-X

AN INTRODUCTION TO ALGEBRAIC STRUCTURES, Joseph Landin. Superb self-contained text covers "abstract algebra": sets and numbers, theory of groups, theory of rings, much more. Numerous well-chosen examples, exercises. 247pp. $5^3/_8 \times 8^1/_2$. 0-486-65940-2

QUALITATIVE THEORY OF DIFFERENTIAL EQUATIONS, V. V. Nemytskii and V.V. Stepanov. Classic graduate-level text by two prominent Soviet mathematicians covers classical differential equations as well as topological dynamics and ergodic theory. Bibliographies. 523pp. $5^3/_8 \times 8^1/_2$. 0-486-65954-2

THEORY OF MATRICES, Sam Perlis. Outstanding text covering rank, nonsingularity and inverses in connection with the development of canonical matrices under the relation of equivalence, and without the intervention of determinants. Includes exercises. 237pp. $5^3/_8 \times 8^1/_2$. 0-486-66810-X

INTRODUCTION TO ANALYSIS, Maxwell Rosenlicht. Unusually clear, accessible coverage of set theory, real number system, metric spaces, continuous functions, Riemann integration, multiple integrals, more. Wide range of problems. Undergraduate level. Bibliography. 254pp. $5^3/_8 \times 8^1/_2$. 0-486-65038-3

MODERN NONLINEAR EQUATIONS, Thomas L. Saaty. Emphasizes practical solution of problems; covers seven types of equations. ". . . a welcome contribution to the existing literature. . . ."—*Math Reviews*. 490pp. $5^3/_8 \times 8^1/_2$. 0-486-64232-1

MATRICES AND LINEAR ALGEBRA, Hans Schneider and George Phillip Barker. Basic textbook covers theory of matrices and its applications to systems of linear equations and related topics such as determinants, eigenvalues and differential equations. Numerous exercises. 432pp. $5^3/_8 \times 8^1/_2$. 0-486-66014-1

LINEAR ALGEBRA, Georgi E. Shilov. Determinants, linear spaces, matrix algebras, similar topics. For advanced undergraduates, graduates. Silverman translation. 387pp. $5^3/_8 \times 8^1/_2$. 0-486-63518-X

MATHEMATICAL METHODS OF GAME AND ECONOMIC THEORY: Revised Edition, Jean-Pierre Aubin. This text begins with optimization theory and convex analysis, followed by topics in game theory and mathematical economics, and concluding with an introduction to nonlinear analysis and control theory. 1982 edition. 656pp. $6^1/_8 \times 9^1/_4$. 0-486-46265-X

SET THEORY AND LOGIC, Robert R. Stoll. Lucid introduction to unified theory of mathematical concepts. Set theory and logic seen as tools for conceptual understanding of real number system. 496pp. $5^3/_8 \times 8^1/_4$. 0-486-63829-4

A TREATISE ON ELECTRICITY AND MAGNETISM, James Clerk Maxwell. Important foundation work of modern physics. Brings to final form Maxwell's theory of electromagnetism and rigorously derives his general equations of field theory. 1,084pp. 5³/₈ x 8¹/₂. Two-vol. set. Vol. I: 0-486-60636-8 Vol. II: 0-486-60637-6

MATHEMATICS FOR PHYSICISTS, Philippe Dennery and Andre Krzywicki. Superb text provides math needed to understand today's more advanced topics in physics and engineering. Theory of functions of a complex variable, linear vector spaces, much more. Problems. 1967 edition. 400pp. 6¹/₂ x 9¹/₄. 0-486-69193-4

INTRODUCTION TO QUANTUM MECHANICS WITH APPLICATIONS TO CHEMISTRY, Linus Pauling & E. Bright Wilson, Jr. Classic undergraduate text by Nobel Prize winner applies quantum mechanics to chemical and physical problems. Numerous tables and figures enhance the text. Chapter bibliographies. Appendices. Index. 468pp. 5³/₈ x 8¹/₂. 0-486-64871-0

METHODS OF THERMODYNAMICS, Howard Reiss. Outstanding text focuses on physical technique of thermodynamics, typical problem areas of understanding, and significance and use of thermodynamic potential. 1965 edition. 238pp. 5³/₈ x 8¹/₂. 0-486-69445-3

THE ELECTROMAGNETIC FIELD, Albert Shadowitz. Comprehensive under- graduate text covers basics of electric and magnetic fields, builds up to electromagnetic theory. Also related topics, including relativity. Over 900 problems. 768pp. 5⁵/₈ x 8¹/₄. 0-486-65660-8

GREAT EXPERIMENTS IN PHYSICS: FIRSTHAND ACCOUNTS FROM GALILEO TO EINSTEIN, Morris H. Shamos (ed.). 25 crucial discoveries: Newton's laws of motion, Chadwick's study of the neutron, Hertz on electromagnetic waves, more. Original accounts clearly annotated. 370pp. 5³/₈ x 8¹/₂. 0-486-25346-5

EINSTEIN'S LEGACY, Julian Schwinger. A Nobel Laureate relates fascinating story of Einstein and development of relativity theory in well-illustrated, nontechnical volume. Subjects include meaning of time, paradoxes of space travel, gravity and its effect on light, non-Euclidean geometry and curving of space-time, impact of radio astronomy and space-age discoveries, and more. 189 b/w illustrations. xiv+250pp. 8³/₈ x 9¹/₄. 0-486-41974-6

THE VARIATIONAL PRINCIPLES OF MECHANICS, Cornelius Lanczos. Philosophic, less formalistic approach to analytical mechanics offers model of clear, scholarly exposition at graduate level with coverage of basics, calculus of variations, principle of virtual work, equations of motion, more. 418pp. 5³/₈ x 8¹/₂. 0-486-65067-7